Visceral Leishmaniasis and Post-kala-azar Dermal Leishmaniasis

This reference book explores various aspects of visceral leishmaniasis (VL) and post-kala-azar dermal leishmaniasis (PKDL), including their pathophysiology, treatment options, and disease control strategies. The initial chapter provides an overview of VL, its global impact, and pathogenesis. It also examines the development of PKDL from VL, outlining the transition from the acute form of the disease to the chronic dermal condition. The book further explores the symptoms and diagnosis of both VL and PKDL. Furthermore, the book delves into the immunopathogenesis of both VL and PKDL, shedding light on the complex interactions between the parasite and the host immune system. It reviews the impact of environmental factors, particularly the contamination of arsenic and antimony in drinking water, on the progression of the disease. Lastly, the book discusses preventive measures for PKDL and highlights the major challenges in effectively implementing these strategies. This book serves as a useful resource for researchers, healthcare professionals, and policymakers seeking to deepen their understanding of these diseases and develop effective strategies for their prevention and control.

Key features

- Provides in-depth coverage of VL and PKDL, including pathogenesis, treatment, and disease control strategies
- Elucidates the symptoms and diagnostic methods for VL and PKDL
- Offers insights into the complex immunopathogenesis of VL and PKDL, exploring the interactions between parasites and host immune system
- Examines treatment options for both VL and PKDL, including current therapeutic approaches and emerging strategies
- Reviews the impact of environmental factors, such as arsenic and antimony contamination, on the disease progression
- Discusses preventive measures and challenges in implementing effective strategies for PKDL prevention and control

Visceral Leishmaniasis and Post-kala-azar Dermal Leishmaniasis

Pathogenesis, Treatment and Disease Control

Edited by Veer Singh and Ashish Kumar

CRC Press
Taylor & Francis Group
Boca Raton London New York

CRC Press is an imprint of the
Taylor & Francis Group, an **informa** business

Designed cover image: Shutterstock Images

First edition published 2025
by CRC Press
2385 NW Executive Center Drive, Suite 320, Boca Raton FL 33431

and by CRC Press
4 Park Square, Milton Park, Abingdon, Oxon, OX14 4RN

CRC Press is an imprint of Taylor & Francis Group, LLC

ISBN: 9781032633053 (hbk)
ISBN: 9781032643007 (pbk)
ISBN: 9781032643120 (ebk)

DOI: 10.1201/9781032643120

Typeset in Times
by Deanta Global Publishing Services, Chennai, India

Dedicated to those who care, conserve, and protect, but do not destroy the beauty and unique characteristics of

Kshit (Earth)

Jal (Water)

Pawak (Fire)

Gagan (Sky)

and

Sameera (Air)

Contents

Preface

Visceral leishmaniasis (VL), also known as kala-azar, and post-kala-azar dermal leishmaniasis (PKDL), are neglected tropical diseases caused by the protozoan parasite *Leishmania*. These diseases pose significant public health challenges, particularly in the Indian subcontinent, East Africa, and parts of Latin America, where they contribute to considerable morbidity and mortality. Despite recent advancements in our understanding of their pathogenesis, treatment, and control strategies, numerous challenges remain.

This book, *Visceral Leishmaniasis and Post-kala-azar Dermal Leishmaniasis: Pathogenesis, Treatment and Disease Control*, aims to provide a comprehensive overview of the current state of knowledge in these areas. It assembles leading experts in the field to discuss various aspects of VL and PKDL, including their epidemiology, pathogenesis, diagnosis, treatment, and control strategies.

The first section of the book examines the epidemiology of VL and PKDL, highlighting the global burden of these diseases and the factors contributing to their spread. It also discusses the challenges in surveillance and control efforts, particularly in resource-limited settings. Understanding the epidemiology of VL and PKDL is crucial for designing effective control programs and implementing targeted interventions.

The book then explores the pathogenesis of VL and PKDL, exploring the interactions between *Leishmania* parasites and the host immune system. It discusses the mechanisms of parasite survival and dissemination within the host, as well as the immunological responses that contribute to disease progression. A better understanding of the pathogenesis of these diseases is essential for the development of novel therapeutic strategies.

Next, the book focuses on the diagnosis of VL and PKDL, reviewing the various methods available for detecting *Leishmania* infection. It highlights the challenges in diagnosing these diseases, particularly in endemic regions with limited access to healthcare facilities. Improving diagnostic tools and techniques is essential for early detection and treatment of VL and PKDL.

The book also explores the current treatment options for VL and PKDL, including the use of antileishmanial drugs and other therapeutic approaches. It discusses the challenges in drug development and the emergence of drug resistance, as well as the need for new treatment strategies. Ensuring access to effective treatment is crucial for reducing the morbidity and mortality associated with VL and PKDL.

The book also focuses on disease control strategies for VL and PKDL, discussing the importance of integrated vector management, community engagement, and other public health interventions. It highlights the progress made in controlling these diseases in some regions and the challenges that remain in achieving sustainable control and elimination.

Overall, this book provides a comprehensive overview of the current state of knowledge in the field of VL and PKDL. It underscores the need for continued research and innovation to address the challenges posed by these diseases and emphasizes the importance of collaborative efforts between researchers, healthcare providers, policymakers, and communities. The editors and authors hope that this book will serve as a valuable resource for researchers, healthcare professionals, and policymakers working to control and eliminate VL and PKDL.

AIM AND SCOPE

Visceral Leishmaniasis and Post-kala-azar Dermal Leishmaniasis: Pathogenesis, Treatment and Disease Control aims to provide a comprehensive and up-to-date overview of the current understanding of VL and PKDL. These tropical diseases, caused by the *Leishmania* parasite, pose significant public health challenges in endemic regions. This book brings together leading experts in the field to discuss various aspects of these diseases, including their epidemiology, pathogenesis, diagnosis, treatment, and control strategies. By providing a multidisciplinary perspective on VL and PKDL, the book seeks to advance our understanding of these diseases and contribute to the development of effective control strategies.

Acknowledgements

It gives us immense pleasure to acknowledge the ICMR – Rajendra Memorial Research Institute of Medical Sciences, Agamkuan, Patna. Veer Singh acknowledges the Department of Health Research (DHR), Ministry of Health and Family Welfare, Government of India, for their support through the Young Scientist Fellowship Grant, R.12014/37/2022 HR. Ashish Kumar thanks the Indian Council of Medical Research (ICMR), Department of Health Research, Ministry of Health and Family Welfare, Government of India, for providing workplace and necessary research support.

The editors of this book are also grateful to Dr Krishna Pandey, Director, ICMR-RMRIMS, for his guidance and support during the completion of this book. It is also our great pleasure to acknowledge and express our enormous gratitude to all the contributors who have provided their quality material to prepare this book.

We are grateful to our beloved family members, who joyfully supported and stood by us during many hours of our absence while we worked to finish this book project. We are also grateful to our friends and colleagues who offered their support and encouragement throughout the writing process.

We would also like to acknowledge the publishing team at CRC Press for their professionalism, enthusiasm, and belief in this project. Their expertise in design, production, and marketing has been essential in bringing this book to life.

Editors

Dr Veer Singh is currently working as a Young Scientist (DHR) at the Department of Biochemistry, ICMR-Rajendra Memorial Research Institute of Medical Sciences, Patna, India. He obtained his PhD from the Indian Institute of Technology (Banaras Hindu University), Varanasi. He has authored more than 15 publications, and his area of work involves environmental contamination and toxicity, the role of heavy metal contamination in drinking water in visceral leishmaniasis (VL), and post-kala-azar dermal leishmaniasis (PKDL) development.

Dr Ashish Kumar is currently working as a Scientist-C and Head of the Department of Biochemistry, ICMR-Rajendra Memorial Research Institute of Medical Sciences, Patna, India. He has published more than 40 articles in international and national journals. He has received several awards and secured funding (sponsored projects) from many national and international agencies. Currently, four doctoral and two postdoctoral students are working under his supervision. Dr Kumar's work is in the interdisciplinary area of life sciences, mainly focused on VL and PKDL diagnosis, treatment, and environmental impacts on disease development.

Contributors

Kumar Abhishek
Department of Molecular Biology
ICMR-Rajendra Memorial Research Institute of Medical Sciences
Patna, Bihar, India

Ghufran Ahmed
Department of Biochemistry
ICMR-Rajendra Memorial Research Institute of Medical Sciences
Patna, Bihar, India

Dayakar Alti
Department of Immunology
ICMR-Rajendra Memorial Research Institute of Medical Sciences
Patna, Bihar, India

Mahesh S. Chivate
School of Biochemical Engineering
IIT (BHU)
Varanasi, Uttar Pradesh, India

Pradeep Das
ICMR-National Institute of Cholera and Enteric Diseases
Kolkata, West Bengal, India

Sapna Desai
Department of Pharmacology
Pioneer Pharmacy College
Vadodara, Gujarat, India

Himanshu Gaur
Department of Food Technology, School of Life Sciences
ITM University
Gwalior, Madhya Pradesh, India

John J. Georrge
Department of Bioinformatics
University of North Bengal
Raja Rammohunpur, West Bengal, India

Balashankar K.
Faculty of Bioinformatics
Sathyabama University
Chennai, Tamil Nadu, India

Rishikesh Kumar
Department of Clinical Medicine
ICMR-Rajendra Memorial Research Institute of Medical Sciences
Patna, Bihar, India

Shobha Kumari
Department of Biochemistry
ICMR-Rajendra Memorial Research Institute of Medical Sciences
Patna, Bihar, India

Vikash Kumar
Department of Biochemistry
ICMR-Rajendra Memorial Research Institute of Medical Sciences
Patna, Bihar, India

Pinki Kumari
Department of Biochemistry
ICMR-Rajendra Memorial Research Institute of Medical Sciences
Patna, Bihar, India

Moti Lal
Department of Biotechnology, School of Life Sciences
ITM University
Gwalior, Madhya Pradesh, India

Anukrti Mishra
School of Biotechnology
IFTM University
Moradabad, Uttar Pradesh, India

Vishal Mishra
School of Biochemical Engineering
IIT (BHU)
Varanasi, Uttar Pradesh, India

Saurav Kumar Mishra
Department of Bioinformatics
University of North Bengal
Raja Rammohunpur, West Bengal, India

Neelam
Department of Microbiology, School of Life Sciences
ITM University
Gwalior, Madhya Pradesh, India

Himanshu N. Panchal
Department of Pharmacology, PIP
Parul University
Vadodara, Gujarat, India

Krishna Pandey
Department of Clinical Medicine
ICMR-Rajendra Memorial Research Institute of Medical Sciences
Patna, Bihar, India

Gaurav Kumar Pandit
Department of Botany
Patna University
Patna, Bihar, India

Divya Prasad
Department of Zoology
Patna University
Patna, Bihar, India

Bhavani R.
Department of Virology
ICMR-Rajendra Memorial Research Institute of Medical Sciences
Patna, Bihar, India

Ravi Ranjan
Department of Biochemistry
ICMR-Rajendra Memorial Research Institute of Medical Sciences
Patna, Bihar, India

Madhvi Saxena
Department of Computer Science
Vishwakarma University
Pune, Maharashtra, India

Mayank Saxena
Department of Biotechnology and Department of Computer Science and Engineering
IILM University
Greater Noida, Uttar Pradesh, India

Prashant Saxena
Department of Biotechnology Engineering
University Institute of Engineering
Chandigarh University
Mohali, Punjab, India

Swati Sharma
Department of Virology
ICMR-Rajendra Memorial Research Institute of Medical Sciences
Patna, Bihar, India

Isha Singh
Department of Biotechnology and Department of Computer Science and Engineering
IILM University
Greater Noida, Uttar Pradesh, India

Kamal Singh
Department of Virology
ICMR-Rajendra Memorial Research Institute of Medical Sciences
Patna, Bihar, India

Meenakshi Singh
Department of Botany
Patna University
Patna, Bihar, India

Sakshi Singh
Department of Molecular Biology
ICMR-Rajendra Memorial Research Institute of Medical Sciences
Patna, Bihar, India

Vishal Singh
School of Biochemical Engineering
IIT (BHU)
Varanasi, Uttar Pradesh, India

Divya J. Tandel
Department of Pharmacology, PIP
Parul University
Vadodara, Gujarat, India

Pratibha Tiwari
Department of Biotechnology
IIMT College
Greater Noida, Uttar Pradesh, India

Ritesh Kumar Tiwari
Department of Botany
Patna University
Patna, Bihar, India

Gyanendra Tripathi
Department of Bioengineering
Integral University
Lucknow, Uttar Pradesh, India

Sonali Vedika
Department of Biochemistry
ICMR-Rajendra Memorial Research Institute of Medical Sciences
Patna, Bihar, India

Manisha Verma
School of Biochemical Engineering
IIT (BHU)
Varanasi, Uttar Pradesh, India

Kusum Yadav
Department of Biochemistry
University of Lucknow
Lucknow, Uttar Pradesh, India

Priyanka Yadav
Department of Biochemistry
University of Lucknow
Lucknow, Uttar Pradesh, India

1 General Introduction to Visceral Leishmaniasis

*Pinki Kumari, Sonali Vedika, and Ashish Kumar**
ICMR-Rajendra Memorial Research Institute of
Medical Sciences, Patna, Bihar, India
*Corresponding Author: Dr. Ashish Kumar, Scientist, ICMR-Rajendra
Memorial Research Institute of Medical Sciences, Patna, Bihar, India

1.1 INTRODUCTION

Leishmaniasis, a vector-borne disease, is caused by an obligate intra-macrophage protozoan parasite. It is a group of diseases with a diverse and complex nature (Tuon et al., 2008). It has been reported that more than 20 species of *Leishmania* are pathogenic for humans, and 30 sandfly (*Phlebotomus*) species are proven vectors of *Leishmania* species (Thakur, 2006). In the Indian subcontinent, the disease is transmitted anthropogenetically, but in the Mediterranean and Latin America, transmission is mostly zoonotic. A few *Leishmania* species are more associated with human hosts (*Leishmania tropica*), whereas some are classically associated with animal species (*Leishmania major*) and therefore considered zoonoses (Zijlstra et al., 2001). Leishmaniasis is classified as a neglected tropical disease and is listed among the six major health concerns (malaria, schistosomiasis, filariasis, trypanosomiasis, leishmaniasis and leprosy) world-wide bythe World Health Organization/Tropical Disease Research (WHO/TDR), affecting approximately 350 millions people across five vulnerable continents. The disease is prevalent in tropical and sub-tropical regions, span 97 countries in America, Africa, Asia, and Europe (Steverding, 2017). The three different major forms of it are visceral leishmaniasis (VL), cutaneous leishmaniasis (CL), and mucocutaneous leishmaniasis (MCL). The number of cases varies with time, and hence estimating it is challenging. For VL, the number of cases approximately per year may have decreased to 100,000, but previously reported cases were over 400,000. The number of cases for CL has been approximated from about 700,000 to 1.2 million or more (Thakur et al., 2020; Lata et al., 2021).

In the Western world, leishmaniasis is prevalent in the southern and central parts of America and Mexico, while in the Eastern world region, it is found in the Middle East, Africa, Asia, and Southern Europe (Chhabra and Singla, 2014). Uruguay and Chile have not reported cases. Infrequent incidences of CL have been noted in Texas and Oklahoma . Evidence shows that VL spread from regions in Southern India and became prevalent in Indian states like Bihar, Bengal, and Assam until the late 1990s (Singh, 2014). In recent times, the determinants and distribution of VL have been found in West Bengal, Jharkhand, Uttar Pradesh, and Bihar (Thakur et al., 2018). Unfortunately, Bihar continues to be the most affected state, with a caseload of 90%, despite many successful programs targeted at eliminating VL (Olliaro et al., 2018). There have been cases of resurgence of the disease in the states of Assam, Tamil Nadu, and Gujarat, and new regions with distressed conditions have also come to light during the last decade (Dhiman, 2014). Despite ten years of efforts to control and eradicate the disease at local, national, and international levels, the status of leishmaniasis is such that it is designated as one of the "neglected tropical diseases" by the World Health Organization (WHO).

DOI: 10.1201/9781032643120-1

The human-pathogen encounter is more likely if the reservoir host and vector stay in a common surrounding (Lemma et al., 2017). One of the two forms is anthroponotic, which causes the spread of contamination from one human to another, and the second form is zoonotic, responsible for transmission from animals to humans (Alborzi et al., 2006; Dhiman, 2014).

Different species of parasites have evolved to develop different invasion strategies to invade the host macrophages and modulate the secretion of cytokines and chemokines to establish the Th2-type immune system (Lukes et al., 2007).

Various diagnostics have been developed over the years for the diagnosis of leishmaniasis; still, efficiently diagnosing it has always been a major shortcoming due to the resemblance of its clinical features with some other prevalent diseases, such as leprosy, typhoid, and malaria. Microscopy is considered the gold standard for confirming the infection. Many drugs are available to treat the disease; however, due to the emergence of drug resistance, some drugs are no longer in practice or are being used with other drugs in combinatorial therapy. Presently, there is no vaccine available for leishmaniasis, though some potential vaccine candidates have been identified (Lemma et al., 2017).

1.2 HISTORY OF LEISHMANIASIS

The disease leishmaniasis was known by different names in the Indian subcontinent such as kala hazar, Burdwan fever, Assam fever, black fever, and Dum Dum fever. Old World VL, or kala-azar, was first noticed in Jessore (now in Bangladesh) in India in 1824 (Gibson, 1983). The disease was not identified or characterized for a long time. Later, Leishman discovered some peculiar bodies in the pulp of the spleen from a soldier who died of Dum Dum fever at Netley Hospital in Dum Dum, Kolkata (Leishman, 1903). However, the parasite was not correctly identified and was considered a trypanosome, a reported parasite in Africa around that time. Meanwhile, Donovan stumbled upon the parasites in tissue obtained from the spleen and in the blood of a young boy with infection and found the Leishman bodies as the agents causing kala-azar (Donovan, 1903). In 1903, Ronald Ross (1857–1932), who is considered the greatest authority on parasitology, put an end to all the controversy by jointly crediting Leishman and Donovan for the discovery of the parasite and naming the parasite as *Leishmania donovani* (Ross, 1903). Later, in 1922, the first case of post-kala-azar dermal leishmaniasis (PKDL) in a kala-azar patient was documented by Dr U.N. Brahmachari, and the patient was treated with tartar emetic, a trivalent antimony compound (Brahmachari et al., 1922) (Figures 1.1 and 1.2).

1.3 TYPES OF LEISHMANIASIS

The female sandflies of the genera *Phlebotomus* and *Lutzomyia* transmit the disease in the Old World and New World by transferring the protozoan parasite's geographically distinct species. Hence, apparent variations in the clinical presentations of the infected host are found (Cecilio et al., 2022). Three players determine the clinical outcomes of infection: vector, parasite species type, and the host's immune response towards the infecting parasite (Desjeux et al., 2001). For instance, *Leishmania donovani* and *Leishmania infantum* primarily cause VL, whereas *Leishmania major*, *Leishmania mexicana*, *Leishmania braziliensis*, and *Leishmania amazonensis* cause cutaneous lesions.

Leishmaniasis has three form classifications differing in symptoms and outcomes. In all forms of the disease, the presence of amastigotes in monocytes remains the authentication mark of the disease (Hide et al., 2007).

1.3.1 Cutaneous Leishmaniasis

Cutaneous leishmaniasis, the most common form of leishmaniasis, is caused by flagellated parasites *Leishmania tropica*, *L. major*, and *L. aethiopica* in the Old World, and *L. braziliensis* complex and

FIGURE 1.1 The discovery of the causative agent of leishmaniasis (kala-azar) is credited to William Leishman and Charles Donovan (Source: Bari, 2006)

L. mexicana complex in the New World (Perez-Franco et al., 2016). Reportedly, 20 different species of *Leishmania* are known to cause CL (Ahmed et al., 2023; WHO 2010). The New World form of CL is more severe and long-lasting than Old World CL. Algeria, Ethiopia, North Sudan, Iran, Syria, Colombia, Brazil, Costa Rica, Peru (Migone 1913), and Afghanistan account for 70% to 75% of the globally reported CL cases (Alvar et al., 2012). The development of ulcers or nodules on exposed parts of the body, such as the face, legs, and arms, is a common characteristic of the disease. The lesions are usually single and often self-healing; however, multiple ulcers can also appear as a result of multiple bites from the sandfly. Irrespective of the species, CL does not jeopardize the lives of individuals; however, lesions can lead to extreme mutilation and may result in stigma against the individuals in society (Ul Bari et al., 2006) (Figure 1.3).

1.3.2 Mucocutaneous Leishmaniasis (MCL)

Mucocutaneous leishmaniasis (MCL) is the extended infection case of CL in which partial or total destruction of the mucous membranes of the nose, mouth, and throat cavities is seen (David CV et

FIGURE 1.2 U.N. Brahmchari reported first PKDL cases in Indian Subcontinent. (Source: https://www.getbengal.com/details/despite-saving-more-than-50-lakh-lives-from-kalazar-u-n-brahmachari-was-denied-nobel-prize)

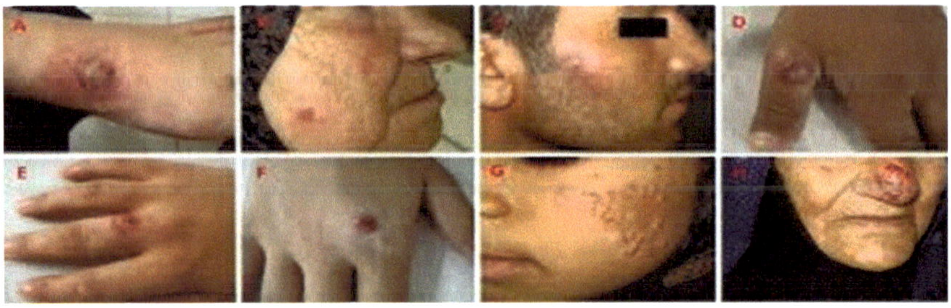

FIGURE 1.3 Skin lesions of Cutaneous leishmaniasis (CL) (Source: Karimi et al., 2021)

al., 2009). Bony parts are not likely to be affected. It is also known as espundia. It is related to infections by the New World's *L. braziliensis* and also of *L. (V.) panamensis*, *L. (V.) guyuanensis*, and *L. (V.) amazonensis* (Herwaldt, 1999). It can be a threat to the survival of the patient if accompanied by malnutrition and acute respiratory pneumonia. Currently, some South American countries are so vulnerable to the disease that 90% of all MCL occurs in Bolivia, Peru, and Brazil (Figure 1.4).

1.3.3 Visceral Leishmaniasis (VL)

VL is a potentially fatal vector-borne disease. It is ranked second in mortality among tropical diseases. According to the WHO, in 2020, approximately 13,000 cases of VL occurred. It is reported as endemic in more than 70 countries, with 200 million people vulnerable across every continent

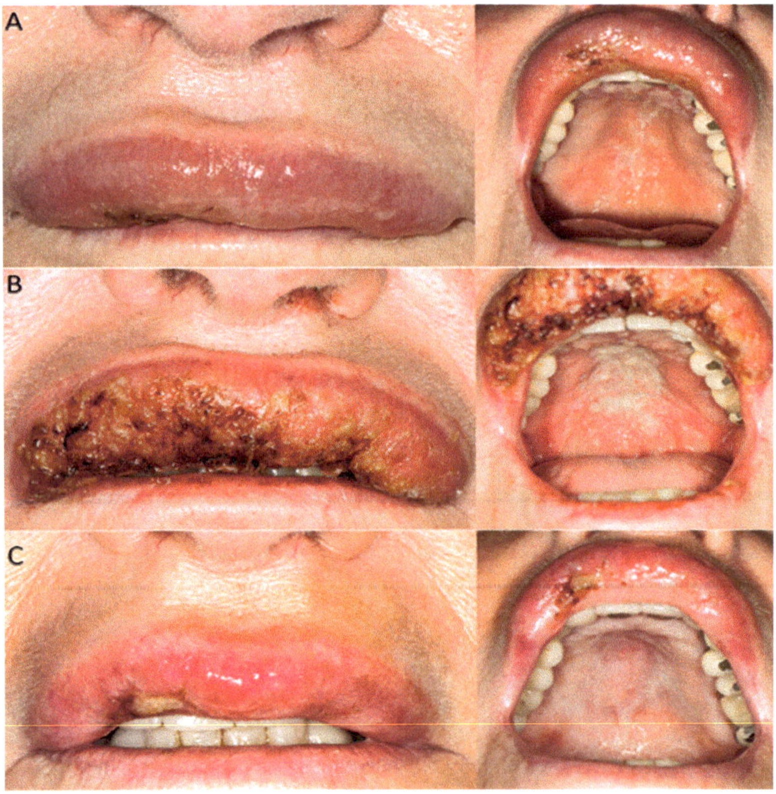

FIGURE 1.4 Skin lesions of Mucocutaneous Leishmaniasis (MCL) (Source: Linse et al., 2022)

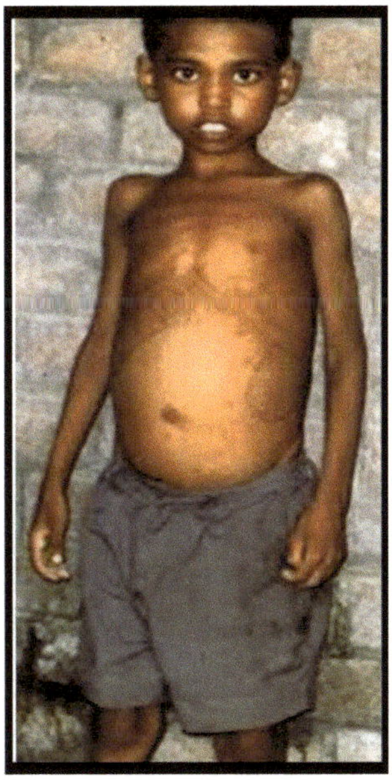

FIGURE 1.5 A typical Visceral Leishmaniasis patient with enlarge with liver and spleen (Source: https://ncvbdc.mohfw.gov.in/)

except Antarctica and Australia. Brazil (Martins-Melo et al., 2014), Ethiopia, India, Kenya, Somalia, South Sudan, and Sudan (Zijlstra et al., 2001) have reported more than 90% of the VL cases worldwide. VL is often underestimated but severe, with some 500,000 new cases and 50,000 deaths occurring annually (Martins-Melo et al., 2014; Migone, 1913).

The main causative agents of VL are *Leishmania donovani* and *Leishmania infantum*, also known as *Leishmania chagasi* in South America. In Southeast Asian countries, particularly India, Bangladesh, and Nepal, and in Eastern Africa, especially Sudan, Ethiopia, Kenya, and Somalia, the disease is caused by *L. Donovani*; (Lukes et al., 2007) however, in the Mediterranean area, the Middle East, Afghanistan, Iran (Alborzi et al., 2006), Pakistan, and Brazil, the species *L. infantum* is considered to be responsible for the infection. Bihar in India is an endemic focus for VL, as are the border districts of Jharkhand, West Bengal, and parts of Uttar Pradesh (Chappuis et al., 2007) (Figure 1.5).

A variation in the incubation period has also been observed. Incubation time ranges from as little as a few weeks up to six months or sometimes many years. VL is usually characterized by an enlarged liver and spleen, fever, and weight loss, and if not treated, can be potentially deadly because of infectious or hemorrhagic complications (Scarpini et al., 2022). In some cases, it may be asymptomatic or latent and appear after years if the person becomes immunodeficient (Perez-Franco et al., 2016; Ramsh et al., 1995).

1.3.4 Post-kala-azar Dermal Leishmaniasis

PKDL is very likely the aggravation of VL caused by *Leishmania donovani*. It is taken as a general health issue in the areas with endemicity of VL, as PKDL is considered to have a crucial role in transmission.

It is characterized by a rash on the skin in the form of macules, papules, or nodules in patients (Brahmachari, 1922; Zijlstra et al., 2003). It may appear in patients who are given medical treatment against VL (Ramsh et al., 1995). There may be chances of reinfection after VL. Some evidence suggests that PKDL is usually confined to following VL caused by *L. donovani* and, therefore, occurs

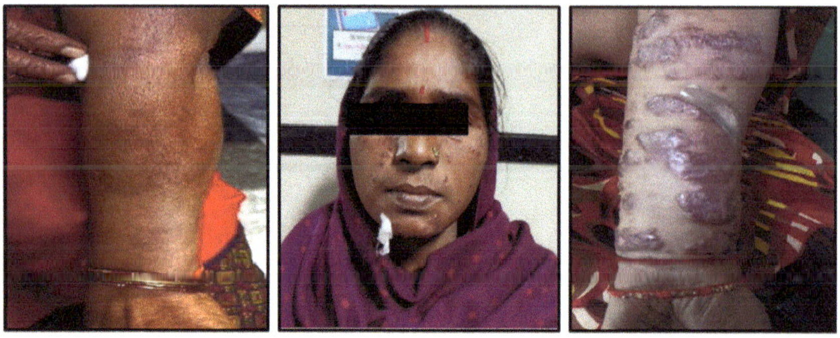

Macular/Hypopigmented Lession **Papular** **Nodular**

FIGURE 1.6 A typical Post kala azar dermal leishmaniasis (PKDL) patients with pigmented macules, papules and nodules (Source: Division of Biochemistry RMRIMS Patna, Bihar, India)

in Africa and Asia. However, significant differences in epidemiological and clinical manifestations have also been reported. The most obvious are the variations in the clinical presentation; for instance, 90% of rashes are papular in Africa, whereas 90% are in macular form in Asia, and the duration between VL and PKDL is 0–13 months in Africa but often 2–3 years in Asia. Surprisingly, patients develop immunity similar to cured VL patients after successful treatment. The immunity has permanent sustenance unless there is a condition of immunosuppression (Brahmachari, 1922; Zijlstra et al., 2003).

There have also been cases where patients present with VL and PKDL at the same time (para-kala-azar leishmaniasis), and surprisingly, around 10% of patients with PKDL didn't have a previous history of VL (Figure 1.6).

1.4 MORPHOLOGY OF THE *LEISHMANIA* PARASITE

Leishmania species depend on two hosts, i.e., a vertebrate and an invertebrate vector. The organism survives in two basic organizations – intracellular non-motile amastigotes inside the vertebrate macrophage and extracellular elongated motile promastigotes in the gut of the sandfly vector (Tuon et al., 2008). Amastigote forms are taken up when the sandfly consumes blood meal from an infected vertebrate host. In the gut of the sandfly, these amastigotes develop into promastigotes, where they multiply by binary fission. The amastigotes undergo a change to procyclic promastigotes in the region of posterior midgut of the sandfly without taking longer than a few hours after the intake of blood meal. This is followed by transformations into nectomonad (elongated, 15–20 µm), haptomonad (attached to stomodeal valve), and metacyclic (highly motile, long flagella, 5–8 µm) promastigote forms sequentially. In this form, the nucleus is located in the center, and the kinetoplast is generally observed as one oriented transversely near the anterior end. The single and delicate flagellum measures 15–28 µm in length. A characteristic feature of the protozoan parasites can be the presence of a kinetoplast, i.e., a granule consisting of DNA placed within the solitary mitochondrion that, in turn, is associated with the flagellar bases. The kinetoplast DNA has thousands of circular DNA molecules that show catenation properties. These DNA grids have two forms: one has a kinetoplast with 25–250 large circles of approximately 30 kb in size and 5,000–10,000 small circles of about 2 kb in size. Also, a distinctive organelle of kinetoplastids called the glycosome is found together with lysosomes in the cytoplasm (Michels et al., 1991).

Amastigotes appear elliptical and lack flagella, measuring 3–5 µm in length and 1.5–3 µm in width. They are observed to have a "flagellar pocket," which has an infolding of the superficial layer

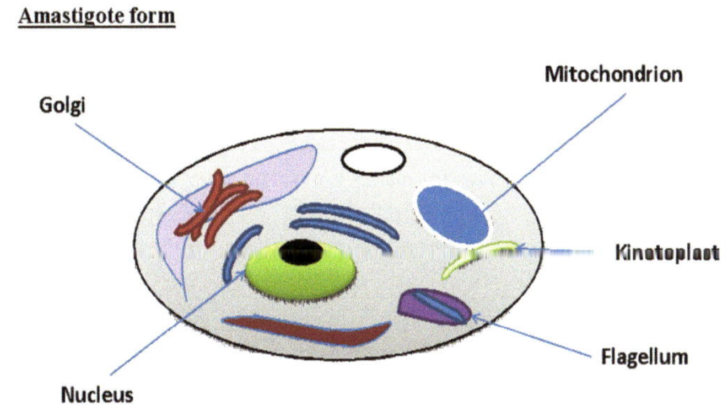

FIGURE 1.7 Morphology: Amastigote form of *Leishmania* parasite

forming an inner volume. Since the flagellum ceases to enlarge on the farther side of the cell body, it is not operative in amastigotes. The pocket offers a position where endocytosis and exocytosis occur (Webster et al., 1993). Leishman-Donovan bodies is another name given to the amastigotes. The distinguished structural feature of the amastigote form is found to be a centrally situated round or oval nucleus, and close to it, a smaller round or rod-shaped kinetoplast (Figure 1.7).

Promastigotes are present in the digestive tract of sandfly vectors and can be cultured artificially. They consist of flagella, and their shape resembles a spindle. They are 15–20 μm in length and 3–5 μm in width. The flagellum is anteriorly located on the body and provides motility. Glycol conjugate lipophosphoglycan (LPG) densely covers the surface of the parasite, including flagellum (Bates, 2008). Procyclics are immature forms of the promastigotes, expressing shorter LPG molecules, but the mature metacyclic infective form expresses higher LPG molecules on their surface. These LPG molecules on the surface of metacyclic promastigotes shield the parasite from complement-mediated killing in the vertebrate host. In vitro conditions also readily allow a transformation from amastigote to promastigote form in 24–48 hours if a commodious culture medium and a temperature of 26–28°C are provided. These promastigotes are not morphologically different from those grown in culture (Figure 1.8).

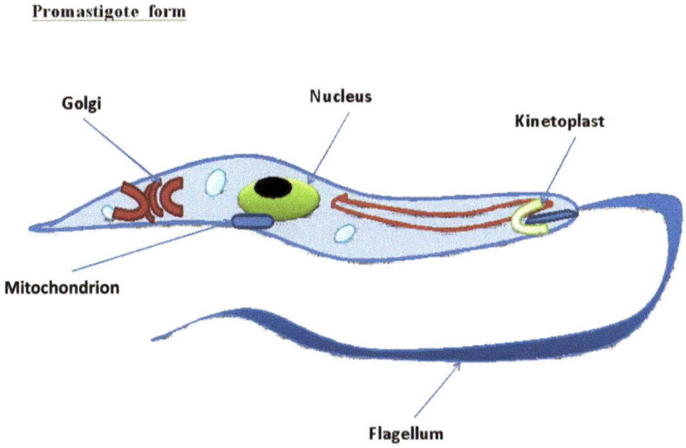

FIGURE 1.8 Morphology: Promastigote form of *Leishmania* parasite

1.5 VECTOR OF *LEISHMANIA*: SANDFLIES

The vector sandfly shares the Psychodidae family with non-biting, non-vector flies. The sub-family Phlebotominae includes vectors of the protozoan parasite *Leishmania* species, the causative agents of leishmaniasis. Surprisingly, in parts of South America, they are known to spread the bacterium *Bartonella bacilliformis*, which causes Carrion's disease, as well as the Pappataci virus (an arbovirus) responsible for sandfly fever (also known as Pappataci fever) (Sharma et al., 2008). Approximately 800 sandfly species have been recognized, and only some 10% of them are reported to act as vectors, with about 30 being medically significant (Sharma et al., 2008).

Like all true Diptera flies, sandflies undergo four stages of complete metamorphosis – from egg to larva to pupa and further to adult. Unlike mosquitoes, they do not need stagnant water for the development of the immature stages but require relatively warm, moist environments. Female adult flies lay 15–80 tiny eggs in a suitable habitat, such as one with a high organic content like animal excreta and soil. Eggs are banana-shaped and measure 0.3–0.5 mm in length. The color of eggs varies in the initial stages from white to light gray depending on the species; however, eggs can turn black or brown within a few hours. On average, 6–17 days are required to hatch of eggs in an incubation temperature-dependent way (Sharma et al., 2008).

Adult sandflies usually have a length comparable to one-third the length of an average mosquito, i.e., less than 3.5 mm in length. Both female and male sandfly adults feed on plant juice for nutritional requirements; however, most female flies also require at least one blood meal to fulfill the need for complete growth and development of a cluster of eggs. Sandflies are very prone to dehydration; therefore, most are active at dawn and dusk. They look for shelter in protected natural habitats with organic content. They are not good at flying; hence, they hop, usually close to the ground (Cecílio et al., 2022). Sandfly species of the Old World live in desert and semi-arid ecosystems, whereas New World species are forest dwellers. Their bodies and small wings are densely covered with hair, and they are found to hold their wings upright in a V-shape during periods of rest (Cecílio et al., 2022).

The early studies performed on the taxonomy of phlebotomine sandflies were extensively based on the of morphological considerations of non-living specimens. The approach to studying sandflies has shifted from observing the morphological features of dead specimens to laboratory rearing and colonization, chromosome analysis, multivariate morphometrics, isoenzyme analysis, molecular and phylogenetic analysis, and, more recently, mass spectrometry. The sandflies that transmit leishmaniasis are classified in two genera: *Phlebotomus* in the Old World and *Lutzomyia* in the New World (Galati et al., 2003; Killick-Kendrick, 1994) (Figure 1.9).

1.6 LIFE CYCLE OF *LEISHMANIA*

The *Leishmania* parasite follows a digenetic life cycle involving two stages without a sexual stage. The developmental stage of the parasite called metacyclic promastigotes is injected into the human host via the proboscis, while the vector female phlebotomine sandfly takes blood as its meal; hence, the promastigotes are designated as the infective stage of the parasite. These promastigotes at the site of the wound are phagocytized by macrophages and, under the developmental process, get transformed into amastigotes within 12–24 hours. These amastigotes increase in number in infected cells and start affecting different tissues in an infection-dependent manner. Binary fission is employed to divide after the amastigotes have grown sufficiently within the phagolysomal compartment of the human host. The heavily populated phagolysomes burst, releasing amastigotes into blood circulation, and new infections of uninfected cells continue. Their course from the blood to the liver, spleen, and bone marrow is observed. The cycle begins again after the parasites are taken by the sandfly to be transmitted to a new host. The sandfly vectors thus act as carriers that spread the infecting amastigotes present in an infected host with infection to healthy hosts belonging to the same species or other species (Galati, 2003; Killick-Kendrick, 1994). The life cycle of *Leishmania donovani* is shown in Figure 1.10.

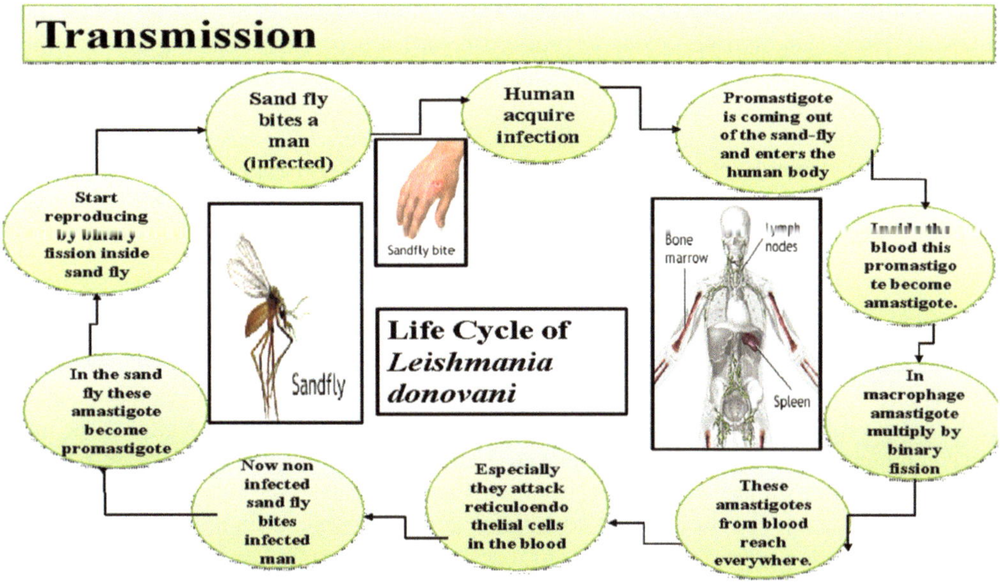

FIGURE 1.9 Diagrammatic representation of transmission of Leishmaniasis by sandfly

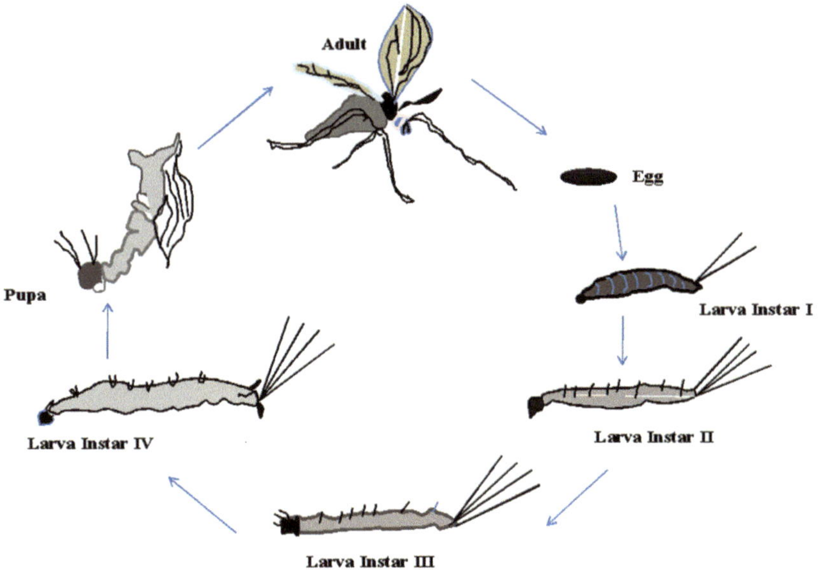

FIGURE 1.10 Diagrammatic representation of the life cycle of *Leishmania*

1.7 DIAGNOSIS OF LEISHMANIASIS

Early diagnosis and treatment are recommended to overcome the challenges associated with leishmaniasis. In addition, an appropriate prognostic marker may prevent treatment failures or limitations. Clinical features of leishmaniasis resemble those of other usual diseases such as malaria, typhoid, tuberculosis, tropical splenomegaly, schistosomiasis, brucellosis, bacterial endocarditis, histoplasmosis, malnutrition, lymphoma, and leukemia, and many of these diseases appear as co-infections with VL (Herwaldt et al., 1999). Clinical characteristics suggest that VL patients suffer from fever, splenomegaly, pancytopenia, hypoalbuminemia, and hypergammaglobulinemia (Berman, 1997; Thakur et al., 2004). Several different methods for visualizing parasites in suspected infected tissues have evolved, including light microscopy of the stained specimen (Reimão et al., 2020), in vitro culturing, animal inoculation, and immunodiagnosis by detection of parasite antigens in tissue, blood, or urine samples, by detection of nonspecific or specific antibodies against *Leishmania* (immunoglobulin), or by assays exploiting the principle of cell-mediated immunity (Berman, 1997).

1.7.1 Microscopic Examination

Microscopic visualization is the most dependable and typical method practiced in the diagnosis of VL. These tests have 100% specificity with variable sensitivity. The amastigotes of *Leishmania* are demonstrated using a light microscope in splenic/bone marrow aspirate or lymph node aspirate, or the buffy coat of peripheral blood (for example, in the case of HIV co-infection). The amastigotes, commonly known as Leishman-Donovan (LD) bodies, can be easily seen in smears prepared from infected tissue stained with Giemsa stain. This staining enables the cytoplasm to conveniently become pale blue and the nucleus and kinetoplast to turn a purple or pink color. The sensitivity of amastigotes in the splenic aspirate smear ranges from 93.1% to 98.7%; it is considered the gold standard in the confirmation of kala-azar. Additionally, the smear from the bone marrow has a sensitivity that varies from 52% to 85%. However, sample collection from bone marrow or the spleen involves significant risk, and involving this invasive procedure in the observation of asymptomatic infection is not dependable. The diagnosis of PKDL is made by validating parasites in the smear or biopsy specimen obtained from the skin of the suspected individual. Parasites may be scant in PKDL; however, the yield from nodules is higher than from macules. The severity of the disease can be measured based on the LD body count in the biopsy smear (Berman, 1997). One way to measure severity is to draw a density score of the parasite based on a scale that ranges from 0 (no parasites per 1,000 oil-immersion fields) to +6 (>100 parasites per field) (Table 1.1).

1.7.2 Parasite Culture

The micoscopic method has typically been used to recognize cases with VL. The parasite is usually confirmed in the aspirate obtained from the spleen or bone marrow and in the buffy coat of peripheral blood (especially in the case of HIV co-infection). Light microscopy of the specimen can be used to identify amastigotes or *Leishmania donovani* bodies produced by staining with Giemsa or Leishman stain. The smear of splenic aspirate gives a sensitivity exceeding 95%. It is the gold standard practice for diagnosing kala-azar. There is also a need for culture so that a sufficient number of

TABLE 1.1
LD Count Designation of *Leishmania* Parasites

LD Body Designation	Number of Parasites per Field
6+	>100 parasites per field
5+	10–100 parasites per field
4+	1–10 parasites per field
3+	1–10 parasites per 19 fields
2+	1–10 parasites per 100 fields
1+	1–10 parasites per 1,000 fields
0	0 parasites per 1,000 fields

organisms can be obtained for an antigen to be used in immunologic diagnosis and further speciation. It can be used for inoculation in animals that are susceptible to it, in experiments, for in vitro drug sensitivity studies, and so on (as a supplement to other methods or in case of failure of other routine methods). The preferred culture media are either monophasic (Schneider's insect medium, M199, or Grace's medium) or diphasic (Novy-MacNeal-Nicolle medium and Tobie's medium). The diphasic medium commonly contains rabbit blood (30%) and agar (70%) coated with RPMI 1640. Inoculation with a volume as small as one to two drops of bone marrow or splenic aspirate to check the growth of parasites in culture tubes, followed by incubation at 25°C in a biological oxygen demand (BOD) incubator and observation under a microscope, is also preferred.

1.7.3 Molecular Diagnosis

1.7.3.1 Polymerase Chain Reaction (PCR)

PCR may be used in the diagnosis and prognosis of disease, strain identification, and detection (Reimão et al., 2020). Several target sequences have been used for PCR, including ribosomal RNA genes, kinetoplast DNA (KDNA), mini-exon derived RNA (med RNA) genes, genomic repeats, the beta-tubulin gene region, the glycoprotein 63 gene locus, and internal transcribed spacer (ITS) regions. The sensitivity of PCR for detecting of *Leishmania* DNA from confirmed VL patients in bone marrow, lymph node aspirate, and blood is respectively 82%, 86.8%, and 70%–99%. PCR is useful for the confirmation of the diagnosis in immunocompromised (HIV and *Leishmania* co-infected) patients. It has been reported that 75% of co-infected patients were PCR-positive (Dixit et al., 2020). The efficacy of specific therapy is evaluated only through clinical parameters because all available antibody-based diagnostic methods fail to distinguish between past and present infection and are unreliable in the case of immunocompromised patients. Although PCR-based methods give higher sensitivity (93.80%), its sensitivity for detecting Leishmanial DNA in the blood of parasitologically proven VL cases was only 70%. In addition, blood (38.2%) and bone marrow (29.9%) samples from non-VL patients were shown to be positive for leishmanial DNA. Hence, an optimized PCR would be of great importance as a diagnostic tool, especially in the prognosis of kala-azar (Lainson, 2010).

1.7.3.2 RT-PCR

Quantitative PCR (Q-PCR) has emerged as an alternative diagnostic tool that addresses many of the limitations of conventional PCR. It requires less amplification time and can accurately quantify parasite load in clinical samples.

It has been demonstrated that real-time PCR is more accurate than traditional PCR in diagnosing leishmaniasis. Several SYBR green and TaqMan chemistry-based single and multiplex real-time PCR methods have shown significant sensitivity and specificity for diagnosing PKDL.

Over the years, its popularity has grown because of its great sensitivity, speed, quantification, high throughput potential, and automation (Hossain et al., 2017). Moreover, because the tubes do not need to be opened for post-PCR analysis, Q-PCR may be applied to a broad variety of samples and it significantly lowers the chance of cross-contamination. It is used quite extensively in referral labs. As with traditional PCR, Q-PCR uses the identical nucleotide sequence but analyzes the fluorescence signal generated by TaqMan, fluorescence resonance energy transfer (FRET), minor groove binder (MGB), or by utilizing intercalating fluorescent dyes like SYBR green.

For example, using kDNA as a target for Q-PCR to diagnose VL and track parasite dynamics has been documented in a number of investigations. With the highest limit of detection up to 5×10^4 parasites per PCR reaction tube, the kDNA target allows for the detection of less than 1 parasite/mL of blood (0.0125 parasites/mL or 12.5 parasites/µl of blood). For the detection of VL utilizing peripheral blood, kDNA-based Q-PCR has been shown to have a sensitivity of 100%. A corresponding splenic score showed a strong correlation with the number of parasite genomes per milliliter of blood (Galluzzi et al., 2018; Reimão et al. 2020).

Melt curve analysis can be used to discriminate between different species of *Leishmania* by using the 18S rRNA sequences. ITS regions with varying sequences are used to type species. Additionally, a large number of protein-coding genes, including heat shock protein 70 (HSP70), can be utilized.

Several studies have indicated that Q-PCR has also shown promising potential for use as a tool to assess cures. Q-PCR can reveal a full recovery or provide early warning signs of treatment failure or recurrence. The high cost of reagents and equipment makes Q-PCR a difficult tool to use in field settings, even with its impressive advancements. As a result, it is still far from being a routine clinical application.

1.7.3.3 Loop-Mediated Isothermal Amplification (LAMP)

LAMP is expected to be a novel and inventive method for amplifying DNA quickly and with high specificity in isothermal circumstances when Bst DNA polymerase, which has a special strand-displacing activity, is present. Neither complex analysis nor high-tech equipment is needed for it. Four to six distinct primers, each intended to detect six to eight sequences, are used in this approach. Moreover, the inhibitory substances found in DNA samples have less impact on sensitivity. Numerous investigations have demonstrated the efficacy of LAMP in leishmaniasis diagnosis. Additionally, the test demonstrated potential utility in evaluating cure in post-treatment phases; of the patients followed up with over a long period, two experienced relapses. Four patients tested positive (Wong et al., 2018).

Additionally, the assay has been effectively validated at two endemic sites in India, augmenting its usefulness for mass surveillance of leishmaniasis. In order to use this assay in field settings for mass screening of *Leishmania* infection in endemic areas, the industry has already seized upon it and developed it into a kit format. In an attempt to avoid the requirement for DNA isolation and complete the assay faster and cheaper, the use of direct crude clinical samples rather than extracted DNA has also been investigated. A 93.06% sensitivity was obtained for diagnosing VL using LAMP based on direct blood lysis. This is encouraging. Comparably, direct sample lysis with slit aspirate for PKDL cases could be investigated and improved as a less intrusive method (Wong et al., 2018).

1.7.3.4 Recombinase Polymerase Amplification

Recombinase polymerase amplification (RPA) is a molecular tool of choice that offers rapid, specific, and cost-effective identification of pathogens. Researchers are facilitated by minimal sample preparation requirements, low operation temperature (37–42°C), and commercial availability of freeze-dried reagents. The method proved effective when applied outside laboratory settings, in remote areas, and, interestingly, onboard automated sample-to-answer microfluidic devices (Daher et al., 2016).

RPA entails two primers and one probe (optional) with simple design requirements. Recombinase enzymes with accessory proteins are used for DNA unwinding and primer annealing. RPA has high specificity and efficiency (10^4-fold amplification in ten minutes) and displays a completely isothermal profile without requiring an additional temperature step for DNA denaturation. The required agents for RPA are available in dried formats; hence, their application becomes efficient in diagnostics. Thus, RPA is undoubtedly a promising isothermal molecular technique for clinical microbiology laboratories and emergency response in clinical settings (Daher et al., 2016).

1.7.4 Immunological Methods of Diagnosis

The immunological method has undergone considerable upgrades in recent years. It is worth appreciating that we have developed a number of tests with desirable specificity and sensitivity. The most practiced techniques are indirect immunofluorescence (IFI), immunosorbent assay (ELISA), multiplex assay, and electroimmunotransference (EIT). In each one of them, a different method is used, but they all employ the common principle of antigen-antibody binding.

1.7.4.1 Antigen Detection Method

The abundance of antibodies, circulating immune complexes, serum amyloid, rheumatoid factor, and self-antibodies happens to shield immunologically important determinants of antigens or competitively prevent the binding of free antigens, and therefore it becomes challenging to detect antigens in the serum of patients. *Leishmania* antigens can possibly be detected by a new latex agglutination test (KAtex) in urine samples of patients who are suspected of having kala-azar. In initial trials, 100% specificity and sensitivity between 65% and 100% (Sarkari et al., 2002; Rijal et al., 2004) have been found. However, the appearance of high false positives was major drawback of this test. Nonetheless, boiling urine or pre-treating urine with dithiothreitol (DTT) could potentially reduce the false positives by 73% as represented by Hatam et al. (2009).

1.7.4.2 Antibody Detection Tests

A number of tests have evolved to detect the presence of antibodies to recognize leishmaniasis. Among these, the 39-amino acid repeated conserved kinesin gene (rK39) sequence-based strip test is, as of now, the most suitable available diagnostic tool for VL being used in far-flung areas. The test gave sensitivity and specificity estimates of 93.9% and 95.3%, respectively (Sundar et al., 1998). However, in East Africa, it is reported that this test is less accurate (Boelaert et al., 1999). Moreover, a significant number of individuals with no symptoms gave positive results with the rK39 strip, and also, the sero-prevalence in people claimed as healthy individuals varies, ranging from less than 10% to up to more than 30%. Therefore, it becomes challenging to distinguish relapse by this strip (Figure 1.11).

1.7.4.3 Direct Agglutination Test (DAT)

The direct agglutination test for VL was developed in East Africa by El Harith et al. in 1986 with high sensitivity (94.8%) and specificity (97.1%) (Harith et al., 1986), and is validated by other laboratories (Singla et al., 1993; Schoone et al., 2001). Samples of plasma and serum can be used to carry out the test. A freeze-dried form or suspension of whole, stained promastigotes is used in the method either way. The presence of antibodies against *Leishmania* in serum samples and blood collected on filter paper can be rapidly detected in less than three hours by the fast agglutination-screening test (FAST) method developed by Schoone et al. The improved version of DAT using freeze-dried antigen (Boelaert et al., 1999) has recently been developed. As far as the accuracy of diagnosis is concerned, the DAT is an excellent candidate, but the lack of standardized antigens makes its use in India very limited. The adaptation of this test from lab to field is not convenient.

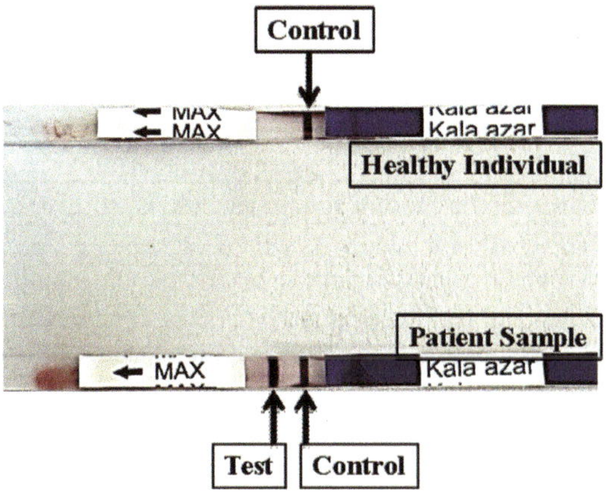

FIGURE 1.11 Serodiagnosis of leishmaniasis using rK39 strip. (Source: Division of Biochemistry RMRIMS Patna, Bihar, India)

TABLE 1.2
Drugs for the Treatment of Leishmaniasis

Drug	History	Dose	Drawback
Sodium antimony gluconate	First-line drugs	Intramuscularly for 28 days at a dosage of 20 mg/kg body weight	Drug resistance; severe side effects
Miltefosine	Earliest potent drug administered orally; came into being as a drug against cancer	For adults weighing more than 25 kg, 100 mg daily as one capsule (50 mg) in the morning and one capsule in the evening after meal for 28 days; for adults weighing less than 25 kg, 50 mg as one capsulr daily in the morning after meal for 28 days; a daily amount of 2.5 mg in pediatric cases if provided for a month	Teratogenicity; gastrointestinal disturbances
Amphotericin-B and its liposomal combinations	Came into being as a drug against fungal infection	10 mg per kg body weight single dose of Ambisome is the standard drug regimen for VL treatment; 0.5 mg per kg body weight per day (on days 1–20, 50–70, and 100–120) for PKDL treatment	Nephrotoxicity; prolonged treatment duration

Also, it is time-consuming (requires 16–18 hours' incubation), and test results must be evaluated by the naked eye, therefore requiring expertise, which adds to its drawback (Table 1.2).

1.8 TREATMENT

Over the years, various drugs are being developed and some are re-purposed to be used in treatment of leishmaniasis. These include Sodium antimony gluconate, Amphotericin B-deoxycholate, liposomal Amphotericin B and Miltefosine to name a few. Amphotericin B (AmB) became the first choice drug with the increasing unresponsiveness to antimony. Amphotericin B acts on perdominantly expressing membrane sterols such as ergosterol and episterol of parasites. Miltefosine is considered to be the earliest potent drug administered orally for leishmaniasis. However, these drugs come with challenge of relapse and unresponsiveness over the years. Some commonly practised drugs with their prescribed doses have been tabulated in Table 1.2.

1.9 CONCLUSION

Leishmaniasis continues to be a challenging infection. There is a chance of relapse, particularly in cases where the immune system is suppressed. This is why secondary prophylaxis is required in this condition. We face cases where the appearance of leishmaniasis with other diseases complicates the approach to tackling them. In endemic countries struggling with drug abuse, drug users who use injections create a potential for higher incidence rates. Thus, human-to-human transmission is a major risk factor. Since no efficacious vaccine is available, and since climate change has likely led towards endemicity, the challenges have worsened. Therefore, the disease must be addressed by potential drug treatments, extensive research, and periodic awareness among people, especially those who are vulnerable. However, new vaccine agents with the potential to tackle the issue of leishmaniasis may open new doors and change the way the disease is handled and managed (Desjeux, 2001; David et al., 2009).

ACKNOWLEDGMENT

The authors are thankful to the ICMR-Rajendra Memorial Research Institute of Medical Sciences, Patna, for its necessary support during this study.

REFERENCES

Ahmed G, Jamal F, Tiwari RK, Singh V, Rai SN, Chaturvedi SK, Pandey K, Singh SK, Kumar A, Narayan S, Vamanu E. Arsenic exposure to mouse visceral leishmaniasis model through their drinking water linked to the disease exacerbation via modulation in host protective immunity: a preclinical study. *Scientific Reports* 2023; 13(1): 21461. https://doi.org/10.1038/s41598-023-48642-z

Alborzi A, Rasouli M, Shamsizadeh A. Leishmania tropica-isolated patient with visceral leishmania in southern Iran. *Am. J. Trop. Med. Hyg.* 2006; 74(2): 306–307.

Alvar J, Velez I, Bern C, Herrero M, Desjeux P, Cano J, Jannin J, Boer M. Leishmaniasis worldwide and global estimates of its incidence. *PLoS One*. 2012; 31(5). doi: 10.1371/journal.pone.0035671

Bates PA. Leishmania sand fly interaction: progress and challenges. *Curr. Opin. Microbiol.* 2008; 11(4): 340–344. doi: 10.1016/j.mib.2008.06.003

Berman JD. Human leishmaniasis: clinical, diagnostic, and chemotherapeutic developments in the last 10 years. *Clin. Infect. Dis.* 1997; 24: 684.

Boelaert M, El Safi S, Jacquet D, De Muynck A, Van Der Stuyft P Le Ray D. Operational validation of the direct agglutination test for diagnosis of visceral leishmaniasis. *The American Journal of Tropical Medicine and Hygiene*. 1999; 60: 129–134.

Brahmachari U. A new form of cutaneous leishmaniasis—dermal leishmanoid. *Indian Med. Gazette.* 1922; 57(4): 125–127.

Cecílio P, Cordeiro-Da-Silva A, Oliveira F.Sand flies: Basic information on the vectors of leishmaniasis and their interactions with Leishmania parasites. *Communications Biology* 2022; *5*(1): 305. https://doi.org/10.1038/s42003-022-03240-z

Chappuis F, Sundar S, Hailu A, Ghalib H, Rijal S, Peeling RW, Boelaert M. Visceral leishmaniasis: what are the needs for diagnosis, treatment and control? *Nat. Rev. Microbial.* 2007; 5:873–882.

Chhabra MB, Singla LD. Leishmaniasis. In: Garg SR, editor. Zoonosis: parasitic and mycotic diseases. New Delhi: Daya Publishing House; 2014. pp. 134–147.

Daher RK, Stewart G, Boissinot M, Bergeron MG. Recombinase polymerase amplification for diagnostic applications. *Clin. Chem.* 2016; 62(7): 947–958. doi: 10.1373/clinchem.2015.245829

David CV, Craft N. Cutaneous and mucocutaneous leishmaniasis. *Dermatol. Ther.* 2009; 22(6): 491–502.

Desjeux P. The increase in risk factors for leishmaniasis worldwide. *Trans. R. Soc. Trop. Med. Hyg.* 2001; 95: 239–243.

Dhiman RC. Emerging Vector-Borne Zoonoses: Eco-Epidemiology and Public Health Implications in India. *Frontiers in Public Health* 2014; 2: 168. https://doi.org/10.3389/fpubh.2014.0016:

Dixit KK, Singh R, Salotra P. Advancement in molecular diagnosis of post kala-azar dermal leishmaniasis. *Indian J. Dermatol.* 2020; 65(6): 465–472. doi: 10.4103/ijd.IJD_311_19

Donovan C. Memoranda: on the possibility of the occurrence of trypanomiasis in India. *Br. Med. J.* 1903; 196:201–202.

Galati, Eunice Aparecida Bianchi. "Classificação de phlebotominae." *Flebotomíneos no Brasil.* 2003; 23–52.

Galluzzi L, Ceccarelli M, Diotallevi A, Menotta M, Magnani M. Real-time PCR applications for diagnosis of leishmaniasis. *Parasit. Vectors.* 2018; 11(1): 273. doi: 10.1186/s13071-018-2859-8

Gibson ME. The identification of kala-azar and the discovery of *Leishmania donovani. Med. Hist.* 1983; 27: 203–213.

Harith AE, Kolk AHJ, Kager PA. A simple and economical direct agglutination test for sero-diagnosis and sero-epidemiological studies of visceral leishmaniasis. *Trans. R. Soc. Trop. Med. Hyg.* 1986: 80(4): 583–587. doi: 10.1016/0035-9203(86)90149-5

Hatam GR, Ghatee MA, Hossini SM, Sarkari B. Improvement of the newly developed latex agglutination test (Katex) for diagnosis of visceral lieshmaniasis. *J. Clin. Lab. Anal.* 2009; 23(4): 202–205.

Herwaldt BL. Leishmaniasis. *Lancet.* 1999; 354(9185): 1191–1199. doi: 10.1016/S0140-6736(98)10178-2.

Hide M, Bras-Goncalves R, Banuls AL. specific CPB copies within the *Leiahmania donovani* complex: evolutionary interpretations and potential clinical implications in humans. *Parasitol.* 2007; 134(03): 379–389.

Hossain F, Ghosh P, Khan MAA, Duthie MS, Vallur AC, Picone A, Howard RF, Reed SG, Mondal D. Real-time PCR in detection and quantitation of *Leishmania donovani* for the diagnosis of visceral Leishmaniasis patients and the monitoring of their response to treatment. *PLoS One.* 2017; 12(9): e0185606. doi: 10.1371/journal.pone.0185606

Karimi T, Sharifi I, Aflatoonian M R, Aflatoonian B, Mohammadi M A, Salarkia E, Babaei Z, Zarinkar F, Sharifi F, Hatami N, Khosravi A, Eskandari A, Solimani E, Shafiee M, Mozaffari M, Heshmatkhah A, Amiri R, Farajzadeh S, Kyhani A, . . . Bamorovat M. A long-lasting emerging epidemic of anthroponotic cutaneous leishmaniasis in southeastern Iran: population movement and peri-urban settlements as a major risk factor. *Parasit. Vectors* 2021; 14(1): 769–775. https://doi.org/10.1186/s13071-021-04619-3

Killick-Kendrick R. A catalogue of old world phlebotomine sandflies (diptera: Psychodidae, phlebotominae). (Occasional papers on Systematic Entomology, no. 8). *Trans. R. Soc. Trop. Med. Hyg.* 1994; 88(6): 717.

Lainson R. The Neotropical Leishmania species: a brief historical review of their discovery, ecology and taxonomy. *Revista Pan-Amazonica de Saude.* 2010; 1(2): 13–32.

Lata S, Kumari S, Das R, Pasi S, Dhinman R. Typical and atypical cutaneous leishmaniasis in Himachal Pradesh (India). *Heliyon.* 2021; 7(6), e07282. doi: 10.1016/j.heliyon.2021.e07282

Leishman WB. On the possibility of the occurrence of trypanomiasis in India. *Br. Med. J.* 1903; 1(2213): 1252–1254.

Lemma W, Bizuneh A, Tekie H, Belay H, Wondimu H, Kassahun A, Shiferaw W, Balkew M, Abassi I, Baneth G, Hailu A. Preliminary study on investigation of zoonotic visceral leishmaniasis in endemic foci of Ethiopia by detecting Leishmania infections in rodents. *Asian Pacific Journal of Tropical Medicine* 2017; *10*(4): 418–422.

Linse K, Bogdan C, Haenssle HA, Toberer F. Mucocutaneous Leishmaniasis due to Leishmania infantum Infection. *Acta Dermato Venereologica.* 2022; 102: 1–2.

Lukes J, Schonian G, Dujardin JC, Soteriadou K, Dedet JP, *et al.* Evolutionary and geographical history of the *Leishmania donovani* complex with a revision of current taxonomy. *PNAS.* 2007; 104: 9375–9380.

Martins-Melo FR, da Silveira Lima M, Ramos Jr AN, Alencar, CH, Heukelbach J. Mortality and case fatality due to visceral leishmaniasis in Brazil: a nationwide analysis of epidemiology, trends and spatial patterns. *PLoS One.* 2014; 9(4): e93770.

Michels P, Opperdoes. The evolutionary origin of glycosomes. *Parasitol Today* 1991; *7*(5): 105–109. https://doi.org/10.1016/0169-4758(91)90167-m

Migone LE. La buba du Paraguay, leishmaniose ameri-caine. *Bull. Soc. Pathol. Exot.* 1913; 6: 210–218.

NCVBDC. What are Signs & Symptoms of Kala-Azar? National Center for Vector Borne Diseases Control, India. https://ncvbdc.mohfw.gov.in/index4.php?lang=1&level=0&linkid=470&lid=3743 (Accessed on June 30, 2024).

Olliaro P, Grogl M, Boni M, Carvalho EM, Chebli H, Cisse M, *et al.* Harmonized clinical trial methodologies for localized cutaneous leishmaniasis and potential for extensive network with capacities for clinical evaluation. *PLoS Negl. Trop. Dis.* 2018; 12(1): e0006141 doi: 10.1371/journal.pntd.0006141

Perez-Franco JE, Cruz-Barrera ML, Robayo ML, Lopez MC, Daza CD, Bedoya A, Echeverry MC. Clinical and parasitological features of patients with American cutaneous leishmaniasis that did not respond to treatment with meglumine antimoniate. *PLoS Negl. Trop. Dis.* 2016; 10(5): e0004739.

Ramsh V, Mukherjee A. Post-kala-azar dermal leishmaniasis. *Int. J. Dermatol.* 1995; 34(2): 85–91.

Reimão JQ, Coser EM, Lee MR, Coelho AC. Laboratory diagnosis of cutaneous and visceral leishmaniasis: current and future methods. *Microorganisms.* 2020; 8(11): 1632. doi: 10.3390/microorganisms8111632

Ross R. Further notes on Leishman's bodies. *Br. Med. J.* 1903; 2: 1401.

Scarpini S, Dondi A, Totaro C, Biagi C, Melchionda F, Zama D, Pierantoni L, Gennari M, Campagna C, Prete A, Lanari M. Visceral leishmaniasis: epidemiology, diagnosis, and treatment regimens in different geographical areas with a focus on pediatrics. *Microorganisms*, 2022; 10(10): 1887.

Sarkari B, Chance M, Hommel M. Antigenuria in visceral leishmaniasis: detection and partial characterisation of a carbohydrate antigen. *Acta Trop.* 2002 June; 82(3): 339–448.

Schoone GJ, Hailu A, Kroon CC, Nieuwenhuys JL, Schallig HD, Oskam L. A Fast Agglutination Screening Test (FAST) for the detection of anti-leishmania antibodies. *Trans. R. Soc. Trop. Med. Hyg.* 2001; 95(4): 400–401.

Singla N, Singh GS, Sundar S, Vinayak VK. Evaluation of the direct agglutination test as an immunodiagnostic tool for kala-azar in India. Trans. R. Soc. Trop. Med. Hyg. 1993; 87(3): 276–278. https://doi.org/10.1016/0035-9203(93)90125-A

Sharma U, Singh S. Insect vectors of Leishmania: distribution, physiology and their control. *J. Vector Borne Dis.* 2008; 45(4): 255–272.

Singh S. Changing trends in the epidemiology, clinical presentation, and diagnosis of Leishmania-HIV co-infection in India. *Int. J. Infect. Dis.* 2014; 29: 103–112.

Steverding D. The history of leishmaniasis. *Parasites Vectors* 2017; 10: 82.

Sundar S, Rosenkaimer F, Makharia MK, Goyal AK, Mandal AK, Voss A, Hilgard P, Murray HW. Trial of oral miltefosine for visceral leishmaniasis. *Lancet.* 1998; 352(9143): 1821–1823.

Thakur C, Dedet J, Narain S, Pratlong F. Leishmania species, drug unresponsiveness and visceral leishmaniasis in Bihar, India. *Transactions of the Royal Society of Tropical Medicine and Hygiene*, 2001; 95(2): 187–189. https://doi.org/10.1016/s0035-9203(01)90160-9

Thakur CP. Leishmaniasis research—the challenges ahead. *Inian. J. Med. Res.* 2006; 123(3): 193–194.

Thakur L, Singh KK, Shanker V, Negi A, Jain A, Matlashewski G, *et al.* Atypical leishmaniasis: a global perspective with emphasis on the Indian subcontinent. *PLoS Negl. Trop. Dis.* 2018; 12(9): e0006659.

Thakur S, Joshi J, Kaur S. Leishmaniasis diagnosis: an update on the use of parasitological, immunological and molecular methods. *J. Parasit. Dis.* 2020; 44(2): 253–272. doi: 10.1007/s12639-020-01212-w

Tuon FF, Neto VA, Amato VS. Leishmania: origin, evolution and future since the Precambrian. *FEMS Immunol. Med. Microbio.* 2008; 54(2): 158–166.

Ul Bari A, Ber Rahman S. Correlation of clinical, histopathological, and microbiological findings in 60 cases of cutaneous leishmaniasis. *Indian J. Dermatol. Venereol. Leprolog.* 2006; 72(1): 28.

Webster P, Russell DG. The flagellar pocket of trypanosomatids. *Parasitology Today* 1993; 9.6: 201–206.

Wong YP, Othman S, Lau YL, Radu S, Chee HY. Loop-mediated isothermal amplification (LAMP): a versatile technique for detection of micro-organisms. *J. Appl. Microbiol.* 2018; 124(3): 626–643. doi: 10.1111/jam.13647

Zijlstra EE, el Hassan AM. Leishmaniasis in Sudan. Visceral leishmaniasis. *Trans. R. Soc. Trop. Med. Hyg.* 2001; 95 (Suppl 1): S27–S58.

Zijlstra EE, Musa AM, Khalil EAG, El Hassan IM. El-Hassan AM. Post-kala-azar dermal leishmaniasis. *Lancet Infect. Dis.* 2003; 3: 87–98.

2 Geographical Distribution of VL and PKDL in India

*Ravi Ranjan, Shobha Kumari, andAshish Kumar**
Department of Biochemistry, ICMR-Rajendra Memorial Research Institute of Medical Sciences, Patna, Bihar, India
*Corresponding Author: Dr. Ashish Kumar, Department of Biochemistry, ICMR-Rajendra Memorial Research Institute of Medical Sciences, Patna, Bihar, India

2.1 INTRODUCTION

Leishmaniasis is a neglected tropical disease caused by protozoan parasites. The disease has three clinical forms: cutaneous leishmaniasis (CL), visceral leishmaniasis (VL), and mucocutaneous leishmaniasis (MCL). Among these, VL is said to be the most severe form of the disease, and it is fatal if left untreated. Visceral leishmaniasis is also named kala-azar due to the appearance of the symptoms in the Asian subcontinent. Another clinical form of leishmaniasis is post-kala-azar dermal leishmaniasis (PKDL), a skin condition that can develop in people who have successfully treated VL. PKDL can appear months or years after VL treatment as a rash or other skin lesions (Ganguly, 2015). While it is not deadly, PKDL can be highly stigmatizing. Leishmaniasis is a significant global health concern, particularly in tropical and subtropical regions. Leishmaniasis affects over one billion people worldwide, and it is most common in Southeast Asia, Eastern Africa, and Brazil. VL is endemic in more than 60 countries, with 90% of cases reported in India, Bangladesh, Brazil, Nepal, and Sudan. PKDL is also concentrated in six countries, with India contributing 75% of the total cases (Alvar et al., 2012; Zijlstra et al., 2017).

To control the cases and spread of VL and PKDL, a proper method for diagnosing and treating PKDL and VL is very much required. The World Health Organization (WHO) launched a strategy in 2005 to eliminate VL in the Indian subcontinent (Picado et al., 2012). This strategy has successfully reduced the number of VL cases in India, but cases are still being reported. The disease-elimination programme faces different challenges, including instability, population movements, and environmental changes. There is also a need for more awareness about the disease and better access to treatment (Ejazi et al., 2013). Despite these challenges, there is still hope for controlling VL. Approximately 200 million people on the Indian subcontinent are at risk for VL, with an annual incidence of about 420,000 cases (Joshi, 2008; Hasker, 2012). India, a diverse nation with varying climates and ecological landscapes, is home to multiple forms of leishmaniasis. Cutaneous leishmaniasis and visceral leishmaniasis are the predominant forms affecting the country. The incidence of VL in India has significantly reduced in recent years. India has made significant progress in reducing the number of VL cases from 32,803 in 2005 to 3,128 in 2019, representing a 90% drop. There were 1,275 cases reported in 2021 (WHO, 2020). However, challenges remain, such as persistent pockets of high endemicity, weak reporting of relapsed cases, increased cases of PKDL, and inadequate tools for vector surveillance. Therefore, this chapter delves into the various aspects of leishmaniasis, with a focus on its endemicity and the distribution of VL and PKDL in India (Figure 2.1).

Understanding the geographical distribution of the disease is important for effective control and prevention strategies (Murray et al., 2015). The endemicity and geographical distribution of

 DOI: 10.1201/9781032643120-2

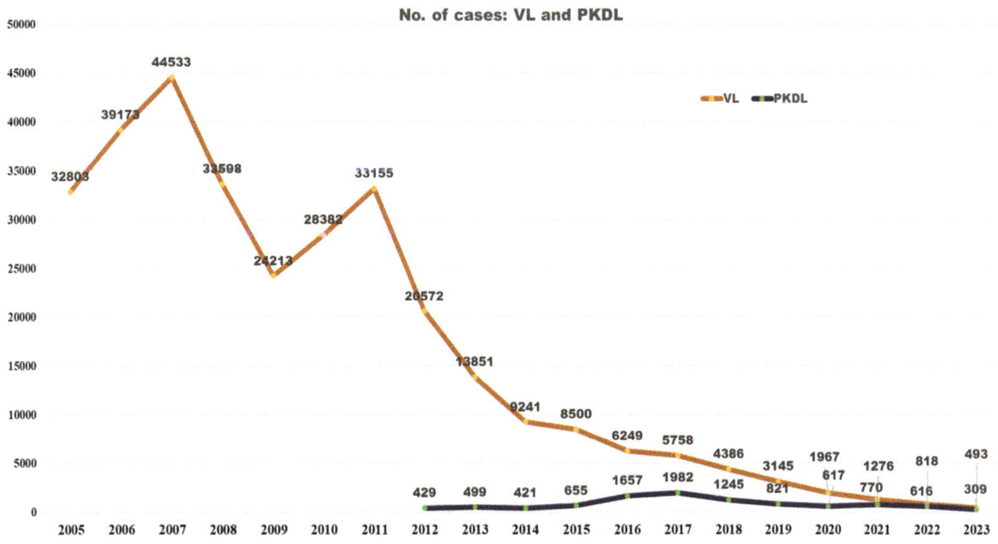

FIGURE 2.1 Distribution of VL and PKDL in India

any vector-associated disease aids in the prediction, prevention, targeted intervention, and control. Similarly, several factors influence the neglected disease VL and PKDL in India, such as environmental, socioeconomic, behavioural, and biological factors (Emery et al., 2020). One factor that contributes to the transmission and persistence of the disease is the presence of vector species, such as the sandfly (*Phlebotomus argentipes*), that breed in soil that is moist and rich in organic matter (Fialho et al., 1995). Other factors that increase people's exposure and susceptibility to sandfly bites are poor housing conditions, a lack of sanitation, overcrowding, and malnutrition. There is also a lack of awareness about the disease among affected communities, as well as the communities' health-seeking behaviours, which may lead to delays in diagnosis and treatment. Even if an individual overcomes all these factors and receives proper treatment, the emergence of drug resistance to amphotericin B, the first-line treatment currently being used, poses a significant challenge. Additionally, there is a lack of effective tools for detecting unsymptomatic VL early and treating PKDL cases that act as reservoirs. Since treatment of leishmaniasis is of long duration and the cost of the available drugs is high, the affected community has mostly been observed to avoid seeking consultation from a specific doctor (Singh et al., 2016).

Knowledge of the region where the disease is most prevalent can be focused on developing more effective preventive measures, which will help identify high-risk areas for proper screening and the early availability of treatment facilities. This can help prevent serious complications and mortality in untreated cases (Murray et al., 2015). Understanding the habitat of sandflies is also important for properly using insecticide spray to make the environment less suitable for the sandfly vector.

2.2 DISTRIBUTION OF THE DISEASE

The distribution of these diseases is closely linked to environmental factors, such as the presence of sandfly vectors and the availability of water. The diseases are also more common in areas where poverty and malnutrition are widespread. Kala-azar remains a significant public health problem in the country as the disease is endemic in 54 districts across four states – Bihar (33 out of 38 districts), Jharkhand (4 out of 24 districts), Uttar Pradesh (6 out of 75 districts), and West Bengal (11 out of 23 districts). Climatic conditions, economic status, and other demographic factors significantly impact the disease's prevalence in the region. According to the number of cases reported in these states, Bihar alone contributes almost 80% of India's total VL cases (Hasker 2019). India has made

TABLE 2.1
VL and PKDL Cases in India

State	VL	PKDL	HIV + VL
Bihar	32,053	3,931	883
Jharkhand	7,071	3,619	44
West Bengal	1,942	1,290	21
Uttar Pradesh	714	253	2
Kerela	20	0	
Sikkim	19	0	
Uttarakhand	11	0	
Assam	2	0	
Punjab	1	0	

significant progress towards achieving the elimination target of less than one case per 10,000 population at the district level by 2020 (Table 2.1).

VL and PKDL cases, including co-infection with HIV, are represented by area in Figure 2.2.

After Bihar, Jharkhand and West Bengal report the most cases, while Uttar Pradesh only reports cases of VL. The endemicity of leishmaniasis is closely associated with poverty, malnutrition, and poor sanitation. These states are characterized by a combination of factors conducive to the proliferation of sandfly vectors, such as poor housing conditions, a lack of proper drainage, and agricultural activities. The distribution of PKDL is also concentrated in the same endemic areas as VL.

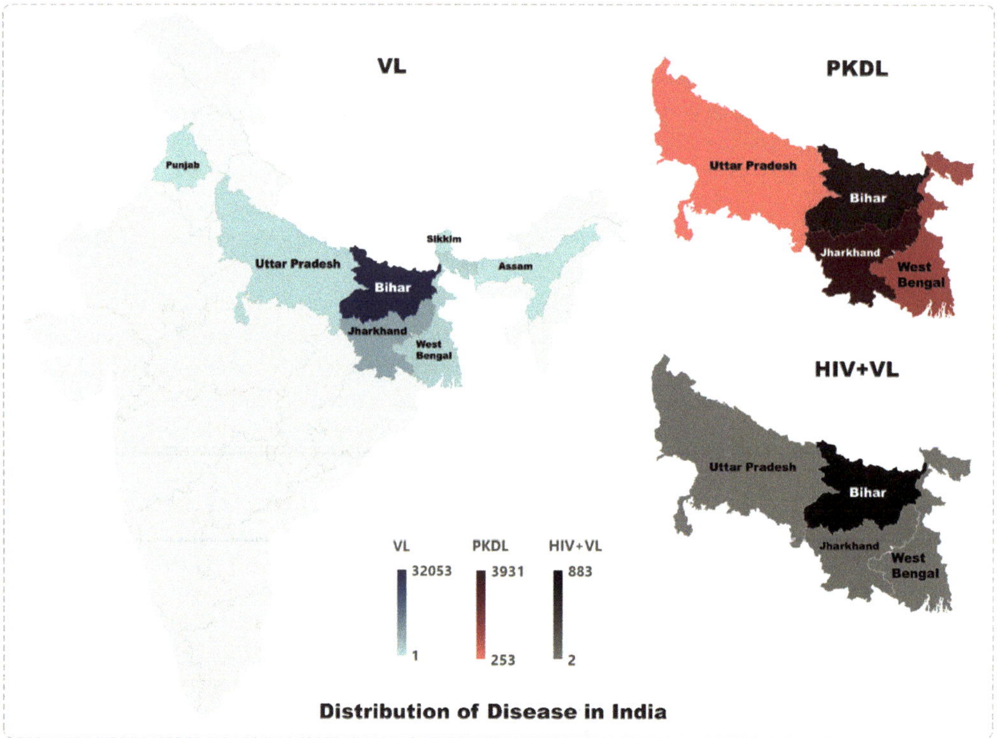

FIGURE 2.2 VL and PKDL cases, including HIV co-infection

TABLE 2.2
The Districts of Bihar, West Bengal, Jharkhand, and Uttar Pradesh Where VL Is Endemic

State	Number	Endemic Districts
Bihar	33	Araria, Arwal, Aurangabad, Banka, Begusarai, Bhagalpur, Bhojpur, Buxar, Darbhanga, East Champaran, Gaya, Gopalganj, Jamui, Jehanabad, Kaimur, Katihar, Khagaria, Kishanganj, Lakhisarai, Madhepura, Madhubani, Munger, Muzaffarpur, Nalanda, Nawada, Patna, Purnia, Rohtas, Saharsa, Samastipur, Saran, Sheikhpura, Sheohar
West Bengal	11	Birbhum, Dakshin Dinajpur, Darjeeling, Hooghly, Howrah, Jalpaiguri, Kolkata, Malda, Murshidabad, Nadia, North 24 Parganas
Jharkhand	4	Garhwa, Latehar, Palamu, Chatra
Uttar Pradesh	6	Bahraich, Ballia, Basti, Deoria, Gonda, Gorakhpur

Bihar, West Bengal, and eastern Uttar Pradesh have reported the highest number of PKDL cases. The cases of visceral leishmaniasis and post-kala-azar dermal leishmaniasis overlap in distribution, which is of significant concern as PKDL patients also act as reservoirs for disease, hampering disease control efforts. According to a WHO report, the micro-stratification surveys conducted in the area reveal that there are 633 blocks in these four states that are endemic (WHO, 2020). The state-wise distribution of VL in India is represented in Table 2.2.

2.2.1 Leishmaniasis in Bihar

VL mainly affects poor rural communities in India. As previously mentioned, Bihar accounts for about eight in every ten cases reported in India. According to reports, the disease is endemic in 33 districts of Bihar (Haskar, 2012). Districts like Vaishali, Muzaffarpur, East Champaran, West Champaran, Saran, Gopalganj, Siwan, Begusarai, and Sitamarhi were highly endemic and reported hundreds of cases annually. However, due to different control measures and routine screening, the number of cases in these regions has reduced significantly, and in 2017, Vaishali was reported to be free of kala-azar. While the overall burden of VL has decreased significantly, the disease persists in certain pockets within the endemic regions and exhibits localized clusters of cases within villages or hamlets.

In the state, the districts of Vaishali, Saran, and Muzaffarpur have more cases of VL and PKDL and are considered highly endemic regions, reporting no less than three cases per 1,000 persons screened (V. Kumar et al., 2020). Different measures were implemented in this area, and Vaishali was selected as a model after successfully implementing strategies for vector control and multi-level tracking of VL and PKDL patients. All 16 blocks of the Vaishali district were selected for the VL elimination target, and the programme was successful by 2016. The integrated VL control strategy helped reduce the number of VL cases from 664 in 2014 to 163 in 2016 and the number of endemic villages from 282 in 2014 to 142 in 2016. The case reduction rate increased from 22.6% in 2014 to 58.8% in 2016 (V. Kumar et al., 2020). Similarly, a study supported by the National Institute of Allergy and Infectious Diseases (NIAID), Maryland, United States, focused on an area comprising 50 villages in the Muzaffarpur district with a population of 81,210. They registered 207 VL cases during the study period. It was concluded from the study that VL cases were strongly clustered at the hamlet level (Bulstra et al., 2018). If we try to understand the clusters or distribution of VL and PKDL in Bihar, some of the common factors that influence endemicity are favourable temperature, humidity, and climate for the proper sustenance and breeding of the sandfly vector (Bhunia et al., 2020). A large cohort study in Bihar identified several additional factors associated with VL, including age, ownership of goats, the presence of bamboo trees, and the type of material used in house walls. Other risk factors include domestic animals, bed net use, and poverty (Haskar, 2012). A study found that a particular caste is disproportionately affected. However, this disparity is not directly related to the caste itself, but rather to the community's lifestyle and sanitation conditions. Overall, factors such as lower daily income and the historical marginalization of certain communities contribute to their increased vulnerability to this disease. This suggests other factors may contribute to the increased risk of VL in this group (Pascual Martinez et al., 2012). Another study found that the annual coverage of indoor residual insecticide spraying in the study area was only 12%, well below the recommended 80% coverage for effective control of VL. It suggests that strengthening vector control efforts, such as indoor residual insecticide spraying, is crucial for VL control. It also suggests that housing plans launched by the union and state government may positively affect VL control. Special attention should be given to hamlets inhabited by the affected population in the area (Faber et al., 2022).

2.2.2 Distribution of Leishmaniasis in Jharkhand

Of the 24 districts in Jharkhand, four – Grhwa, Latehar, Palamu, and Chatra – are endemic for kala-azar. However, cases are also rising in some non-endemic regions, like Gumla district. To eliminate VL and PKDL in the state, the government is working in collaboration with various health organizations and NGOs. Challenges include poverty, lack of awareness about the diseases, inadequate healthcare infrastructure, and difficulties in accessing treatment in remote areas (NVBDC, 2014).

2.2.3 Uttar Pradesh

Uttar Pradesh ranks after Bihar and Jharkhand in terms of the number of cases; the eight endemic districts are Bahraich, Balrampur, Deoria, Gorakhpur, Kushinagar, Siddharthnagar, and Sonbhadra. These districts are in the eastern part of the state and near the border area of Bihar. These regions are considered hotspots for VL and PKDL in Uttar Pradesh. Apart from the main factors influencing the endemicity of a disease, there are other associated factors, like the migration of people for jobs from endemic areas to non-endemic areas. These migrants may carry the parasites with them, serving as reservoirs for the new area and transmitting the diseases.

2.2.4 West Bengal

Since the distribution of VL and PKDL is closely linked to environmental factors, such as the presence of sandfly vectors and the availability of water, the diseases are also more common in areas where poverty and malnutrition are widespread. The hotspots, or endemic districts, for VL and PKDL in West Bengal are in the eastern parts of the state and in the northern parts, close to Bihar. The districts with the highest burden of VL and PKDL are Malda, Murshidabad, Dakshin Dinajpur, Darjeeling, Jalpaiguri, Nadia, Birbhum, Hooghly, Howrah, Jalpaiguri, Kolkata, and North 24 Parganas.

2.2.5 Leishmaniasis in Non-Endemic Regions

Leishmaniasis has also been reported in some of the non-endemic regions of India. However, these cases are mainly cutaneous leishmaniasis (CL) or PKDL. These cases may act as reservoirs for leishmaniasis. Two studies reported the presence of cutaneous and visceral leishmaniasis in the hilly areas of Himachal Pradesh and other areas, including the southern Himalayas (Sharma et al., 2003; Mokhtari and Kumar, 2013). The presence of the disease in these regions suggests a potential shift in the disease's traditional distribution. This further highlights the need for an improved understanding of the disease's transmission and risk factors, particularly in non-endemic areas, to aid in control and eradication efforts.

Gujarat and Rajasthan are two states in India that have a high prevalence of cutaneous leishmaniasis. Also, in a report from Himachal Pradesh, 38 new cases were detected in the Kinnaur district and 161 new cases in the sub-alpine valley along the Sutlej River. A study identified a new focus of VL in the Himalayas, with a concerning increase in cases occurring annually in India (Raina et al., 2009). Additionally, a recent study reported emerging visceral leishmaniasis cases in the Kashmir valley (Basu et al., 2023). These studies highlight the significant burden of leishmaniasis in these regions and the need for further research and control measures.

2.3 FACTORS INFLUENCING THE DISTRIBUTION OF VL AND PKDL IN SPECIFIC AREAS

Several factors, including inadequate hospital facilities, low hygiene awareness, poverty, lifestyle choices, and climatic conditions, contribute to the endemicity of a disease in a particular region. In the context of leishmaniasis in India, these elements significantly impact its prevalence, with mainly climate and related factors being particularly influential.

2.3.1 Climate and Environment

Climatic factors and environmental conditions influence tropical diseases like VL and PKDL. Sandflies, which transmit the parasites, are sensitive to temperature and humidity, and changes in climate patterns can influence the distribution and prevalence of these diseases. As leishmaniasis is transmitted by the sandfly vector, its transmission depends on the climate. The prevalence of leishmaniasis varies across the several Indian states in which it is found, depending on the regions' climatic factors and ecological conditions. Leishmaniasis transmission depends on the climate, as it is transmitted by the sandfly vector and the climate of the region impacts the sandfly's life cycle. The sandfly needs warm temperatures and humidity to breed and survive (Patil and Chatterjee, 2023). Warm temperatures are also conducive to the development of the parasite within the sandfly. These conditions are favourable for the sandfly vectors' breeding and survival. The eastern region of India experiences warm and humid conditions, which are conducive to the proliferation of sandflies and the transmission of the parasite, leading to the spread of leishmaniasis. Thus, the states of Bihar, West Bengal, and Jharkhand are among the most affected regions. Some parts of southern India have a mix of temperature, humidity, and vegetation cover that can support sandfly populations. However, other factors like sanitation, hygiene, and living standards make the population less prone to the disease.

2.3.2 Spread of the Sandfly Vector

The role of insect vectors is very important for the spread of a disease in an area. The distribution of sandflies, the insect vectors of leishmaniasis, closely overlaps with the geographical distribution of VL. Understanding these overlaps is important for targeted vector control efforts and the prevention of this fatal disease. There are 44 species of *Phlebotomus* found in different regions of India, mainly *P. argentipes*, *P. papatasi*, *P. sergenti*, *P. salehi*, *P. hindustanicus*, and *P. major* (Kaul and Jain, 1995). *P. argentipes*, the most common, is spread over the Eastern Gangetic Plains, including Bihar, Jharkhand, West Bengal, and the eastern part of Uttar Pradesh. *P. papatasi* is predominantly found in India and is responsible for some cases of cutaneous leishmaniasis in the regions of Rajasthan, Gujarat, Madhya Pradesh, Kerala, and Tamil Nadu (Poche et al., 2017).

2.3.3 Disease of the Poor?

Treating visceral disease can be expensive, making it less accessible to people with lower incomes. Leishmaniasis disproportionately affects the poorest populations, with factors such as poor housing, lack of personal protective measures, and economic migration contributing to increased infection risk (Alvar et al., 2006). The regions most affected by VL and other forms of leishmaniasis are often socially limited and poor, with limited access to healthcare and resources. Leishmaniasis is therefore also called a disease of poor, as it disproportionately affects the poorest populations. People with low incomes experience factors indirectly associated with financial status, including insufficient nutrition, poor housing, lack of personal protective measures, and economic migration. The population of the area may have weak immune systems due to a lack of proper diet and malnutrition. They also have limited access to the healthcare system; especially in the case of leishmaniasis, the drugs used for treatment are costly and have limited availability at public health centres (Sundar et al., 2010). According to the survey reports of the Ministry of Rural Development and the National Sample Survey (NSSO), Bihar, Jharkhand, Uttar Pradesh, and West Bengal are among the top states in terms of the percentage of population below the poverty line (National Multidimensional Poverty Index, 2015–17). Socioeconomic disparities can lead to limited health education and awareness about diseases like leishmaniasis. Communities with lower levels of education and awareness may not take adequate preventive measures or recognize early symptoms, contributing to the persistence of the disease. An individual's well-being is also linked to the cleanliness and hygiene of their environment. Additionally, the materials used in building house walls play a role, as sandlfies thrive in mud, which is a common construction material in many homes in the region (Mosawi et al., 2019; Hasker et al., 2012).

2.4 CONCLUSION

Visceral leishmaniasis remains a significant public health concern in India. The disease is currently present in clusters across India, particularly in the endemic regions of the Eastern Gangetic Plains. The spread of VL is a concern due to the shifting distribution of the disease. Analyzing the geographical pattern of the disease reveals that it primarily exists in endemic pockets, concentrated in specific regions such as Bihar, Jharkhand, West Bengal, Uttar Pradesh, and some other parts of India. VL is also found in neighboring areas, including some parts of Bangladesh and southern Nepal. The spread of VL is closely associated with ecological factors like climate, temperature, rainfall, and the abundance of vectors in the area, all of which significantly influence its distribution and transmission.

Some studies have also reported that sudden changes in environmental conditions may affect the spread of VL and impact strategies for its control and prevention. While several preventive measures have led to a decrease in cases in recent years, challenges remain, including vector control issues, drug resistance, underreporting, and misdiagnosis. Understanding the geographical distribution is crucial because remote areas often lack proper housing, medical infrastructure, and awareness, which hinders accurate case detection and management. Effective vector control requires community engagement alongside government initiatives.

ACKNOWLEDGEMENT

The authors are thankful to the ICMR-Rajendra Memorial Research Institute of Medical Sciences, Patna, for the necessary support during this study.

REFERENCES

Alvar, J., I. D. Vélez, C. Bern, M. Herrero, P. Desjeux, J. Cano, J. Jannin, M. den Boer, and WHO Leishmaniasis Control Team. "Leishmaniasis worldwide and global estimates of its incidence." *PloS One* 7, no. 5 (2012): e35671.

Alvar, J., S. Yactayo, and C. Bern. "Leishmaniasis and poverty." *Trends in Parasitology* 22, no. 12 (2006): 552–557.

Basu, A., P. Saharan, G. S. Katoch, N. Dash, V. Pandiarajan, and A. Mewara. "Emerging pediatric visceral leishmaniasis in Kashmir valley: a report of three cases." *Tropical Doctor* 53, no. 2 (2023): 317–318.

Bulstra, C. A., E. A. Le Rutte, P. Malaviya, E. Hasker, L. E. Coffeng, A. Picado, O. Prakash Singh, M. Boelaert, S. J. de Vlas, and S. Sundar. "Visceral leishmaniasis: spatiotemporal heterogeneity and drivers underlying the hotspots in Muzaffarpur, Bihar, India." *PLoS Neglected Tropical Diseases* 12 (2018): n.p.

Directorate of National Vector Borne Disease Control Programme. "National roadmap for kala-azar elimination August 2014." (2014) https://ncvbdc.mohfw.gov.in/WriteReadData/l892s/Road-map-KA_2014.pdf.

Ejazi, S. A., and N. Ali. "Developments in diagnosis and treatment of visceral leishmaniasis during the last decade and future prospects." *Expert Review of Anti-Infective Therapy* 11, no. 1 (2013): 79–98.

Emery, D., and D. W. Denning. "The global distribution of actinomycetoma and eumycetoma." *PLoS Neglected Tropical Diseases* 14, no. 9 (2020): e0008397.

Faber, C., C. M. Quiñonez, O. Horstick, K. M. Rahman, and S. Runge-Ranzinger. "Indoor residual spraying for the control of visceral leishmaniasis: A systematic review." *PLoS Neglected Tropical Diseases* 16, no. 5 (2022): e0010391.

Fialho, R. F., and J. J. Schall. "Thermal ecology of a malarial parasite and its insect vector: consequences for the parasite's transmission success." *Journal of Animal Ecology* 64, no. 5 (Sept 1995): 553–562.

Ganguly, S., P. Saha, M. Chatterjee, S. Roy, T. K. Ghosh, S. K. Guha, P. K. Kundu, D. K. Bera, N. Basu, and A. K. Maji. "PKDL—a silent parasite pool for transmission of leishmaniasis in kala-azar endemic areas of Malda District, West Bengal, India." *PLoS Neglected Tropical Diseases* 9, no. 10 (2015): e0004138.

Joshi, A., J. P. Narain, C. Prasittisuk, R. Bhatia, G. Hashim, A. Jorge, M. Banjara, and A. Kroeger. "Can visceral leishmaniasis be eliminated from Asia?" *Journal of Vector Borne Diseases* 45, no. 2 (2008): 105.

Hasker, E., P. Malaviya, V. K. Scholar, P. de Koning, O. Prakash Singh, S. Kansal, K. Cloots, M. Boelaert, and S. Sundar. "Post kala azar dermal leishmaniasis and leprosy prevalence and distribution in the Muzaffarpur health and demographic surveillance site." *PLoS Neglected Tropical Diseases* 13, no. 10 (2019): e0007798.

Hasker, E., S. Prakash Singh, P. Malaviya, A. Picado, K. Gidwani, R. Pratap Singh, J. Menten, M. Boelaert, and S. Sundar. "Visceral leishmaniasis in rural Bihar, India." *Emerging Infectious Diseases* 18, no. 10 (2012): 1662.

Kaul, S. M., and D. C. Jain. "Distribution of Phlebotomine sandflies (Diptera: Psychodidae) according to the physiographic divisions of India." *The Journal of Communicable Diseases* 27, no. 3 (1995): 155–163.

Kumar, A., V. Kumar Singh, R. Tiwari, P. Madhukar, S. Kumar, V. Gautam, C. Engwerda, S. Sundar, and R. Kumar. "Post Kala-azar Dermal Leishmaniasis (PKDL) in the Indian subcontinent: challenges and strategies for elimination." *Frontiers in Immunology* 14 (2023): 1236952.

Kumar, V., R. Mandal, S. Das, S. Kesari, D. S. Dinesh, K. Pandey, V. R. Das, R. K. Topno, M. P. Sharma, R. K. Dasgupta, and P. Das. "Kala-azar elimination in a highly-endemic district of Bihar, India: a success story." *PLoS Neglected Tropical Diseases* 14, no. 5 (2020): e0008254.

Leishmaniasis – India (who.int): https://www.who.int/india/health-topics/leishmaniasis.

Mokhtari, M., and P. Vijayananda Kumar. "Visceral leishmaniasis-associated hemophagocytosis: a single center experience." *Archives of Iranian Medicine* 16, no. 8 (2013): 471–473. https://journalaim.com/PDF/40_aug13_007.pdf.

Mosawi, S. H., Z. Zarei, M. Shams, K. Mohammadi, and S. A. Sajjadi. "Environmental health and leishmaniasis by indication on Afghanistan: a review." *Encyclopedia of Environmental Health* (2019): 458–465. https://colab.ws/articles/10.1016%2FB978-0-12-409548-9.11740-3.

Murray, K. A., N. Preston, T. Allen, C. Zambrana-Torrelio, P. R. Hosseini, and P. Daszak. Global biogeography of human infectious diseases. *Proceedings of the National Academy of Sciences* 112, no. 41 (2015): 12746–12751.

Pascual Martinez, F., A. Picado, P. Roddy, and P. Palma. "Low castes have poor access to visceral leishmaniasis treatment in Bihar, India." *Tropical Medicine & International Health* 17, no. 5 (2012): 666–673.

Patil, R. R., and P. K. Chatterjee. "Epidemiology of visceral leishmaniasis in India." *Leishmania Parasites – Epidemiology, Immunopathology and Hosts.* IntechOpen. (2023) doi:10.5772/intechopen.112444.

Picado, A., A. P. Dash, S. Bhattacharya, and M. Boelaert. "Vector control interventions for visceral leishmaniasis elimination initiative in South Asia, 2005–2010." *The Indian Journal of Medical Research* 136, no. 1 (2012): 22.

Poche, D. M., R. M. Poche, S. Mukherjee, G. A. Franckowiak, L. N. Briley, D. J. Somers, and R. B. Garlapati. "Phlebotomine sandfly ecology on the Indian subcontinent: does village vegetation play a role in sandfly distribution in Bihar, India?." *Medical and Veterinary Entomology* 31, no. 2 (2017): 207–213.

Prakash Singh, O., B. Singh, J. Chakravarty, and S. Sundar. "Current challenges in treatment options for visceral leishmaniasis in India: a public health perspective." *Infectious Diseases of Poverty* 5, no. 2 (2016): 1–5.

Raina, S., D. M. Mahesh, R. Kaul, K. S. Satindera, D. Gupta, A. Sharma, and S. Thakur. "A new focus of visceral leishmaniasis in the Himalayas, India." *Journal of Vector Borne Diseases* 46, no. 4 (2009): 303–306.

Sharma, R. C., V. K. Mahajan, N. L. Sharma, and A. Sharma. "A new focus of cutaneous leishmaniasis in Himachal Pradesh (India)." *Indian Journal of Dermatology, Venereology and Leprology* 69 (2003): 170.

Sundar, S., R. Arora, S. P. Singh, M. Boelaert, and B. Varghese. "Household cost-of-illness of visceral leishmaniasis in Bihar, India." *Tropical Medicine & International Health* 15 (2010): 50–54.

World Health Organization. "Eliminating visceral leishmaniasis: India takes decisive steps to overcome last-mile challenges." 2020 (2020). https://www.who.int/news/item/05-03-2020-VL-India-takes-decisive-steps-overcome-last-mile-challenges#:~:text=India%20recently%20carried%20out%20a,(kala%2Dazar)%20strategy.&text=%E2%80%9CChallenges%20need%20to%20be%20identified,Bekedam%2C%20WHO%20Representative%20to%20India.

Zijlstra, E. E., F. Alves, S. Rijal, B. Arana, and J. Alvar. "Post-kala-azar dermal leishmaniasis in the Indian subcontinent: a threat to the South-East Asia region Kala-azar elimination programme." *PLoS Neglected Tropical Diseases* 11, no. 11 (2017): e0005877.

3 Visceral Leishmaniasis as a Public Health Problem in India
Action Toward its Elimination

Sakshi Singh[1], *Veer Singh*[1], *and Kumar Abhishek*[1,2]*
[1]ICMR-Rajendra Memorial Research Institute of Medical Sciences, Patna, Bihar, India
[2]DHR-Model Rural Health Research Unit, Kurhani, Muzaffarpur, Bihar, India
*Corresponding Author: Dr. Kumar Abhishek, ICMR-Rajendra Memorial Research Institute of Medical Sciences, Patna, Bihar, India, and DHR-Model Rural Health Research Unit, Kurhani, Muzaffarpur, Bihar, India

3.1 INTRODUCTION

Tropical countries have a higher prevalence of diseases that are protozoan parasitic in nature (Sundar et al. 2016). Visceral leishmaniasis (VL) or kala-azar is one of the most fatal diseases, but is often neglected. It is characterized by prolonged fever (for more than two weeks), weight loss, anemia, and splenomegaly (Bhattacharya and Dash, 2016). Two to four million new cases occur every year, primarily in six countries: Bangladesh, Sudan, India, Ethiopia, Brazil, and South Sudan (Singh et al. 2016a). Of the 676 districts in India, VL is only prevalent in 54, among which 34, or nearly 70%, are in the state of Bihar (Hasker et al. 2018). Previously, there were limited ways to control, prevent, and treat the disease, which mainly entailed giving injections of sodium stibogluconate and performing indoor residual spraying (DDT-50%). Because only one drug was on the market, resistance gradually grew, and fatalities were more frequent. Since 2003, the National Vector-Borne Disease Control Programme has been the primary initiative providing central support to all endemic states as part of the national health mission. The disease primarily affects impoverished populations who lack proper access to healthcare and live in unsanitary conditions. Enhancing awareness through healthcare specialists and conducting campaigns in endemic areas would help ensure that affected individuals can access information and treatment more easily.

Despite the effective measures taken, several challenges remain. These include a lack of proactive health-seeking behavior among local populations in endemic areas, difficulties arising from geographical locations, inadequate housing, poor living conditions, and unsatisfactory hygiene practices. Additionally, there are obstacles in implementing the kala-azar program as planned. Other contributing factors include population mobility and environmental and climatic variations. However, thanks to ongoing government efforts, there has been a noticeable decline in cases, highlighting the importance of this issue on the public health agenda aimed at elimination. If these efforts continue in the right direction, it is possible to achieve complete eradication of the disease in the near future (Vélez et al. 2017).

Significant improvements have been made in addressing the challenges associated with kala-azar, but many hurdles remain. Favorable conditions that support kala-azar epidemics include rural areas located below 600 meters above sea level, heavy annual rainfall, humidity levels above 70%,

DOI: 10.1201/9781032643120-3

temperature variations between 15°C and 38°C with daily fluctuations of more than 7°C, abundant vegetation, subsoil water, and alluvial soil (Singh and Singh, 2019). Deforestation caused by increasing agriculture, livestock raising, and urbanization significantly impacts the ecological niches of thousands of phlebotomine species. More than a billion poor people worldwide are affected by tropical diseases, which are normally neglected. Pharmaceutical industries often give less priority to developing drugs for these diseases due to the lower likelihood of substantial returns and profits.

VL has several causes, some of which remain unknown. While it is suggested that VL is transmitted exclusively through humans, this has not been conclusively proven; humans are often targeted due to factors such as the impactful role of post-kala-azar dermal leishmaniasis (PKDL), the potential presence of latent human carriers, and domestic animals. Currently, transmission can occur through both clinical VL and subclinical VL, highlighting the need to understand the dynamics of asymptomatic VL infections. Individuals who are infectious but asymptomatic pose additional challenges to achieving a cure for the disease.

The absence of an animal reservoir has been suggested based on positive results from polymerase chain reaction (PCR) tests and the detection of antibodies in domestic animals. Sandflies exhibit opportunistic feeding behavior, considering domestic mammals as potential blood sources. If transmission through animals is confirmed, it would present another challenge even after addressing the human parasite reservoir (Singh et al. 2016b).

Prevention and control strategies for the disease are developed based on evidence-based practices that align with the World Health Organization's guidelines and region-specific methods. These strategies include early diagnosis, comprehensive case management, vector control through indoor residual spraying, proper monitoring, and program management. Since 1992, the number of cases has decreased by around 97%, and the death rate has fallen to less than 5%, with only 37 reported deaths in 2020. The strategies aim to reduce delayed symptoms and ensure complete management of cases. As the burden of disease increases, robust surveillance is essential for monitoring PKDL and HIV-VL coinfections, along with effective planning (Sundar et al. 2018, Burza et al. 2022). Additionally, immunochromatographic serological tests can be used for rapid and cost-effective diagnosis (Bansal and Jain, 2023).

3.2 ELIMINATION OF VISCERAL LEISHMANIASIS

Central, state, and district levels, along with partner organizations, have come together to work toward the eradication of the neglected disease of VL. Previously, the disease did not receive serious attention, resulting in many deaths. However, advancements in technology and improved diagnostics and treatment strategies – particularly the introduction of drugs like liposomal amphotericin B – have sparked renewed interest in finding a cure. Planning has been undertaken at the national, state, district, and subdistrict levels. Resources are available as needed, multiple partnerships are providing assistance, diagnostic methods have become more accessible, there is political support at every level, and drug donations from the World Health Organization (WHO) are contributing to efforts aimed at making VL completely curable. The recurrence of VL cases can be predicted through population-based antibody surveys (Coffeng et al. 2021) (Figure 3.1).

If left untreated, VL can be fatal to more than 90% of affected patients. India is one of the countries most impacted by kala-azar. The potential to eliminate kala-azar is promising due to several factors: the limited spread of the disease, humans being the only reservoir (with no other animal reservoirs), VL's single vector (the sandfly, *Phlebotomus argentipes*), and new impactful treatments along with rapid diagnostic tests. Public health measures currently in practice include financial support, professional human resources, micro-level stratification, case-specific supervision, maintaining electronic health records for easy access, single-dose treatment with liposomal amphotericin B, constructing concrete houses in endemic areas, technical advisory support through regular program updates, collaboration with diverse partners. Several vaccine development strategies are under consideration, including recombinant peptides, DNA vaccines, killed whole parasites, and genetically

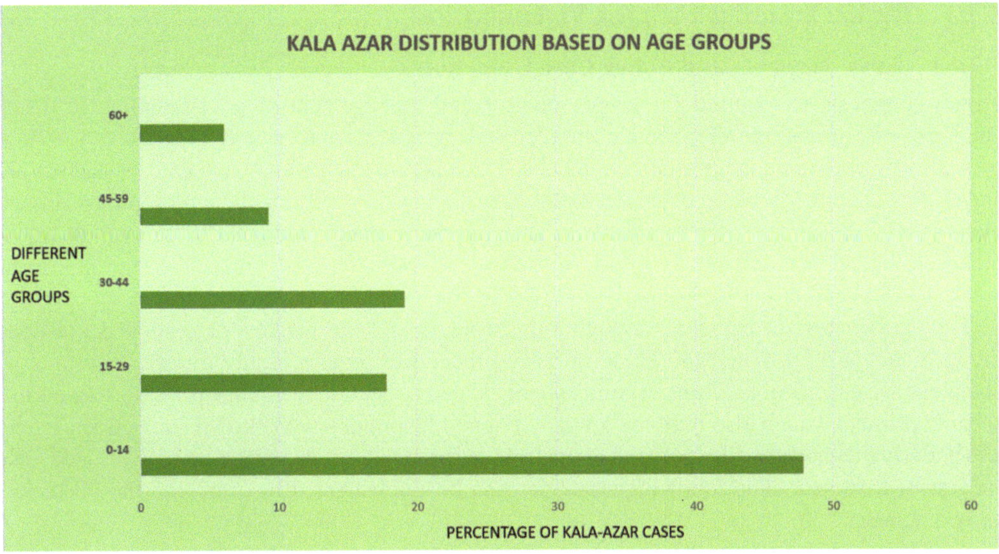

FIGURE 3.1 Distribution of kala-azar based on different age groups and percentage of kala-azar cases

modified live-attenuated parasites (Hendrickx et al. 2018). Researchers have also called for a data-sharing platform for VL clinical trials to facilitate informative analyses of developments and results in case management for better disease control (Selvapandiyan et al. 2019).

Cattle treated with fipronil-based drugs can effectively control sandfly populations since the drug permeates the cattle's blood and feces, targeting adult sandflies through the blood and sandfly larvae through the feces (Poche et al. 2016). Improvements in understanding the pathogenesis of VL have led to the development of various drugs targeting different biochemical pathways. Combination therapies using multiple drugs have also proven effective. The goal is to eliminate the disease or significantly reduce its incidence while maintaining this progress over the coming years. It is essential to conduct annual surveillance of case numbers and maintain comprehensive data. One of the best ways to tackle the disease is by raising public awareness about current developments through organized campaigns. Active volunteers should spread awareness through door-to-door outreach in villages to empower residents to make informed decisions. The WHO has classified leishmaniasis as a Category I disease (emerging and uncontrolled), and the World Health Assembly (WHA) Resolution 43.18 recognizes it as a major public health concern (Hirve et al. 2017).

In 2005, the Ministers of Health of Bangladesh, India, and Nepal affirmed their strong political commitment to eliminating VL by 2015 through inter-country cooperation and cross-border collaboration aimed at reducing incidence to less than 1 per 10,000 population at the upazila (administrative block) and district levels in each country (Hasker et al. 2018). The VL elimination framework was updated in 2012 to identify five key strategies for achieving this goal: early diagnosis and comprehensive case management; effective disease and vector surveillance; social mobilization and partnership building; and clinical and operational research (Singh and Singh, 2019). A more recent initiative led by the WHO aims to define a roadmap for the prevention, control, elimination, and eradication of 17 neglected tropical diseases, including VL, by 2030 as part of achieving the Sustainable Development Goals (Sundar and Agarwal, 2018). Despite considerable progress and a declining trend in VL incidence rates, 90 out of 456 endemic blocks (20%) in India remain highly endemic for VL. Meanwhile, new ecological niches for focal indigenous transmission have emerged in previously non-endemic hilly areas of Nepal and Bhutan as well as in Bangladesh and Thailand. Since 2005, the Special Program for Research and Training in Tropical Diseases (TDR) has coordinated and financed research aimed at developing innovative tools to support the VL elimination initiative in the Indian subcontinent.

3.2.1 The Key Strategies Used for VL Elimination

3.2.1.1 Early Diagnosis and Active Case Detection

Timely diagnosis is required for proper control and management of a communicable disease. Diagnosis of a susceptible VL patient is initiated with symptoms of fever and splenomegaly, followed by screening with rapid diagnostic kits (RDTs). The RDTs available for VL diagnosis, such as rK39 and rK26, are antibody-based detection methods that are well-suited for endemic settings. However, they are not confirmatory due to sensitivity and specificity issues (Manisha Vaish et al. 2012). The other serological tests suitable for mid-level laboratory settings are enzyme-linked immunosorbent assay (ELISA) and direct agglutination test (DAT), but these are also non-confirmatory.

The examination of parasites through either splenic or bone marrow aspiration is considered the gold standard for diagnosis. These invasive procedures require highly trained personnel with comprehensive medical facilities. Another sophisticated molecular technique is PCR using parasite-specific primers, which also requires a well-equipped laboratory with skilled staff (Boelaert et al. 2014). Early diagnosis of VL is crucial to control the spread, lower its impact on patients, and ensure successful treatment. The following strategies can be implemented to overcome the obstacles to early diagnosis:

(a) **Active case detection:** Conduct house-to-house surveys in endemic areas between the months of March and August to coincide with peak case incidence.
(b) **Screening contacts:** Test family members, household companions, and neighbors of the recently diagnosed VL or PKDL patients.
(c) **Medical camps:** Organize medical camps in highly endemic areas where cases are reported regularly, focusing on people presenting with fever.

Rural villages in endemic areas need healthcare tools and techniques to facilitate early detection of VL for timely treatment. Diagnosis typically involves serological tests, indirect fluorescence antibody (IFA) tests, ELISA, DAT, and the rK39-based immunochromatographic test (ICT) (Sundar and Agarwal, 2018). The rK39-based test has sensitivity in a variable range and has restricted use for PKDL and HIV-VL coinfections. Therefore, other tests like the loop-mediated isothermal amplification (LAMP) assay and ELISA are being considered for these conditions (Goyal et al. 2020). Enhancing district-level healthcare infrastructure and training staff in pathological testing methods for VL will enable the diagnosis of skin samples through microscopy. These improvements can provide villagers with affordable access to testing. If proper microscopy techniques and sampling protocols are followed, only about 30% of patients may require further testing for PCR confirmation. An effective long-term solution involves developing non-invasive sampling methods and diagnosis techniques suitable for field conditions (Goyal et al. 2020).

3.2.1.2 Complete Treatment

Previously, sodium stibogluconate was used as treatment, but as resistance emerged around 2002, miltefosine was adopted as an alternative. However, miltefosine was found to be less effective over time, showing higher relapse rates. Combination therapy is recommended to protect against drug resistance. Nowadays, liposomal amphotericin B (AmBisome) treatment is seen to be working effectively and has been adopted as the drug of choice. Additionally, a combination of miltefosine and paromomycin has been shown to be effective.

Monitoring drug sensitivity and resistance is necessary for managing current treatment options and developing new drug discovery models (Hendrickx et al. 2018). It is crucial to provide effective leishmania treatments at minimal costs in endemic areas, ensuring proper dosage and a full course of medication, especially since many infected individuals are poor. Easy access to effective drugs and routine efficacy check-ups should be standard practice. Funding agencies, political stakeholders, and wealthy individuals should invest in initiatives aimed at elimination. Collaborative efforts among various organizations are necessary to review and improve strategies regularly.

The therapeutic drugs available for VL treatment – either alone or in combination – are pentavalent antimonials, amphotericin B (and its liposomal formulation AmBisome), miltefosine, and paromomycin (Van Griensven et al. 2023). The pentavalent antimonials, which were the main first-line drug for more than 50 years, are no longer used in the Indian subcontinent because of parasite resistance. However, they are still used in other parts of the world for treatment.

Amphotericin B, which is basically an antifungal drug, is considered a second line option for the treatment of VL. The lipid formulation AmBisome has emerged as the first-line treatment in the Indian subcontinent due to its high efficacy and is included as a standard treatment in the government's kala-azar elimination program. Miltefosine is currently the only oral drug available for VL treatment; it was incorporated into the VL elimination program after clinical trials in 2002. Although miltefosine is an anti-cancer drug with potential side effects, it is currently used for PKDL treatment in the Indian subcontinent (Park et al. 2021). Paromomycin has also been registered for VL treatment after clinical trials. Both miltefosine and paromomycin are seldom used in combination with AmBisome.

Like diagnosis, the treatment of VL is also a challenging task. A single administration of AmBisome is the standard treatment adopted for VL in the Indian subcontinent, boasting a cure rate of over 95%. However, reports of VL relapse following AmBisome treatment have raised concerns (Mahajan et al. 2022). Cases of VL with other coinfections like HIV or tuberculosis further complicate treatment, often leading to unresponsiveness. Combination therapy using two or more anti-leishmanial drugs – either in co-administration or in a sequential manner, and often including AmBisome – is being trialed for managing unresponsive cases at some tertiary treatment centers.

Adhering to standard guidelines for drug therapy is essential for successful treatment of patients and should be complemented by follow-ups lasting six months. The patients will be referred to the tertiary centers for combination therapy to manage unresponsive cases. Effective treatment of VL is crucial to prevent patients from becoming reservoirs for the disease and to interrupt the transmission cycle (Hendrickx et al. 2018) (Figure 3.2).

3.2.1.3 Training Social Health Workers and Doctors

In 2013, India created a national health mission to eliminate VL as a public health issue. Accredited social health activists (ASHAs) serve as volunteers at the village level, with one ASHA available for every 1,000 people (Fathima et al. 2015). The focus should be on developing an action plan for disease elimination and its sustainability. Social health workers should arrange proper awareness programs and campaigns to educate villagers about preventative measures and appropriate actions to take if symptoms arise. The central domestic budget allocates funds to appoint kala-azar treatment supervisors at both state and district levels, ensuring that health facilities, laboratory technicians, and professionals are available even in small districts for early diagnosis, treatment, and follow-up care. The strategy for kala-azar elimination is well-planned at all levels, incorporating best practices for vector control. Adequate resources should be provided to doctors and nurses knowledgeable in VL elimination strategies to ensure that they can effectively manage patient cases. Continuous dissemination of educational materials, support, and training should be prioritized.

3.2.1.4 Follow-Up with Patients for Timely Diagnosis of PKDL

The way vectors move from one region to another is not known. People infected with VL may develop PKDL weeks or months later, presenting as small flat spots on the skin, known as maculae and papular lesions. These patients can act as significant reservoirs for the infection. In East Africa, nearly 50% of VL patients show PKDL symptoms; however, data on its prevalence in the Indian subcontinent is limited. Sandflies show a high rate of infection when exposed to nodular PKDL lesions, highlighting an urgent need to establish reports on PKDL occurrences and infection rates

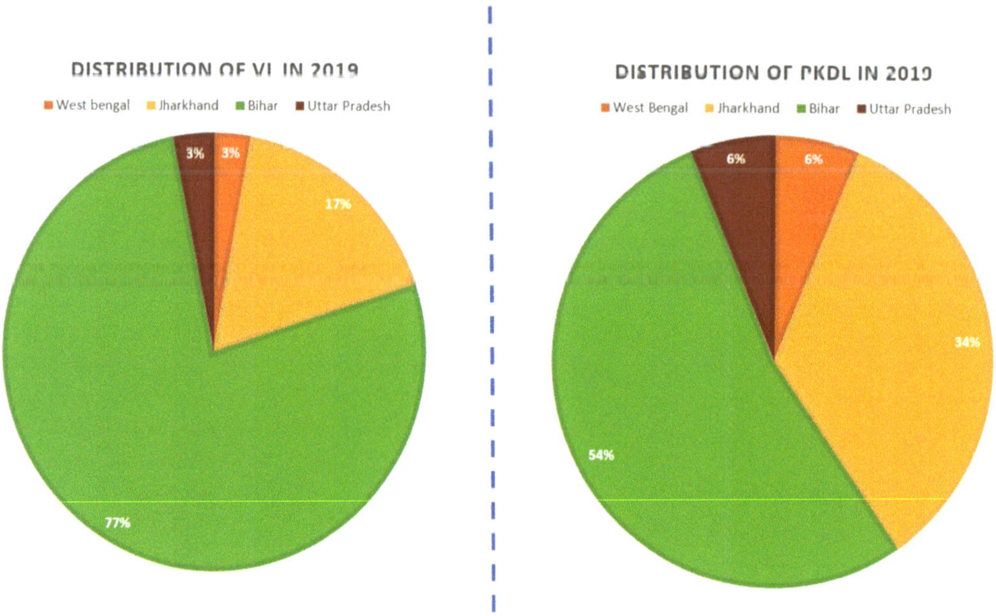

FIGURE 3.2 Distribution of VL and post-kala-azar dermal leishmaniasis in different Indian states in 2019

in endemic regions. Utilizing xenodiagnosis can help assess the infectious potential of sandflies (Bourdeau et al. 2020).

Challenges such as the limited availability of animal models, a lack of treatment options for PKDL, and a low number of reported cases hinder research and development efforts. According to the National AIDS Control Organization, India has the second-largest global burden of HIV, which increases the risk of asymptomatic VL infection, making this condition highly fatal. The coinfection of HIV and VL increases the risk of transmission through blood contact and may reactivate symptomless VL infections. However, such coinfections are rare, with India reporting about 5.6% of VL cases in HIV-positive patients.

3.3 STRENGTHENING THE CAPACITY OF HUMAN RESOURCES IN HEALTH

3.3.1 Training Social Health Workers

VL is associated with poverty, poor nutritional health, and illiteracy and is often referred to as a "disease of the poorest of the poor." The disease primarily affects poor people living in remote locations lacking basic facilities and proper knowledge and support. People in these regions have less access to quality healthcare and often rely on limited local facilities. This is where local social workers, such as ASHAs, auxiliary nurse midwives (ANMs), and Anganwadi workers (AWWs), play a vital role in supporting health facilities and informing people about government schemes. Indeed, these social activists are among the first to approach patients in such areas. Training these social workers to identify potential VL cases is crucial for achieving elimination. Regular training sessions for ANMs and ASHAs should focus on identifying VL symptoms, referring patients to primary health centers or community health centers for proper treatment, maintaining records of referrals, and conducting post-treatment follow-ups for therapy status.

PKDL develops in more than 80% of VL cases, with the macular form being the most common in the Indian subcontinent. This form shows hypopigmented white patches on the skin, which can be mistaken for other skin infections (Salotra and Singh, 2006). PKDL initially shows no critical symptoms, leading patients to consult physicians late, making them a sustainable reservoir for infection despite lower transmission rates. Therefore, social workers should be trained to identify PKDL and distinguish it from other skin disorders (Zijlstra, 2013). Hence, strengthening the capabilities of rural social workers is essential for the elimination of kala-azar.

3.3.2 Training Medical Healthcare Personnel

Diagnosing VL and managing its treatment can be challenging, necessitating adherence to guidelines for proper treatment to avoid toxic effects from drug therapy and ensure complete recovery. Regular training for medical officers and nurses should be conducted at primary health centers or district-level facilities, especially in VL-endemic areas. Training should extend beyond government healthcare staff to include private practitioners and other healthcare stakeholders, equipping them to identify VL suspects and refer them for treatment. Their clinics or workplaces should display definitions of VL suspects to help raise awareness among visitors. Additionally, local practitioners who are more accessible in remote locations should be trained to identify and refer potential cases to the nearest treatment center. A standard operating procedure (SOP) for proper treatment should be circulated during each training session. This collaborative approach involving both public and private healthcare personnel is essential for achieving elimination (Mondal et al. 2016).

3.3.3 Integrated Vector Management: Principles and Components

The key components of integrated vector management (IVM) include chemical control, environmental management, biological control, personal protection measures, and community participation.

3.3.3.1 IVM Strategies for VL Control

Several important IVM strategies can be useful for controlling VL:

1. An overview of IVM strategies implemented in endemic regions.
2. Case studies of successful IVM programs in VL control.
3. Challenges and limitations of current IVM approaches.

Recent advancements in IVM strategies incorporate techniques such as remote sensing and GIS-based approaches (Reference). Additionally, socioeconomic and environmental factors impact the effectiveness of IVM (Huda et al. 2012).

3.3.3.2 Biannual Spray of Insecticide: Indoor Residual Spray

Sandflies have a short lifespan, and their populations can rapidly rebound after insecticide application. Biannual spraying helps maintain a sustained reduction in vector populations throughout the year. The seasonal nature of sandfly activity, with peaks during certain times of the year, necessitates regular indoor residual spraying (IRS) to coincide with periods of high transmission risk (Lawyer et al. 2017). Biannual IRS helps to prevent insecticide resistance by rotating or alternating between different classes of insecticides, as recommended by the WHO.

Implementation of biannual IRS involves:

- Identifying high-risk areas for VL transmission based on epidemiological data and vector surveillance.
- Training and deploying personnel to conduct IRS using approved insecticides and equipment.
- Engaging the comm"unity through awareness campaigns to ensure participation and compliance with IRS activities.
- Monitoring and evaluating IRS effectiveness through entomological surveillance and VL case data (Huda et al. 2012).

As vectors tend to adapt in favorable environments, India has expanded its vector-control regions to include areas that reported active cases over the past three years, conducting two rounds of IRS in these regions. Due to the resistance of sandflies to DDT, synthetic pyrethroids have been introduced for IRS. This campaign has protected more than 35 million people from kala-azar, significantly reducing disease occurrences. Integrated management of vectors and multisectoral actions are essential for controlling transmission. All four endemic states must report cases monthly to the national vector-borne disease control program under the Ministry of Health (Huda et al. 2012).

Female sandflies feed on human blood to produce eggs, and crowded housing conditions or habits like sleeping outdoors increase the likelihood of infection. While countries use deltamethrin and cypermethrin for IRS, further studies on their effectiveness are still needed across various regions. Also, since IRS is only effective indoors, those who sleep outside or in open spaces remain at risk for infections. There is a need to look for further outdoor protection options, such as topical creams, so that individuals can apply in endemic areas. VL control in high-risk areas of India can be achieved through early detection, treatment, and vector control. Currently, IRS is conducted using alpha-cypermethrin 5% powder dissolved in water, which kills sandflies that land on treated surfaces (Mondal et al. 2016).

Studies have shown that biannual IRS can significantly reduce sandfly densities and VL incidence in endemic areas. A study in Bihar, India, found that biannual IRS with DDT reduced the indoor resting density of *Phlebotomus argentipes* by 93.3% and decreased VL incidence by 79.6% compared to areas without IRS. Another study in Bangladesh reported a 67% reduction in VL cases after implementing biannual IRS with alpha-cypermethrin (Huda et al. 2012).

Operational challenges include logistical constraints, access to remote areas, and community acceptance of IRS programs. Environmental and health concerns related to insecticide use highlight the importance of using WHO-approved insecticides and adhering to safety guidelines. Sustainable funding and support for biannual IRS programs are necessary to ensure long-term success in VL elimination efforts (Mondal et al. 2016; Huda et al. 2012).

3.4 CONCLUSION

Significant progress has been made toward eradicating visceral leishmaniasis due to the unwavering commitment and collaborative efforts of government agencies, funding organizations, healthcare officials, and social health workers throughout India. Early diagnosis and treatment are critical for averting fatalities associated with this debilitating disease. India has made remarkable strides in eliminating this malady over the years; however, overcoming challenges related to testing, diagnosis, healthcare provisions, research, and development remains essential. Understanding the role of the vectors responsible for the transmission of the disease is crucial, as is understanding VL's prevalence in other animal species.

Complete eradication of VL can be achieved through improved equipment, up-to-date data, and professional healthcare assistance. It is essential to devise innovative tools and methodologies that are not only user-friendly but also affordable for patients from impoverished rural backgrounds. Human surveillance alone is insufficient; therefore, tools enabling continuous monitoring of sandflies upon contact with humans must be developed. Our efforts toward attaining the ultimate objective of the complete elimination of VL (kala-azar) can be enhanced through an incremental approach that considers the disease's inherent lethality.

ACKNOWLEDGMENT

The authors are thankful to the ICMR-Rajendra Memorial Research Institute of Medical Sciences, Patna, for its necessary support during this study.

REFERENCES

Bansal, N. and Jain, A., 2023. Diagnosing visceral leishmaniasis. *BMJ, 383*, pp. 1–11.

Bhattacharya, S.K. and Dash, A.P., 2016. Treatment of visceral leishmaniasis: options and choice. *The Lancet Infectious Diseases, 16*(2), pp. 142–143.

Boelaert, M., Verdonck, K., Menten, J., Sunyoto, T., van Griensven, J., Chappuis, F. and Rijal, S., 2014. Rapid tests for the diagnosis of visceral leishmaniasis in patients with suspected disease. *Cochrane Database of Systematic Reviews*, (6), pp. 1–104.

Bourdeau, P., Rowton, E. and Petersen, C., 2020. Impact of different Leishmania reservoirs on sand fly transmission: Perspectives from xenodiagnosis and other one health observations. *Veterinary Parasitology, 287*, p.109237.

Burza, S., Mahajan, R., Kazmi, S., Alexander, N., Kumar, D., Kumar, V., Lasry, E., Harshana, A., de Lima Pereira, A., Das, P. and Verma, N., 2022. AmBisome monotherapy and combination AmBisome–Miltefosine therapy for the treatment of visceral leishmaniasis in patients coinfected with human immunodeficiency virus in India: a randomized open-label, parallel-arm, Phase 3 trial. *Clinical Infectious Diseases, 75*(8), pp. 1423–1432.

Coffeng, L.E., Le Rutte, E.A., Munoz, J., Adams, E. and De Vlas, S.J., 2021. Antibody and antigen prevalence as indicators of ongoing transmission or elimination of visceral leishmaniasis: a modeling Study. *Clinical Infectious Diseases, 72*(Supplement_3), pp. S180–S187.

Fathima, F.N., Raju, M., Varadharajan, K.S., Krishnamurthy, A., Ananthkumar, S.R. and Mony, P.K., 2015. Assessment of 'accredited social health activists'—a national community health volunteer scheme in Karnataka State, India. *Journal of Health, Population, and Nutrition, 33*(1), p.137.

Goyal, V., Das, V.N.R., Singh, S.N., Singh, R.S., Pandey, K., Verma, N., Hightower, A., Rijal, S., Das, P., Alvar, J. and Bern, C., 2020. Long-term incidence of relapse and post-kala-azar dermal leishmaniasis after three different visceral leishmaniasis treatment regimens in Bihar, India. *PLoS Neglected Tropical Diseases, 14*(7), p. e0008429.

Hasker, E., Kwete, J., Da Luz, R. I., Mpanya, A., Bebronne, N., Makabuza, J., Claeys, Y., Ilunga, J., Lejon, V., Ngoyi, D. M., Büscher, P., Boelaert, M., & Miaka, E. M. 2018. Innovative digital technologies for quality assurance of diagnosis of human African trypanosomiasis. *PLoS Neglected Tropical Diseases, 12*(9), e0006664. https://doi.org/10.1371/journal.pntd.0006664

Hirve, S., Kroeger, A., Matlashewski, G., Mondal, D., Banjara, M.R., Das, P., Be-Nazir, A., Arana, B. and Olliaro, P., 2017. Towards elimination of visceral leishmaniasis in the Indian subcontinent—Translating research to practice to public health. *PLoS Neglected Tropical Diseases, 11*(10), p.e0005889.

Hendrickx, S., Guerin, P.J., Caljon, G., Croft, S.L. and Maes, L., 2018. Evaluating drug resistance in visceral leishmaniasis: the challenges. *Parasitology, 145*(4), pp. 453–463.

Lawyer, P., Killick-Kendrick, M., Rowland, T., Rowton, E. and Volf, P., 2017. Laboratory colonization and mass rearing of phlebotomine sand flies (Diptera, Psychodidae). *Parasite, 24*, pp. 1–39..

Mahajan, R., Owen, S.I., Kumar, S., Pandey, K., Kazmi, S., Kumar, V., Adams, E. R., Harshana, A., and Burza, S. 2022. Prevalence and determinants of asymptomatic Leishmania infection in HIV-infected individuals living within visceral leishmaniasis endemic areas of Bihar, India. *PLoS Neglected Tropical Diseases, 16*(8), e0010718. https://doi.org/10.1371/journal.pntd.0010718

Manisha Vaish, M.V., Singh, O.P., Jaya Chakravarty, J.C. and Shyam Sundar, S.S., 2012. rK39 antigen for the diagnosis of visceral leishmaniasis by using human saliva. *American Journal of Tropical Medicine and Hygiene, 86*(4), pp. 598–600.

Mondal, D., et al., 2016. Efficacy and safety of single-dose liposomal amphotericin B for visceral leishmaniasis in a rural public hospital in Bangladesh: a feasibility study. *The Lancet Global Health, 2*, pp. 51–57..

Park, S.Y., Kim, J.H., Choi, J.H., Lee, C.J., Lee, W.J., Park, S., Park, Z.Y., Baek, J.H. and Nam, J.S., 2021. Lipid raft-disrupting miltefosine preferentially induces the death of colorectal cancer stem-like cells. *Clinical and Translational Medicine, 11*(11), p. e552.

Poche, D.M., Grant, W.E. and Wang, H.H., 2016. Visceral leishmaniasis on the Indian subcontinent: modelling the dynamic relationship between vector control schemes and vector life cycles. *PLoS Neglected Tropical Diseases, 10*(8), p. e0004868.

Salotra, P. and Singh, R., 2006. Challenges in the diagnosis of post kala-azar dermal leishmaniasis. *Indian Journal of Medical Research, 123*(3), pp. 295–310.

Selvapandiyan, A., Croft, S.L., Rijal, S., Nakhasi, H.L. and Ganguly, N.K., 2019. Innovations for the elimination and control of visceral leishmaniasis. *PLoS Neglected Tropical Diseases, 13*(9), p. e0007616.

Singh, N.S. and Singh, D.P., 2019. A review on major risk factors and current status of visceral leishmaniasis in north India. *American Journal of Entomology, 3*(1), pp. 6–14.

Singh, O.P., Hasker, E., Boelaert, M. and Sundar, S., 2016a. Elimination of visceral leishmaniasis on the Indian subcontinent. *The Lancet Infectious Diseases*, *16*(12), pp. e304–e309.
Singh, O.P., Singh, B., Chakravarty, J. and Sundar, S., 2016b. Current challenges in treatment options for visceral leishmaniasis in India: a public health perspective. *Infectious Diseases of Poverty*, *5*(02), pp. 1–15.
Sundar, S. and Agarwal, D., 2018. Visceral leishmaniasis–Optimum treatment options in children. *The Pediatric Infectious Disease Journal*, *37*(5), p.492.
Sundar, S. and Singh, A., 2016. Recent developments and future prospects in the treatment of visceral leishmaniasis. *Therapeutic Advances in Infectious Disease*, *3*(3–4), pp. 98–109.
Sundar, S., Singh, O.P. and Chakravarty, J., 2018. Visceral leishmaniasis elimination targets in India, strategies for preventing resurgence. *Expert Review of Anti-Infective Therapy*, *16*(11), pp. 805–812.
Van Griensven, J., Dorlo, T.P., Diro, E., Costa, C. and Burza, S., 2023. The status of combination therapy for visceral leishmaniasis: an updated review. *The Lancet Infectious Diseases, 24*(1), pp. 36–46.
Vélez, I.D., Carrillo, L.M., Muskus, C. and Claborn, D., 2017. *Application of the Eco-Epidemiological Method in the Study of Leishmaniasis Transmission Foci* (pp. 31–56). London: InTec.
Zijlstra, E.E. and Alvar, J., 2013. The Post Kala-Azar dermal Leishmaniasis (PKDL) atlas. *The Lancet Infectious Diseases*, *3*(2), pp. 87–98.

4 Pathogenesis of Visceral Leishmaniasis

*Vikash Kumar[1], Divya Prasad[2], Ghufran Ahmed[1], and Ashish Kumar[1]**

[1]ICMR-Rajendra Memorial Research Institute of Medical Sciences, Patna, Bihar, India
[2]Department of Zoology, Patna University, Patna, Bihar, India
*Corresponding Author: Dr. Ashish Kumar, Scientist, ICMR-Rajendra Memorial Research Institute of Medical Sciences, Patna, Bihar, India

4.1 INTRODUCTION

Visceral leishmaniasis (VL), commonly known as kala-azar, is primarily induced by *Leishmania donovani* and *L. infantum* parasites, transmitted through bites from phlebotomine sandflies. VL is the most severe form of leishmaniasis and is characterized by continuous fever, anemia, enlargement of the spleen and liver, weight loss, and potential fatality when left untreated. Presently, there are no specialized antileishmanial drugs or vaccines available for treating leishmaniasis. Nonetheless, several chemotherapeutic options exist. These include antimonial compounds (sodium stibogluconate and meglumine antimoniate), paromomycin, miltefosine, and amphotericin B and its different formulations. Unfortunately, their effectiveness has been compromised by drug resistance, particularly in hyperendemic regions of India.[1] In contrast, liposomal amphotericin B has emerged as a notably effective choice with a reduced risk of resistance development. Although amphotericin B is recognized for its efficacy, it also carries a significant risk of adverse effects.[2,3]

The treatment options for VL are limited and often fail to meet satisfactory standards due to various challenges such as safety concerns, efficacy issues, stability problems, adverse effects, increasing drug resistance, high treatment costs, and the need for hospitalization to complete treatment regimens.[4] One notable issue is that the drugs used to treat leishmaniasis are not specifically designed to target *Leishmania* parasites; instead, they belong to broader categories like antibacterial, antifungal, and anticancer agents.

This introduction aims to offer an in-depth exploration of the intricate mechanisms underlying the pathogenesis of VL while setting the stage for a detailed examination of the molecular intricacies underlying host–parasite interactions, focusing on the role of thiol metabolism. As we delve into the complexities of VL, it becomes apparent that a comprehensive understanding of the pathogenesis is not only essential for unraveling the mysteries of this neglected tropical malady but is also critical for developing targeted therapeutic interventions that can alleviate its burden on global public health.

4.2 PATHOGENESIS OF VISCERAL LEISHMANIASIS

VL pathogenesis refers to the mechanisms by which the disease develops and progresses within the human body during the course of its development. The pathogenesis of VL encompasses a complex series of events that occur within the human body as the disease progresses. This includes the initial infection by the parasite, its spread throughout the body, the host's immune response, tissue

DOI: 10.1201/9781032643120-4

damage, and ultimately the clinical manifestations of the disease.[5,6] Understanding the pathogenesis of VL can be achieved by employing several approaches, which are described below.

4.2.1 Life Cycle of the Parasite

Leishmania's life cycle begins when metacyclic promastigotes are injected into the host by an infected female sandfly during blood-feeding. Within minutes, phagocytic cells like neutrophils and macrophages engulf these promastigotes. Due to their short lifespan, neutrophils give way to macrophages as the final host cells for *Leishmania*'s proliferation. Once inside macrophage phagolysosomes, promastigotes transform into non-motile amastigotes, dividing through binary fission, rupturing macrophages, and infecting nearby ones. When another sandfly bites an infected host, it ingests amastigotes that then differentiate into mature, infectious promastigotes in the sandfly's gut. These infective organisms migrate to the sandfly's mouthparts for transmission during another host's subsequent blood meal.[7]

4.2.2 Causative Agents and Clinical Manifestations

VL is caused by two distinct strains of parasites in different parts of the world. In Asia and Eastern Africa, the disease is caused by a strain called *Leishmania donovani*, where humans act as the primary reservoirs for the parasite. On the other hand, in Latin America and the Mediterranean, the disease is caused by a strain called *Leishmania infantum*, where dogs act as the primary hosts.[8] Clinical manifestations of VL can vary widely based on the host's immune response, ranging from mild to severe.[9–11] The incubation period for VL can vary from 10 days to 24 months, with a typical duration of 2–6 months. The classic symptoms of VL include fever, weight loss, splenomegaly (enlarged spleen, which is seen in most patients), hepatomegaly (enlarged liver), pallor (caused by severe anemia), leukopenia (low white blood cell count), thrombocytopenia (low platelet count), and weight loss. Patients may also experience respiratory or gastrointestinal symptoms such as vomiting and diarrhea. If the disease is severe, malnutrition and lower limb edema may occur, which can progress to anasarca (extreme generalized edema). Other significant signs of VL include hemorrhage, bleeding from the nose or mouth, disseminated intravascular coagulation (DIC), hemophagocytic lymphohistiocytosis (HLH), jaundice, and fluid buildup in the abdomen. Bacterial infections or bleeding are the usual causes of death in these patients.

In some instances, VL may present asymptomatically or as a latent infection that manifests years later with the development of any form of immunodeficiency. There is a possibility that severe disease may be complicated by disseminated intravascular coagulation (DIC) or hemophagocytic lymphohistiocytosis (HLH). After recovery from VL, some individuals may develop post-kala-azar dermal leishmaniasis (PKDL) months or years later – a dermatological condition characterized by macular, papular, or nodular skin lesions on the face, trunk, or limbs. The lesions may be hypopigmented or hyperpigmented and may persist for several months or years. It is important to note that PKDL is not contagious and can be treated with medication.[8,12,13]

4.2.3 Parasite Entry and Invasion: Host–Parasite Interaction

When the parasites are first injected by a sandfly, neutrophils and macrophages are immediately mobilized to the site of infection. The interaction of these cells with the parasites plays a key role in determining the disease outcome. Interestingly, research suggests that more neutrophils are recruited to the infection site when infections are caused by sandfly bites; these neutrophils capture parasites very efficiently, but as neutrophils are transient host cells, they may also serve as "Trojan horses," allowing parasites to covertly infiltrate macrophages and evade cellular activation. It is a macrophage that ingests both free parasites and infected neutrophils after recruitment, serving as the definitive host for parasite replication as well as the effector cell to eliminate parasites.[14]

The phagocytosis of *Leishmania* promastigotes by macrophages is a multifaceted process driven by the interplay of various molecular interactions. The process starts when macrophages recognize pathogen-associated molecular patterns (PAMPs) like lipophosphoglycan (LPG) or gp63 on promastigote surfaces. Glycoinositol phospholipids (GIPLs) are also abundantly present on both promastigotes and amastigotes, where they actively interact with macrophages.[11] The macrophage, on the other hand, contains a number of receptors, including complement receptor 1 (CR1), complement receptor 3 (CR3), fibronectin receptors, and the mannose-fucose receptors. These receptors play an important role in binding promastigotes.[15,16] Opsonins such as complement components C3b/iC3b, galectins, and mannan-binding lectin (MBL) can bind to promastigote surface components like LPG, gp63, and proteophosphoglycans (PPGs) to facilitate their recognition and engulfment.[11,15–17] It has also been shown that toll-like receptors (TLRs) play contrasting roles in the immune response during *Leishmania* infection; TLR-2 activation by LPG promotes parasite survival within macrophages, while TLR-9 activation appears to aid the body in fighting infection by bolstering a host-protective response.[18–21] Following the promastigote-macrophage interaction, the formation of a membrane-bound vesicle known as a phagosome, which encapsulates the promastigotes, occurs around the parasite. Proteins such as Rab GTPases and SNARE proteins facilitate the fusion of these vesicles and the modification of the phagosome's contents. This process aids in the formation and structure of the parasitophorous vacuole, a specialized compartment within the macrophage where the parasite resides.[20]

4.2.4 Parasite Replication and Dissemination

Once inside macrophages, promastigotes differentiate into amastigote forms, which are smaller, rounded, and non-motile versions of the parasites adapted for reproduction within the acidic phagolysosomes of the macrophages. In these phagolysosomes, amastigotes replicate through binary fission, dividing into multiple daughter cells and thereby increasing the parasite population. During this process, the parasites may interfere with the host cell's immune signaling pathways to avoid detection and destruction. As the infection progresses, macrophages harboring amastigote forms of the parasite may encounter an overwhelming load of amastigote particles within their cells. As a result, these infected macrophages could rupture, releasing amastigotes into surrounding tissues such as the bone marrow, spleen, and liver, where they can interact with other macrophages. Various receptors, including the mannose receptor and the Fc receptor, are believed to facilitate the re-entry of amastigotes into host macrophages. Additionally, an infected sandfly can spread the infection by feeding on the blood of a host already harboring amastigotes. During this blood meal, the sandfly can ingest amastigotes alongside the host's blood, which are then absorbed in its stomach[22].

4.2.5 Immunopathogenesis and Cytokine Responses

The host's immune response to *Leishmania* infection is multifaceted and varies based on factors like the specific parasite species involved, the host's vulnerability, and its capacity to combat the infection. The two types of immunity, innate and adaptive, work together to defend against parasites. The immune response to *Leishmania* infection is primarily initiated by innate immune cells like neutrophils, macrophages, and dendritic cells (DCs), which are specialized cells that can recognize and clear invading pathogens. These cells are the first line of defense against parasites and play a critical role in the immune response to *Leishmania* infection. Other cells that contribute to resistance or susceptibility to *Leishmania* infection include natural killer cells, inflammatory monocytes, and mast cells. Effector molecules like cytokines (such as interferon and interleukin), components of the complement system, nitric oxide, and reactive oxygen species (ROS) also assume distinct roles in the host's defense against the infection.[22]

The immune response begins at the precise location where the pathogen enters the body. When promastigotes penetrate the dermis, they trigger the complement cascade, leading to the efficient eradication of over 90% of parasites. *Leishmania* promastigotes are more susceptible to complement-induced lysis compared to metacyclic forms, while amastigotes display higher resistance.[23–26] Furthermore, the complement system opsonizes parasites using fragments like iC3b and C3b, which facilitate their entry into macrophages and neutrophils via complement receptors (CR2).[27] Upon parasite entry, neutrophils are the initial immune cells to migrate to the infection site.[27–29] These cells utilize various mechanisms to eliminate the pathogen, including engulfing and breaking them down within phagolysosomes, while producing ROS and microbicidal agents.[24] Neutrophils also generate a spectrum of cytokines and chemokines, activate diverse immune cell types – including DCs – and supervise T-cell proliferation; they play a critical role in shaping subsequent T-cell-dependent immunity.[30–32]

Macrophages function as the main host for the *Leishmania* parasite and are pivotal in determining disease outcomes. When a parasite enters through a sandfly bite, macrophages can respond to signals from both the parasites and the inflammatory environment around them to display diverse phenotypes. These include M1 macrophages, which are pro-inflammatory, and M2 macrophages, which are anti-inflammatory. How macrophages react – whether they take on an M1 or M2 phenotype – depends on factors such as the species of *Leishmania* involved and the host's immune response. Understanding these cells' fundamental contribution to lesion development enhances our comprehension of M1 and M2 profiles as valuable tools for studying leishmaniasis pathogenesis. The M1 phenotype produces pro-inflammatory cytokines along with ROS and reactive nitrogen species (RNS), aiding in parasite clearance. In contrast, M2 macrophages produce anti-inflammatory cytokines, which support parasite survival.[33–37]

DCs also contribute to shaping disease outcomes.[38,39] Upon infection with *Leishmania*, DCs produce the IL-12p70 subunit, a pivotal player in directing T-helper (Th1) cell polarization. This response bolsters protective immunity by activating natural killer cells, which subsequently produce IFN-γ and engage in cytotoxic activity.[40–43]

In addition to innate immunity, adaptive immune responses involving B-cells and T-cells significantly influence the progression of *Leishmania* infection and the clinical symptoms of the disease. B-cells and T-cells carry specific receptors that identify pathogen-derived proteins attached to major histocompatibility complexes located on the surface of antigen-presenting cells. When they encounter these specific signals, they become activated and differentiate into effector cells. B-cells differentiate into plasma cells, responsible for the production of antibodies and thereby contributing to the humoral response, while T-cells differentiate into either Th1 or Th2 types that contribute to cell-mediated immunity. Th1 responses contribute to the pro-inflammatory reactions that aid in parasite clearance; in contrast, Th2 responses lead to anti-inflammatory reactions that support parasite survival and promote disease progression.[41,43]

4.3 MOLECULAR MECHANISMS OF IMMUNE EVASION

4.3.1 Antigenic Variation in *Leishmania*

Leishmania employs antigenic variation as a strategy to evade the host's immune surveillance, facilitating the parasite's prolonged persistence and survival within hosts. This phenomenon involves altering the surface molecules targeted by the host immune system – such as glycoproteins and lipophosphoglycans – to avoid recognition and clearance by immune cells. One well-studied example is the modulation of LPG, a principal surface glycoconjugate in *Leishmania* parasites whose structure can vary between species, strains, and even within the same strain during infection, contributing to immune evasion. Changes in surface glycoproteins, such as gp63 and PPGs, also occur, affecting antigenicity and immune recognition. Antigenic variation poses a significant challenge to developing effective vaccines and therapeutics against leishmaniasis.[44]

4.3.2 Inhibition of Nitric Oxide Production

The *Leishmania* parasite is known to evade the host's immune response by inhibiting the production of nitric oxide (NO), a key effector molecule involved in killing intracellular pathogens. NO is produced by host macrophages in response to infection and is essential for controlling *Leishmania* replication. However, *Leishmania* has developed mechanisms to counteract NO production, enabling its survival and replication within host cells.[45,46]

One major mechanism employed by *Leishmania* to inhibit NO production involves the modulation of host cell signaling pathways. For example, *Leishmania* can activate protein kinase C (PKC) in host macrophages, leading to the downregulation of inducible nitric oxide synthase (iNOS), the enzyme responsible for NO production. Additionally, *Leishmania* can interfere with the MAPK pathway, which in turn inhibits NO production. Understanding how *Leishmania* inhibits NO production is crucial for developing new therapeutic strategies to combat leishmaniasis.[45,47]

4.3.3 Immunomodulatory Molecules Secreted by *Leishmania*

Leishmania secretes various immunomodulatory molecules that interfere with host signaling pathways, dampening the immune response and promoting its survival. These molecules can interfere with the host's immune signaling pathways, facilitating infection establishment and persistence. One example is LPG, a principal surface molecule in Leishmania, which can modulate the host immune response by impeding macrophage activation and pro-inflammatory cytokine production, thereby leading to a suppressed immune response.[48]

Another example is the secretion of PPGs, which can also modulate the host immune response by inhibiting the activation of macrophages and dendritic cells and by promoting the production of anti-inflammatory cytokines. These immunomodulatory molecules play a critical role in the pathogenesis of leishmaniasis by allowing the parasite to evade host immune defenses and establish chronic infections.[48,49]

4.4 ROLE OF THIOL METABOLISM IN *LEISHMANIA* PATHOGENESIS

It is widely accepted that every living organism has the capability to produce ROS. These molecules are generated either as natural byproducts of aerobic respiration or in response to contact with various noxious substances. ROS can negatively impact cellular components such as nucleic acids (DNA/RNA), proteins, lipids, and membranes. Therefore, to maintain redox homeostasis, aerobic organisms (including the intracellular *Leishmania* parasite, which uses antioxidant defense mechanisms) have developed numerous effective antioxidant defense mechanisms.

The *Leishmania* parasites are exposed to oxidative stress throughout their digenetic life cycle. It has also been reported that Trypanosomatids lack several key enzymes, including glutathione reductase, catalase, thioredoxin reductase, and selenocysteine-containing glutathione peroxidase.[50] To combat the effects of oxidative stress, they utilize superoxide dismutase (SOD), glutathione (GSH), tryparedoxin peroxidase (TXNPx), tryparedoxin (TXN), ascorbate peroxidase (APx), and trypanothione.[4,51] The dithiol trypanothione (T[SH]2) is synthesized by conjugating glutathione with spermidine and is involved in a variety of cellular functions, including the neutralization of ROS (which shields the parasite against oxidative stress), metal detoxification, xenobiotic detoxification, redox homeostasis, and drug resistance within the cells.[52,53] This particular group of intermediate enzymes plays a crucial role in safeguarding the *Leishmania* parasite against oxidative stress caused by ROS during infection by macrophages.[54] In addition to trypanothione, other factors function concurrently to eliminate ROS. Among them is TXN, which detoxifies ROS effects in the presence of tryparedoxin peroxidase. Another factor that works in parallel with trypanothione is APx, classified as heme peroxidase, which plays a vital role in protecting *Leishmania* species against oxidative stress induced by macrophages, thereby aiding parasite survival inside host macrophages.[51,55]

4.5 TISSUE DAMAGE AND ORGAN INVOLVEMENT

VL is characterized by the potential for substantial tissue damage and the engagement of various organs. The dynamic interplay between the parasite and the host's immune system frequently leads to pathological alterations that can culminate in systemic organ dysfunction. In the context of VL, the parasites invade internal organs, namely the spleen, liver, and bone marrow, targeting cells within these vital structures and resulting in hepatosplenomegaly. The bone marrow, which plays a pivotal role in the production of blood cells, also undergoes significant impairment in VL.[56] The invasion by *Leishmania* parasites disrupts normal bone marrow function, contributing to the development of anemia, thrombocytopenia, and leukopenia. These hematological complications further exacerbate the overall impact on the host's health.

In cases of severe VL – especially when left untreated or inadequately treated – the risk of significant organ dysfunction increases markedly. The enlargement of the liver and spleen not only compromises their physiological functions but also adversely affects blood filtration and metabolism. The intricate involvement of these critical organs underscores the severity of VL and highlights the potential for long-term health complications in individuals grappling with this form of leishmaniasis.[8]

4.6 CHALLENGES IN UNDERSTANDING LEISHMANIASIS PATHOGENESIS

Understanding the pathogenesis of leishmaniasis presents several challenges due to the complex interactions between the parasite (*Leishmania* spp.), the host immune response, and the vector (sandfly). The following are some of the key challenges.

4.6.1 Diversity of *Leishmania* Species

There are over 20 species of *Leishmania* that can cause disease in humans, each exhibiting different clinical manifestations and immune responses. Understanding the differences in pathogenesis between these species is crucial yet challenging.[57]

4.6.2 Host Immune Response

The host immune response plays a critical role in determining the outcome of infection. However, the immune response to *Leishmania* is complex and varies depending on factors such as the host's genetic background and previous exposure to the parasite.[58] *Leishmania* parasites have evolved mechanisms to evade the host's immune response, including passive protection, suppression of reactive oxygen or nitrogen intermediates, and modulation of cytokine responses.[59]

4.6.3 Role of the Sandfly Vector

The sandfly vector plays a crucial role in the transmission of *Leishmania* and can influence both the biology and pathogenesis of the parasite. Understanding the interactions between the parasite, vector, and host is essential but challenging due to the complexity of the vector's biology.[60]

4.6.4 Clinical Variability

Leishmaniasis can manifest in several clinical forms – including cutaneous, mucocutaneous, and visceral – making it challenging to understand the factors that determine the clinical outcomes of infection due to variability in clinical presentation and disease progression.[61]

4.6.5 Drug Resistance

The emergence of drug-resistant strains of *Leishmania* poses a significant challenge to the treatment and control of leishmaniasis. Understanding the mechanisms behind drug resistance and developing new treatment strategies are crucial yet difficult tasks.[62]

4.6.6 Lack of Suitable Animal Models

Animal models are essential for studying leishmaniasis pathogenesis. However, finding suitable animal models that accurately mimic human disease is challenging, which limits our understanding of this condition.

Addressing these challenges requires a multidisciplinary approach that includes molecular biology, immunology, epidemiology, and vector biology to gain a comprehensive understanding of leishmaniasis pathogenesis.[61]

4.7 CONCLUSION

VL is caused by *Leishmania donovani* and *Leishmania infantum* and is characterized by parasite dissemination to visceral organs, particularly the spleen, liver, and bone marrow. The pathogenesis of VL involves a complex interplay between the parasite and the host immune response. Upon infection, *Leishmania* parasites are phagocytosed by macrophages, where they replicate within phagolysosomes, evading host immune recognition. This intracellular survival leads to chronic infection and systemic dissemination.

The host immune response plays a critical role in VL pathogenesis. Early in infection, a Th1-type immune response is initiated, characterized by the production of pro-inflammatory cytokines such as IFN-γ. However, *Leishmania* parasites can subvert this response by inducing anergy in T-cells and inhibiting macrophage activation, which allows for parasite persistence. As the infection progresses, there is a shift towards a Th2-type response characterized by the production of anti-inflammatory cytokines such as IL-10. This shift contributes to disease progression and immunopathology, leading to clinical manifestations such as hepatosplenomegaly, pancytopenia, and hypergammaglobulinemia.

In conclusion, the pathogenesis of VL is multifaceted and involves a complex interplay between the parasite and the host immune response. A better understanding of these interactions is crucial for developing effective treatments and vaccines against this devastating disease.

ACKNOWLEDGMENT

The authors are thankful to the ICMR-Rajendra Memorial Research Institute of Medical Sciences, Patna, for its necessary support during this study.

REFERENCES

1. Vanaerschot, M. *et al.* Treatment failure in leishmaniasis: drug-resistance or another (epi-) phenotype? *Expert Rev Anti Infect Ther* **12**, 937–946 (2014).
2. Hamill, R. J. Amphotericin B formulations: a comparative review of efficacy and toxicity. *Drugs* **73**, 919–934 (2013).
3. Dupont, B. Overview of the lipid formulations of amphotericin B. *Journal of Antimicrobial Chemotherapy* **49**, 31–36 (2002).
4. Prakash Singh, O., Singh, B., Chakravarty, J. & Sundar, S. Current challenges in treatment options for visceral leishmaniasis in India: a public health perspective. *Infect Dis Poverty* **5**, 1–15 (2016).
5. Yurdakul, P. Immunopathogenesis of Leishmania infections. *Mikrobiyol Bul* **39**, 363–381 (2005).

6. Poulaki, A., Piperaki, E.-T. & Voulgarelis, M. Effects of visceralising Leishmania on the spleen, liver, and bone marrow: a pathophysiological perspective. *Microorganisms* **9**, 759 (2021).
7. Leta, S., Dao, T. H. T., Mesele, F. & Alemayehu, G. Visceral leishmaniasis in Ethiopia: an evolving disease. *PLoS Negl Trop Dis* **8**, e3131 (2014).
8. Scarpini, S. *et al.* Visceral leishmaniasis: epidemiology, diagnosis, and treatment regimens in different geographical areas with a focus on pediatrics. *Microorganisms* **10**, 1887 (2022).
9. Hong, A., Zampieri, R. A., Shaw, J. J., Floeter-Winter, L. M. & Laranjeira-Silva, M. F. One health approach to leishmaniases: understanding the disease dynamics through diagnostic tools. *Pathogens* **9**, 809 (2020).
10. dos Santos Meira, C. & Gedamu, L. Protective or detrimental? Understanding the role of host immunity in leishmaniasis. *Microorganisms* **7**, 695 (2019).
11. Yasmin, H., Adhikary, A., Al-Ahdal, M. N., Roy, S. & Kishore, U. Host–pathogen interaction in leishmaniasis: immune response and vaccination strategies. *Immuno* **2**, 218–254 (2022).
12. da Silva, G. A. R. *et al.* First case of autochthonous human visceral leishmaniasis in the urban center of Rio de Janeiro: case report. *Rev Inst Med Trop Sao Paulo* **56**, 81–84 (2014).
13. Pedrosa, C. M. Clinical manifestations of visceral leishmaniasis (American visceral leishmaniasis). *The Epidemiology and Ecology of Leishmaniasis,* Claborn, D., Ed., IntechOpen: London, UK, 17–30 (2017).
14. Carreira, J. C. A. & da Silva, A. V. M. The role of neutrophils in the interaction with Leishmania: far beyond a simple trojan horse? *Open J Anim Sci* **11**, 399–421 (2021).
15. Ueno, N. & Wilson, M. E. Receptor-mediated phagocytosis of Leishmania: implications for intracellular survival. *Trends Parasitol* **28**, 335–344 (2012).
16. Liu, D. & Uzonna, J. E. The early interaction of Leishmania with macrophages and dendritic cells and its influence on the host immune response. *Front Cell Infect Microbiol* **2**, 83 (2012).
17. Elmahallawy, E. K. & Alkhaldi, A. A. M. Insights into Leishmania molecules and their potential contribution to the virulence of the parasite. *Vet Sci* **8**, 33 (2021).
18. Halliday, A., Bates, P. A., Chance, M. L. & Taylor, M. J. Toll-like receptor 2 (TLR2) plays a role in controlling cutaneous leishmaniasis in vivo, but does not require activation by parasite lipophosphoglycan. *Parasit Vectors* **9**, 1–14 (2016).
19. Srivastava, S., Pandey, S. P., Jha, M. K., Chandel, H. S. & Saha, B. Leishmania expressed lipophosphoglycan interacts with Toll-like receptor (TLR)-2 to decrease TLR-9 expression and reduce anti-leishmanial responses. *Clin Exp Immunol* **172**, 403–409 (2013).
20. Moradin, N. & Descoteaux, A. Leishmania promastigotes: building a safe niche within macrophages. *Front Cell Infect Microbiol* **2**, 121 (2012).
21. Bhattacharya, P. *et al.* Toll-like Receptor-9 (TLR-9) signaling is crucial for inducing protective immunity following immunization with genetically modified live attenuated Leishmania parasites. *Pathogens* **12**, 534 (2023).
22. Handman, E. & Bullen, D. V. R. Interaction of Leishmania with the host macrophage. *Trends Parasitol* **18**, 332–334 (2002).
23. Von Stebut, E. Cutaneous Leishmania infection: progress in pathogenesis research and experimental therapy. *Exp Dermatol* **16**, 340–346 (2007).
24. Regli, I. B., Passelli, K., Hurrell, B. P. & Tacchini-Cottier, F. Survival mechanisms used by some Leishmania species to escape neutrophil killing. *Front Immunol* **8**, 1558 (2017).
25. Kolářová, I. & Valigurová, A. Hide-and-seek: a game played between parasitic protists and their hosts. *Microorganisms* **9**, 2434 (2021).
26. Pereira Filho, A. A. *et al.* Evasion of the complement system by Leishmania through the uptake of factor H, a complement regulatory protein. *Acta Trop* **224**, 106152 (2021).
27. Nylen, S. & Gautam, S. Immunological perspectives of leishmaniasis. *J Glob Infect Dis* **2**, 135 (2010).
28. Toepp, A. J. & Petersen, C. A. The balancing act: immunology of leishmaniosis. *Res Vet Sci* **130**, 19–25 (2020).
29. van Gisbergen, K. P. J. M., Sanchez-Hernandez, M., Geijtenbeek, T. B. H. & van Kooyk, Y. Neutrophils mediate immune modulation of dendritic cells through glycosylation-dependent interactions between Mac-1 and DC-SIGN. *J Exp Med* **201**, 1281–1292 (2005).
30. Bennouna, S., Bliss, S. K., Curiel, T. J. & Denkers, E. Y. Cross-talk in the innate immune system: neutrophils instruct recruitment and activation of dendritic cells during microbial infection. *The Journal of Immunology* **171**, 6052–6058 (2003).
31. Gasperini, S., Calzetti, F., Bazzoni, F. & Cassatella, M. A. The neutrophil as a cellular source of chemokines. *Immunol Rev* **177**, 195–203 (2000).

32. Tecchio, C., Micheletti, A. & Cassatella, M. A. Neutrophil-derived cytokines: facts beyond expression. *Front Immunol* **5**, 508 (2014).
33. Costa-da-Silva, A. C. *et al.* Immune responses in leishmaniasis: an overview. *Trop Med Infect Dis* **7**, 54 (2022).
34. Arango Duque, G. & Descoteaux, A. Macrophage cytokines: involvement in immunity and infectious diseases. *Front Immunol* **5**, 491 (2014).
35. Zanluqui, N. G., Wowk, P. F., Pinge-Filho, P. & others. Macrophage polarization in Chagas disease. *J Clin Cell Immunol* **6**, 1–6 (2015).
36. Tomiotto-Pellissier, F. *et al.* Macrophage polarization in leishmaniasis: broadening horizons. *Front Immunol* **9**, 2529 (2018).
37. Almeida, F. S. *et al.* Leishmaniasis: Immune Cells Crosstalk in Macrophage Polarization. *Trop Med Infect Dis* **8**, 276 (2023).
38. von Stebut, E. & Tenzer, S. Cutaneous leishmaniasis: Distinct functions of dendritic cells and macrophages in the interaction of the host immune system with Leishmania major. *International Journal of Medical Microbiology* **308**, 206–214 (2018).
39. Gutiérrez-Kobeh, L. & Wilkins-Rodríguez, A. A. GK1 improves the immune response induced by dendritic cells of BALB/c mice infected with *Leishmania mexicana* promastigotes. *Acta Parasitol* **65**, 27–35 (2020).
40. Rossi, M. & Fasel, N. How to master the host immune system? Leishmania parasites have the solutions! *Int Immunol* **30**, 103–111 (2018).
41. Schleicher, U. *et al.* NK cell activation in visceral leishmaniasis requires TLR9, myeloid DCs, and IL-12, but is independent of plasmacytoid DCs. *J Exp Med* **204**, 893–906 (2007).
42. Liese, J., Schleicher, U. & Bogdan, C. TLR9 signaling is essential for the innate NK cell response in murine cutaneous leishmaniasis. *Eur J Immunol* **37**, 3424–3434 (2007).
43. Rivera-Fernández, I., Argueta-Donohué, J., Wilkins-Rodríguez, A. A. & Gutiérrez-Kobeh, L. Effect of two different isolates of *Leishmania mexicana* in the production of cytokines and phagocytosis by murine dendritic cells. *J Parasitol* **105**, 359–370 (2019).
44. Akopyants, N. S. *et al.* Expression profiling using random genomic DNA microarrays identifies differentially expressed genes associated with three major developmental stages of the protozoan parasite Leishmania major. *Mol Biochem Parasitol* **136**, 71–86 (2004).
45. Mukhopadhyay, D. *et al.* M2 polarization of monocytes-macrophages is a hallmark of Indian post kala-azar dermal leishmaniasis. *PLoS Negl Trop Dis* **9**, e0004145 (2015).
46. Olivier, M., Gregory, D. J. & Forget, G. Subversion mechanisms by which Leishmania parasites can escape the host immune response: a signaling point of view. *Clin Microbiol Rev* **18**, 293–305 (2005).
47. Bogdan, C. Nitric oxide and the immune response. *Nat Immunol* **2**, 907–916 (2001).
48. Zhang, S., Kim, C. C., Batra, S., McKerrow, J. H. & Loke, P. Delineation of diverse macrophage activation programs in response to intracellular parasites and cytokines. *PLoS Negl Trop Dis* **4**, e648 (2010).
49. Olivier, M., Atayde, V. D., Isnard, A., Hassani, K. & Shio, M. T. Leishmania virulence factors: focus on the metalloprotease GP63. *Microbes Infect* **14**, 1377–1389 (2012).
50. Flohe, L., Hecht, H. J. & Steinert, P. Glutathione and trypanothione in parasitic hydroperoxide metabolism. *Free Radic Biol Med* **27**, 966–984 (1999).
51. Kumar, A. *et al.* Ascorbate peroxidase, a key molecule regulating amphotericin B resistance in clinical isolates of *Leishmania donovani*. *Antimicrob Agents Chemother* **58**, 6172–6184 (2014).
52. Fairlamb, A. H. & Cerami, A. Metabolism and functions of trypanothione in the Kinetoplastida. *Annu Rev Microbiol* **46**, 695–729 (1992).
53. Equbal, A. *et al.* Stage-dependent expression and up-regulation of trypanothione synthetase in amphotericin B resistant *Leishmania donovani*. *PLoS One* **9**, e97600 (2014).
54. Iyer, J. P., Kaprakkaden, A., Choudhary, M. L. & Shaha, C. Crucial role of cytosolic tryparedoxin peroxidase in *Leishmania donovani* survival, drug response and virulence. *Mol Microbiol* **68**, 372–391 (2008).
55. Adak, S. & Pal, S. Ascorbate peroxidase acts as a novel determiner of redox homeostasis in Leishmania. *Antioxid Redox Signal* **19**, 746–754 (2013).
56. Molina, R. *et al.* Role of asymptomatic and symptomatic humans as reservoirs of visceral leishmaniasis in a Mediterranean context. *PLoS Negl Trop Dis* **14**, e0008253 (2020).
57. Kaye, P. & Scott, P. Leishmaniasis: complexity at the host–pathogen interface. *Nat Rev Microbiol* **9**, 604–615 (2011).
58. Reiner S. L., Richard M. L. The regulation of immunity to Leishmania major." Annual review of immunology 13 (1995): 151-177.

59. Bogdan, C. & Röllinghoff, M. The immune response to Leishmania: mechanisms of parasite control and evasion. *Int J Parasitol* **28**, 121–134 (1998).
60. Kedzierski, L. & Evans, K. J. Immune responses during cutaneous and visceral leishmaniasis. *Parasitology* **141**, 1544–1562 (2014).
61. McCall, L.-I., Zhang, W.-W. & Matlashewski, G. Determinants for the development of visceral leishmaniasis disease. *PLoS Pathog* **9**, e1003053 (2013).
62. Sacks, D. & Noben-Trauth, N. The immunology of susceptibility and resistance to Leishmania major in mice. *Nat Rev Immunol* **2**, 845–858 (2002).

5 A Current Scenario of Drug Discovery Against Visceral Leishmaniasis

Kamal Singh, Bhavani R.,and Swati Sharma*
ICMR- Department of Virology, Rajendra Memorial Research Institute of Medical Sciences, Patna, Bihar, India
*Corresponding Author: Dr. Kamal Singh, ICMR-Rajendra Memorial Research Institute of Medical Sciences, Patna, Bihar, India

5.1 INTRODUCTION

Visceral leishmaniasis, also known as kala-azar, is a neglected equatorial disease that has affected almost 12 million people globally, with over 350 million people at high risk of developing it. According to the World Health Organization, between 700,000 and 1 million new cases are reported each year. This infection is characterised by anaemia, enlarged liver and spleen, intermittent fever and weight loss. The regions most frequently exhibiting cases of leishmaniasis include East Africa, India and Brazil (World Health Organization, Leishmaniasis fact sheets).

Currently, the only medications available for treating leishmaniasis are pentavalent antimonials (Pentostam), amphotericin B (AmBisome), paromomycin, miltefosine, sitamaquine and other azole derivatives. These first- and second-line medications frequently exhibit a number of negative side effects related to toxicity, teratogenicity, lack of selectivity and ignorance of the targets and their correct processes (Ritmeijer et al. 2001, Berman 2003). Moreover, significant challenges in drug design include the differing pharmacokinetics requirements for the various *Leishmania* species, the lack of specificity of the drug for all 17 species infecting humans, depleted immune response alongside resistance to standard drugs and dissatisfaction with the clinical trial process due to both short-term and long-term toxicity. Therefore, greater focus should be placed on developing novel medications with higher efficacy and specificity while minimising or eliminating toxicity in both in vivo and clinical research (Saporito et al. 2013).

The drug discovery process is cumbersome and involves extensive analysis and studies. The phases of discovery and development, along with ongoing preclinical and clinical investigations, are essential for creating new medications. These phases are then submitted to the US Food and Drug Administration (FDA) for approval and followed by post-marketing surveillance.

The general drug discovery process follows two approaches: ligand-based drug design and structure-based drug design. The choice between these approaches depends on the type of target structure available and is related to the quantitative structure-activity relationship of different chemical structures. This entirely depends on the knowledge of the structure availability for the targeted protein molecule. In order to satisfy the ideal conditions necessary to develop the hit molecule into a successful drug candidate, hit and lead identification must adhere to a set of criteria (Patel et al. 2021).

Different approaches to drug design, such as drug repositioning along with combination therapy, have shown improved working mechanisms against leishmaniasis infection. Repurposing drugs typically involves less time and cost since these medications are already FDA-approved. Additionally,

DOI: 10.1201/9781032643120-5

creating analogues of drug candidates that have been abandoned throughout various stages of clinical trials can be reconfigured into potential therapeutic molecules by modifying functional groups or atoms to reduce current side effects.

Moreover, phenotype-based drug discovery can be pursued when the precise mechanism of action for a disease is unspecified. This approach utilizes assays with the help of stem cells, cell lines and other ex vivo environments to detect early toxicity of small molecules under treatment. It also employs faster imaging methods, biomarker analysis and other multi-parametric phenotypic assessments, obviating the necessity for expertise in target proteins (Comley 2015).

Various omics approaches, such as genomics, transcriptomics and proteomics, are increasingly used to detect biomarkers, important target molecules and differentially expressed genes related to disease conditions. Traditional methods for identifying targeted proteins linked to *Leishmania* infection include MALDI-TOF and 2D-GE. Modern techniques like mass spectrometry, single/multiple reaction monitoring, mass spectrometry, Sequential Window Acquisition of All Theoretical Mass Spectra (SWATH-MS) and label-free quantification procedures allow for selective quantification of small molecules within compound mixtures (Lin et al. 2023).

Many novel drugs are designed based on the different approaches mentioned above. DDD01305143 and DDD853651 are two newly developed drugs that target proteasomes and cdc2-related kinase 12 (CRK12), respectively. Advancements in genomics and sequencing have facilitated the analysis of different targets responsible for pathogen association in the host system, as entire sequences can be collected with the help of Next Generation Sequencing (NGS) methods (Salloum et al. 2021). This chapter mainly focusses on the current landscape of the drug design process and the novel approaches developed to guarantee high levels of safety and effectiveness in drug candidates.

5.2 CURRENT TREATMENT OPTIONS

Protozoan research has traditionally focussed on identifying and characterising cellular targets while addressing medication resistance in leishmaniasis. Recent developments in creative parasitic and recombinant gene constructions in animal models have enabled rapid high-throughput drug screening techniques for both in vitro and in vivo studies. However, many side effects remain associated with conventional anti-leishmanials currently available on the market, which are presented in Figure 5.1.

Therefore, the focus should be on developing new medication candidates that target various leishmanial species while minimising or eliminating harsh effects on the host system. The following are the anti-leishmanial drugs currently available.

5.2.1 Current Drugs for Leishmaniasis

5.2.1.1 Pentavalent Antimonials

Conventionally, the treatment strategies for visceral leishmaniasis (VL) have relied on pentavalent antimonials, particularly generic sodium stibogluconate. These strategies were instigated in the 1940s by Albert David Ltd in India (Berman 2003). This medication was shown to have a good therapeutic index compared to other generic drugs. A study conducted in Sudan revealed that 95.9% of 250 patients were healed with generic sodium stibogluconate, while 91.3% were cured with Pentostam (Veeken et al. 2000). Another study conducted with Ethiopian patients resulted in a VL cure rate of 43.5% for VL-HIV coinfected patients and 92.1% for HIV-negative patients treated with Pentostam (n = 104) and generic sodium stibogluconate (n = 95) (Ritmeijer et al. 2001). These drugs are administered through intramuscular injections at a dosage of 20 mg/kg/day for 30 days (Esfandiarpour and Alavi 2002). The primary mechanisms involved in the destruction of the pathogen include apoptosis, DNA fragmentation, adenosine diphosphate phosphorylation and fatty acid beta-oxidation (Sereno et al. 2001; Sudhandiran and Shaha 2003).

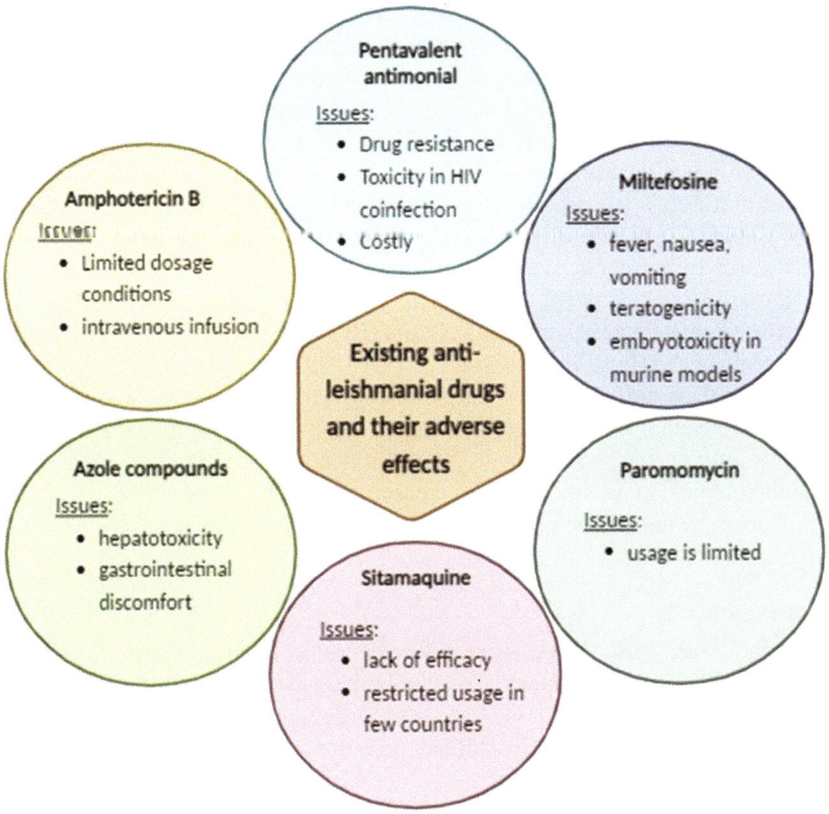

FIGURE 5.1 The standard anti-leishmanial drugs available on the market and their adverse effects

However, these medications have significant drawbacks. They are unaffordable and not cost-effective. Moreover, the patients treated with these drugs may develop resistance, along with some serious health issues such as cardiotoxicity, toxicity and pancreatic enzyme activation (Chulay et al. 1985; Maltezou 2010). Pentavalent antimonials also prevent the production of trypanothione reductases, an important enzyme that protects host reactive nitrogen and oxygen species from the pathogen (Wyllie et al. 2004).

5.2.1.2 Amphotericin B

Amphotericin B is mainly used to treat fungal infections and VL. AmBisome is made of amphotericin B conjugated with liposomes for the active drug delivery mechanism. It is a type of macrocyclic antifungal antibiotic polyene that is given in short doses to increase efficacy following different injection dosages. The liposomal emulsion of the drug helps increase the cure rate to 93% (Sundar et al. 2001).

The major adverse effects include the induction of fever and chills in 87% of treated patients. Its high cost range limits its use in poor countries (Sundar et al. 2000). A randomised study involving children under the age of 14 demonstrated that both daily and alternate-day dosing regimens of amphotericin B resulted in no relapse of VL; however, the daily dosage was more cost-effective (Singh et al. 2010).

5.2.1.3 Miltefosine

Miltefosine is an alkylphosphocholine drug initially used as an anti-cancer agent but later repurposed for treating various bacterial, fungal and parasitic infections. It is a synthetic analogue of lysolecithins that lacks a glycerol backbone (Dorlo et al. 2012). Impavido is the licenced product used to treat human VL; it contains microcrystalline cellulose, titanium dioxide, lactose monohydrate, silicon dioxide and other minor constituents (Authorities 2008). Developed by Zentaris/ASTA Medica, this drug underwent extensive testing to ensure safety and efficacy (Sindermann et al. 2006). A randomised clinical trial involving 299 VL patients showed no disease relapse following six months of oral drug administration. Ninety-four percent of patients showed no recurrence, confirming the oral drug's efficacy and safety (Sundar et al. 2002). Miltefosine's method of action against the *Leishmania* parasite is based on the metazoan apoptosis of cell death. Treatment with miltefosine has led to nuclear condensation of *Leishmania donovani* promastigotes, which confirms that apoptosis-like cell death occurs. Extensive DNA fragmentation was also observed in miltefosine-treated amastigotes (Verma et al. 2004). The drug also disrupts the phosphatidylcholine biosynthesis mechanism (Haase et al. 1991).

Most adverse effects are gastrointestinal, with frequent symptoms of vomiting, diarrhoea and nausea. There have been insufficient clinical studies with pregnant women; therefore, it is better to avoid using this drug during pregnancy. In rats, treatment with miltefosine at a dosage of 1–2 mg/kg in the embryological stage has shown increased teratogenicity and embryotoxicity (Soto and Soto 2006). Currently, only a few countries, such as India, Germany and Colombia, have licenced the drug.

5.2.1.4 Paromomycin

Discovered in the 1960s, paromomysin is an aminoglycoside antibiotic administered in intramuscular, intravenous and topical applications. It acts as both an anti-leishmanial and antibacterial agent. Paromomycin sulphate (PM) is a Class III drug according to the biopharmaceutical classification system (Matos et al. 2020). It primarily inhibits protein synthesis and affects mitochondria-based degradation. In 2006, India approved the usage of PM for the treatment of VL. Studies indicated a 95% cure rate when administered at a dosage of 11 mg/kg/day for 21 days (Sundar et al. 2007).

A Phase IIIb trial conducted in Bangladesh provided key findings regarding the safety and efficacy of the drug. Out of 120 subjects, a 94.2% success rate was observed; however, 28.6% of the subjects experienced pain at the injection site as an adverse effect (Jamil et al. 2015). Paromomycin injections inhibit macromolecular and protein synthesis in *L. donovani* promastigotes, resulting in growth arrest and decreased lipid metabolism and membrane fluidity (Maarouf et al. 1997). Since its usage is limited, it does not create any resistance and thus has fewer side effects compared to other drug treatments.

5.2.1.5 Sitamaquine

Sitamaquine is an 8-aminoquinoline compound effective against VL infections. The drug was synthesised and developed by the Walter Reed Army Institute and GlaxoSmithKline for oral administration. A Phase II clinical trial demonstrated promising results in VL patients; however, a few subjects showed mild adverse effects such as abdominal aches, dyspepsia and glomerulonephritis (Wasunna et al. 2005). The mechanism of action involves sitamaquine binding to positively charged lipids in the parasite's membrane, leading to the lysis of the parasite's cell membrane (Loiseau et al. 2011).

While reports of drug resistance are currently absent, concerns about efficacy and restriction in further phases of clinical trials have led to its avoidance in some countries (Singh et al. 2012).

5.2.1.6 Azole Compounds

Some of the azole compounds, such as itraconazole, fluconazole and ketoconazole, inhibit cytochrome P450-based methylation of lanosterol in fungi and block ergosterol biosynthesis. This mechanism is similar to that observed in leishmanial sterol biosynthesis, particularly in *L. mexicana*, which is relevant for the treatment of cutaneous leishmaniasis (CL) (Beach et al. 1988). An open-controlled study conducted in India concluded that out of ten CL patients receiving oral itraconazole, seven were clinically cured based on urinalysis and liver function tests (Dogra and Saxena 1996). Despite its effectiveness in treating CL, itraconazole can cause hepatotoxicity and gastrointestinal discomfort in some cases (Alrajhi et al. 2002).

5.2.1.7 Diamine Pentamidine

Pentamidine, an aromatic diamine available as isethionate and methanesulfonate salts, is primarily used for treating VL. Although the precise mechanism of action is unclear, it is assumed that the medicine enters *L. donovani* promastigotes through the membrane proteins by utilising arginine and polyamine transporter proteins (Pinheiro and de Souza 2022). A clinical study revealed a differential cure rate for pentamidine of 35% in Peru compared to 96% in Colombia. This disparity is attributed to varying efficacy criteria and differences in leishmanial species organisation (Andersen et al. 2005).

The high toxicity of pentamidine prevents it from being used to treat leishmaniasis; it can cause myocarditis, hypoglycaemia, nephrotoxicity, diabetes mellitus and other side effects. Pentamidine is processed by cytochrome P450 enzymes in human liver microsomes, and *Leishmania* parasites may reduce systemic levels of pentamidine in patients as a result (Pinheiro and de Souza 2022).

5.2.2 Combination Therapy

The repurposing of drugs, along with combination therapy, has opened new avenues for scientists to identify treatments with higher efficacy and reduced toxicity to the host system. The aim of delaying the resistance of the parasite to the drug, cost-effectiveness, increasing drug half-life and enabling slow and targeted release of drug compounds was achieved using combination therapy. The combination of allopurinol with a half dose of pentamidine has been shown to be more effective than the single regimen. Additionally, lower dosages of pentamidine combined with allopurinol showed no indication of hyperglycaemia, whereas a single regimen of pentamindine resulted in irreversible hyperglycaemia in the subjects (Das et al. 2001).

5.2.3 Challenges in the Drug Discovery Approach

The chemotherapy currently used to treat two types of leishmaniasis caused by *L. donovani* in humans and *L. infantum* in other animals relies on drugs that have lost their effectiveness because of the resistance developed by the parasites. Leishmaniasis is a neglected tropical disease found in nearly 100 countries, and diagnosis occurs in only 15% of these countries. Major challenges include complex host-parasite interactions, complicated vector-reservoir complex, ineffective targeting, drug resistance, invasive diagnosis and treatment procedures and cost-related issues.

Certain specific biomolecular structures can be targeted along with the pathways used by the parasite during invasion. This approach is known as molecular therapy. Components such as cells, tissue, organs, molecular pathways and physiological sites can be targeted in such cases. However, even targeted therapy faces many hindrances, such as difficulties in identifying target-specific lead compounds, a lack of understanding of the immunocharacteristics of the pathogen associated with the disease, challenges in passing the in vitro assays and issues with acceptance through clinical trials. The development of resistance due to the genome plasticity of *Leishmania* is also an important factor to consider (Gupta et al. 2023).

5.2.3.1 Blockade in the Cellular Mechanism

Leishmania species show a higher degree of flexibility in residing within different cells in the host organism, making it difficult to target specific infection sites. This leads to reduced drug accessibility and increases the time required for treating infections. Most cases are asymptomatic, and the pathogen can invade different body locations such as cartilage, the central nervous system, placenta and testes (Pacheco-Fernandez et al. 2023). There is evidence showing the localisation of

Leishmania species in the bone marrow, resulting in latent infections. Experiments conducted in mouse models showed a decrease in platelet count due to granulomatous inflammation in the bone marrow, splenectomised mouse models exhibited little or no signs of thrombocytopenia (Rani et al. 2021).

5.2.3.2 Cons of the Existing Anti-Leishmanial Drugs

Pentavalent antimonials such as sodium stibogluconate (Pentostam) and meglumine antimoniate (Glucantime) are primary medications for leishmaniasis. Second-line medications include pentamidine, azole derivatives and amphotericin B; however, these options are toxic and necessitate prolonged parenteral administration. A novel oral medication called miltefosine, initially created to treat cancer, has been given the go-ahead in India to treat VL. The 8-aminoquinoline derivative sitamaquine and the aminoglycoside antibiotic paromomycin are two novel medications currently undergoing clinical testing (Handman et al. 2008).

5.2.3.3 Selectivity and Safety Issues Related to Lead Molecule Identification

A group of scientists from the Bill and Melinda Gates Foundation, along with the authorities from the Medicines for Malaria Venture, Drugs for Neglected Diseases Initiative and TB Alliance, has come up with general criteria satisfying the hit and lead identification for four tropically neglected diseases, including VL. Finalising a lead molecule based on potency alone is difficult because it may not be very specific to the target. The process of hit identification and finalisation depends on two essential categories: physicochemical properties and biological profile. The former focusses on selectivity, specificity and potency, while the latter mainly depicts the basis of drug mechanisms inside the body, such as absorption, distribution, metabolism, excretion and toxicity (ADMET) profiling (Brown 2016). Orthogonal in vitro assays should be approached as early as possible if the mechanism of action is completely unpredictable or unknown. Moreover, early detection of the dose-response curve against the drug over relative mammalian cell lines is critical in reducing dosage-related toxicity (Balana-Fouce et al. 2019).

5.2.3.4 Complications in Host-Parasite Interactions

Both immunopathology and antiparasitic resistance depend heavily on macrophages. Through their innate immunological capabilities and promotion of a protective type 1 T-helper cell (Th1) response, these sentinel cells actively contribute to containing and eliminating *Leishmania*. Intricate and dynamic relationships have coevolved between intracellular *Leishmania* and their host cells. *Leishmania* species have developed ways to resist both adaptive and innate immune responses, leading to immunological dysregulation and diseases typical of CL and VL. This allows the parasite to proliferate and remain infected within the mammalian host for an extended period. The parasite subverts the host cells' immune system while gaining nutrients for its development and proliferation (Lamotte et al. 2017). This hinders the process of drug discovery; thus, cells of the host system can be targeted in such cases.

5.3 EMERGING DRUGS AGAINST LEISHMANIASIS

Treatment strategies for VL have largely failed with existing drugs such as pentavalent antimonials, amphotericin B, miltefosine, paromomycin, sitamaquine, azole compounds, diamine pentamidine, etc. The side effects associated with these drugs include toxicity, drug resistance, in vitro and in vivo assay testing, costly synthesis processes and limited availability in developing countries. Table 5.1 depicts the different investigational drugs and their critical experimental results predicted in both in vivo and in vitro assays. The results of the pharmacokinetic studies conducted with human subjects are also provided.

Single-regime drug use did not exhibit much disease specificity or selectivity. In light of these facts, it was determined that employing drug combinations in conjunction with the repurposing

TABLE 5.1
List of Emerging Investigational Leishmaniasis Drugs and Their Effectiveness in the In Vitro and In Vivo Models

Derivative	Compounds	Species Used	In Vitro Study	In Vivo Study	Human Trials	References
Quinoline derivatives	4-Aminoquinaldine compounds	*L. donovani* promastigotes and amastigotes	PP-9 and PP-10 showed better efficacy rates with half maximal inhibitory concentration (IC_{50}) values ranging from 0.94 to 127μM.	Effective in BALB/c mice infected with sodium gluconate-sensitive and resistant strains of *L. donovani*.	Human testing is yet to be conducted.	(Palit et al. 2012)
	2-substituted quinoline	*L. donovani*	Halogen substitution improved the metabolic stability of the lead compound with an IC_{50} value of 0.2 μM.	An 84.26 % inhibition rate was observed in the hamster model with infection.	Clinical trials are to be conducted with the *Leishmania*-HIV coinfected patients.	(Nakayama et al. 2005; Gopinath et al. 2013)
	8-Aminoquinoline (WR-6026)	*L. donovani*	Oral test for VL.	The hamster model showed the compound to be 708 times more effective than meglumine antimoniate.	In India, a randomised trial with 106 patients at a dosage of 1.5 mg/kg/day showed a success rate of 87% with no relapse of the disease.	(Kinnamon et al. 1978; Wasunna et al. 2005)
Nitroimidazole compounds	Fexinidazole	*L. donovani* (strain LdBOB)	Fexinidazole had half maximal effective concentration (EC_{50}) values of 5.6 ± 0.2 μM and 2.8 ± 0.1 μM against promastigotes and amastigotes, respectively.	With five single doses of 200 mg/kg/day administered to BALB/c mice, *L. donovani*'s in vivo sensitivity to the medication was outstanding, suppressing infection by 98.4%.	A Phase II proof-of-concept study is ongoing to assess the effectiveness of fexinidazole at 1800 mg/day for four days, with another regimen of 1200 mg per day for six days in Sudanese VL patients currently seeking patients.	(ClinicalTrials.gov; Wyllie et al. 2012)
	PA-824	*L. donovani* (strain LdBOB)	The lead compound was tested against promastigotes and intracellular amastigotes in peritoneal macrophages.	After four hours, mice were given either (S)- or (R)-PA-824 at a dosage of 50 mg/kg, reaching maximal blood concentrations of 11,600 or 10,500 ng/ml, respectively; blood levels for (R)-PA-824 dropped to undetectable levels between 8 and 24 hours later.	In a Phase I clinical investigation, (S)-PA-824 exposure was elevated when given with a high-calorie, high-fat meal compared to fasting conditions.	(Ginsberg et al. 2009; Patterson et al. 2013)

(Continued)

TABLE 5.1 (CONTINUED)
List of Emerging Investigational Leishmaniasis Drugs and Their Effectiveness in the In Vitro and In Vivo Models

Derivative	Compounds	Species Used	In Vitro Study	In Vivo Study	Human Trials	References
Naphthaquinones	Buparvaquone and its derivatives	*L. donovani*	Tested on macrophages infected with the *Leishmania* species, showing high activity against the parasites.	The compound was less active in the BALB/c mice model.	Clinical trials are to be conducted.	(Croft et al. 1992)
Imipramine	N-(dimethylaminopropyl)-iminodibenzyl HCl	*L. donovani*	Both IC_{50} and EC_{50} values were similar in sodium stibogluconate-resistant (SSG-R) strains.	A dosage of 5 mg/kg/day for four weeks completely eliminated the parasites in both SSG-R and sodium stibogluconate-sensitive (SSG-S) hamsters.	The FDA approved the oral formulation of the drug after treatment in infected hamsters.	(Mukherjee et al. 2012)

technique is a helpful protocol in the urgent search for innovative, fast, secure and affordable therapies for leishmaniasis. Combinatorial therapy is therefore essential in managing the symptoms brought on by the disease conditions. Since miltefosine is the sole oral medication that is already widely known, it can be used in combination with nifuratel to improve treatment efficacy and specificity. This nifuratel-miltefosine combination has shown positive results in reducing viral load in the thymus and bone marrow cells, as experimentally proven in 6- to 8-week-old female BALB/c mice. Similarly, dosage prediction was also achieved using the in vitro and ex vivo models (Melcon-Fernandez et al. 2023).

Another study aimed at identifying compounds with leishmanicidal activity demonstrated the effective activity of laurinterol, aplysin, debromolaurinterol, isolaurinterol and laurequinone, all of which exhibited better IC_{50} values against promastigotes. Among these lead compounds, laurequinone was as the most promising, with a very low IC_{50} value of 1.87 μM. This study made its conclusion by validating the levels of reactive oxygen species production, cell shrinkage, mitochondrial membrane disruption and dysregulation of membrane permeability against *L. amazonensis* promastigotes after treatment (García-Davis et al. 2023).

The field of drug development is a continuous process that has significantly impacted the evaluation of potential novel treatments for VL and related diseases. New chemical entities are being developed by the Drugs for Neglected Diseases initiative (DNDi), which ensures improved pharmacokinetic approaches and safety profiling for newer drugs.

These novel chemical entities include two compounds identified as proteasome inhibitors: one from the University of Dundee-GlaxoSmithKline collaboration (GSK3494245/DDD01305143) and another from Novartis, along with GSK3186899/DDD853651, which evidence suggests inhibits CRK12 (LXE408). These substances block the proteasome subunit's "chymotrypsin-like" active site (Wyllie et al. 2018, 2019; Thomas et al. 2021).

5.4 DRUG DISCOVERY APPROACHES

5.4.1 Repurposing of Existing Drugs

Drug repurposing, also called drug repositioning, involves identifying new therapeutic uses for existing drugs. Usually, the discovery of a new drug compound takes 10–17 years, encompassing all in vitro and in vivo assays, preclinical and clinical trials and pharmacovigilance criteria after the drug's release into the market. This process costs approximately $1.8 billion and has a very low success rate of 10%. Long-term side effects are a major reason for pulling accepted drugs from the market. In contrast, drug repurposing requires only 3–12 years for the entire process to be completed. Nearly 1–1.5 years is focussed on the compound identification process, with up to another 1.5 years for compound acquisition procedures. Typically, 6–8 years are dedicated to the preclinical and clinical processes to validate the drug molecule. By following these steps, the cost can be reduced to 20–30%, making the approach more potent and effective.

Alternative tactics are now required, and the practice of drug repositioning, or repurposing, has become more popular. Developing a new chemical entity involves several phases; however, with drug repositioning, candidates already passed numerous checkpoints and possess information about their pharmacokinetic and safety profiles, which significantly shortens the adverse effects of adopting this method. This trend is reflected in the growing success of businesses examining their libraries for potential repositioning candidates and finding new applications for drugs outside of their initial prescription (Ashburn and Thor 2004).

Some anti-cancer drugs can also be repurposed as anti-leishmanials, such as kinase inhibitors. One such drug is miltefosine, which was used to treat breast cancer. The targeted pathway in this case is the PI3K/Akt/PkB pathway (Dorlo et al. 2012). Fexinidazole, a nitroaromatic compound, advanced into clinical trials as a therapy for VL based on these encouraging results. In Phase II research conducted in 2013, fexinidazole was administered to 14 VL patients, all of whom demonstrated clinical improvement; however, relapses occurred, making it currently unavailable for VL treatment (ClinicalTrials.gov). Nonetheless, combinatorial treatment with miltefosine seemed to be effective in clinical trials conducted in Eastern Africa.

Successfully repositioned antimicrobial and antifungal drugs include paromomycin and pentamidine. Fluconazole, an antifungal drug, showed good levels of response against *L. major* in Phase II clinical trials but was terminated in Phase III due to low efficacy and species dependency (ClinicalTrials.gov).

Tucaresol, an antiviral drug compound previously used as an anti-HIV drug, now shows potential as a lead compound for reducing amastigote loads in the liver. The optimal oral dosage of the drug is 5 mg/kg, resulting in a depletion of 44–62% of amastigotes (Smith et al. 2000).

Synthesis of analogues related to cinnarizine is feasible since it has shown promising results as a calcium channel blocker against intracellular *L. infantum* amastigotes, with IC_{50} values ranging from 13 to 84 μM (Pinto et al. 2014).

Recent studies indicate that statins may also be effective in working against leishmaniasis infection. One example is mevastatin, used to treat heart diseases, which now can act upon the recombinant *L. donovani* 3-hydroxy-3-methylglutaryl coenzyme A with a reasonable IC_{50} value of 42.2 ± 3 μM (Dinesh et al. 2015).

5.4.2 Phenotype-Based and Target-Based Drug Discovery Processes

Identifying hit compounds is essential for finding lead chemicals that could become future medicines. Numerous assays are needed for the in silico screening of small compounds and the optimisation of these molecules to obtain prospective therapeutic candidates. Understanding the disease-specific medication targets and their modes of action is crucial. Any campaign to find new anti-leishmanial drugs must first decide between target-based drug discovery (TDD) and phenotype-based drug discovery (PDD). Despite being founded on opposing concepts, they complement one another, boosting the likelihood of finding and creating new medications with novel mechanisms of action.

The PDD process is based on the involvement of an external agent in a cell, tissue, organ or the entire organism, independent of the mechanism of action. Identifying new modes of action in both the host and the parasite is of great importance; however, this approach can be challenging in its preclinical stages due to high resource demands (Balana-Fouce et al. 2019). Consequently, following up on PDD can be more difficult than on TDD.

According to TDD, a specific molecular target (protein) is key to the emergence of illness. However, the molecular processes necessary for *Leishmania* survival and disease progression are still being established, resulting in a shortage of dependable drug targets. The introduction of various databases showcasing data on vectors, pathways, pathogens and host informatics resources based on the OrthoMCL algorithm has been beneficial. Databases performing these functions include VEuPathDB, TriTrypDB, ClinEpiDB, etc. (Amos et al. 2022).

5.4.3 Criteria for Hit and Lead Identification

Leishmania species primary PDD starts at doses greater than 10 mM, which is considered an inclusive benchmark. The screening platform tests and compares libraries with a selection of active and inactive compounds. A second phase of screening (dose-response) evaluates the potency log through successive dilutions (inhibitory concentration IC_{50} [M]). Compounds with the highest rates of inhibition (e.g., 70–100%) are included. The optimal Hill slope coefficient for hits ranges from 0.5 to 1.8, with a sigmoidal concentration-growth inhibition curve achieving a maximum efficacy of 100%. Using a mammalian cell line, a hit must exhibit a selectivity index of at least tenfold. Given the potential emergence of resistant strains, it is advisable to investigate the potency of a potential hit using a panel of current clinical isolates alongside a variety of drug sensitivity assays.

A lead compound is defined as a substance that, following no more than five doses of oral administration of 50 mg/kg body weight once or twice per day, results in a reduction of liver parasite load by more than 70% in vivo and has an in vitro IC_{50} of less than 10 mM against an intracellular parasite. A potential lead should also not exhibit promiscuous bioactivity and must complete at least one early safety evaluation, such as the Ames test, investigations into cardiotoxicity or in vivo tolerance tests using the specified doses (Katsuno et al. 2015).

5.4.4 Molecular Docking and Molecular Dynamics Simulation

In the modern era of drug development, molecular docking has improved. It is a method of molecular modelling that displays the relationship between the protein (the target) and the small molecule (ligand). Molecular dynamics simulations are used to further evaluate hit compounds with high negative binding affinities during virtual screening and to better understand interatomic interactions before and after ligand binding with the protein molecule. This confirms whether the binding stability is satisfactory; anti-leishmanial compounds are then subjected to final screening.

Teles et al. performed a virtual screening approach for potential small molecules against *Leishmania infantum* methionyl-tRNA synthetase (LiMetRS) using the $ZINC_{15}$ database and molecular docking with Genetic Optimisation for Ligand Docking (GOLD) version 5.8.1. Compounds with very low binding affinity were selected for further screening through molecular dynamics simulations using the Groningen Machine for Chemical Simulations (GROMACS). The final selected compound was validated by studying its ADMET properties and can thus be further synthesised to observe its effects in vitro and in vivo (Teles et al. 2022).

5.4.5 Omics Approaches in Leishmaniasis Drug Discovery

Proteomics, transcriptomics and genomics are the various omics techniques used for target identification and candidate drug modification processes. Every biomolecular process in our body involves proteins, making it crucial to consider multiple facets of structural, functional and gene regulation/expression data incorporated into proteomics investigations. Classical methods of proteomics include 2-dimensional gel electrophoresis (2D-GE) and matrix-assisted laser desorption ionisation–time of flight (MALDI-TOF) mass spectrometry (MS). The foundation of 2D-GE lies in separating proteins by relative molecular weight in one direction and isoelectric pH (pI value) in the other. By changing the pI settings, this method can analyse more than 10,000 protein spots in a single gel run. However, these techniques have some drawbacks, including inaccurate identification of extremely basic or acidic proteins, inaccurate detection of proteins that are too small or large and inaccurate dynamic analysis.

Some cutting-edge mass spectrometry techniques rely on peptide fragmentation and evenly charge distribution for protein separation based solely on mass. The use of selective reaction monitoring (SRM), selected ion monitoring (SIM) and multiple reaction monitoring (MRM) technologies is another aspect of contemporary proteomics work (Lin et al. 2023). Due to its robustness, data-independent acquisition and high sensitivity, the novel SWATH-MS approach outperforms several other modern mass spectrometry-based techniques. Protein SWATH offers quantitative consistency and precision compared to gel-based or label-based techniques (Guo et al. 2023) (Figure 5.2).

5.4.6 Proteomics Approach

Understanding the representation of structural and molecular proteins is crucial. Proteomic research plays a key role in identifying the many proteins involved in host-parasite interactions within cells. These investigations open the door to the examination of diverse disease-related biomolecules and pathways, aiding both the discovery of prophylactic medications and the selection of qualified vaccine candidates.

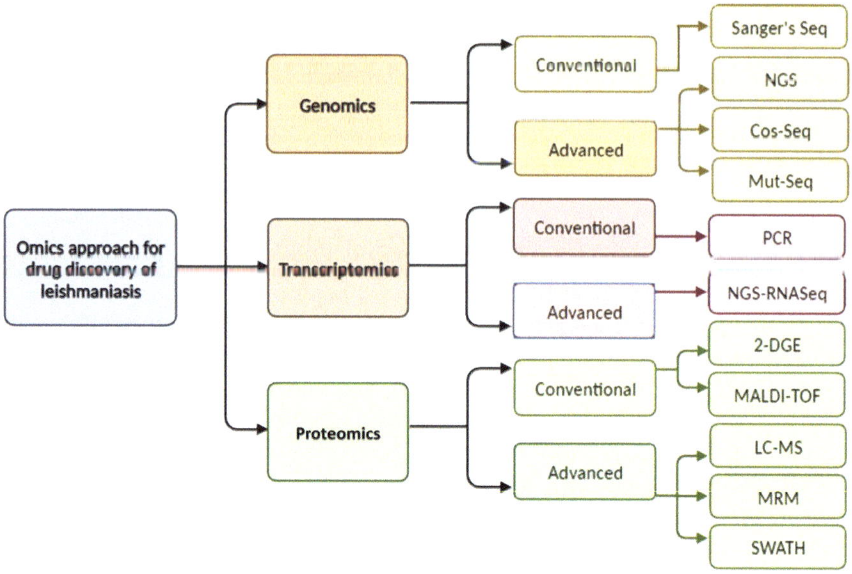

FIGURE 5.2 Omics approaches, including genomics, transcriptomics and proteomics studies used in the drug discovery process

Numerous parasitological and genetic techniques are used to diagnose VL. Additionally, protection against VL is mediated by Th1 cells. Interferon-gamma (IFN-γ), secreted by the Th1 subset of T cells, signals macrophages to begin killing parasites. However, most treated individuals show resistance to an infection-related increase in the Th1 cell antigenic response. Therefore, research on Th1 cells against the *Leishmania* parasite should take precedence for vaccine development. Currently, novel methods such as proteomics offer promising approaches for locating new leishmaniasis vaccine targets. Techniques used in proteomic analysis include 2D-GE, protein elution, high-performance liquid chromatography-mass spectrometry (HPLC-MS) and utilisation of a protein database (Rashidi et al. 2018) (Figure 5.3).

L. donovani, which frequently causes VL, was the subject of proteomics research by Gupta et al. The F2-soluble proteins from the promastigote form of *L. donovani* were found to induce a Th1 immune cell response in mononuclear blood cells from leishmaniasis patients who had recovered, indicating a potential target for therapy. To understand the nature of the F2 proteins, the researchers used a number of proteomics techniques, including 2D-GE and MALDI-TOF MS. This

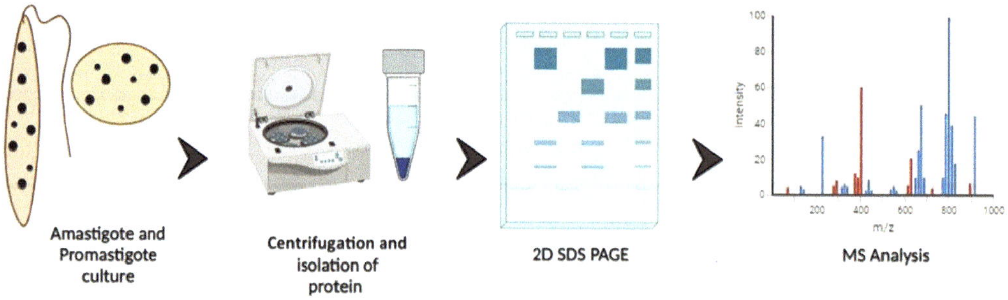

FIGURE 5.3 Proteomics approach in identifying new drug targets. The process involves isolating protein samples from amastigotes and promastigotes of *Leishmania* species, performing sodium dodecyl-sulphate polyacrylamide gel electrophoresis (SDS-PAGE) to analyse the proteins and validating the protein spots using mass spectrometry

investigation identified 33 proteins, including six novel ones, among a total of 63 resolved proteins. They examined known immunostimulatory proteins as potential therapeutic targets, including P450, heat shock protein (HSP) 70, HSP83, fructose 1–6 biphosphate aldolase, elongation factor 2, enolase, triosephosphate isomerase, calreticulin and kinesin-like protein (Gupta et al. 2007).

Another study conducted by Krobthong et al. resulted in the identification of 6099 proteins from three different species: *L. martiniquensis*, *L. donovani* and *L. orientalis*. From these proteins, 1065 antigenic proteins were further analysed using various in silico approaches. Ultimately, eight proteins demonstrated potential antigenic properties and were identified as future therapeutic targets. The principal component approach was used for protein profiling strategies and proved effective for finding novel target proteins analysed alongside drug candidates (Krobthong et al. 2022).

Stable isotope labelling with amino acids in cell culture (SILAC) is a quantitative proteomics-based approach followed by Smirlis et al. in their study of host-pathogen interactions. In this investigation, BALB/c macrophages from primary bone marrow infected for 72 hours with the promastigote stage of *L. donovani* underwent proteomic changes. Strong anion exchange (SAX) fractions were used for analytical purposes, leading to the identification of 86 differentially expressed genes – 35 upregulated and 51 downregulated – as a result of *L. donovani* infection. This study established new approaches for proteomics enrichment analysis with ex vivo protocols, providing novel mechanisms for understanding host-pathogen interaction (Smirlis et al. 2020).

Scientists also employ additional cutting-edge methods, including MRM, SRM, LC-MS and SWATH, to identify novel protein targets and highly effective protein pathways. Quantitative proteomics with sound reasoning is critical for concluding any kind of biomolecular study. West and his colleagues estimated the thermodynamic quantification alongside establishing drug targets, leading to the discovery of validated protein targets for treatment with anti-leishmanial drugs (West et al. 2010).

5.4.7 Genomic and Transcriptomic Approach

Leishmania's biology is studied via genomic and transcriptome research, which sheds light on the intricate interactions occurring within the parasite-host-vector group. Recent advances in gene expression and transcriptome analysis of *Leishmania* species enable comparisons of the parasite's lifecycle stages, facilitating the evaluation of different strains and species, as well as their ecological niches. Furthermore, researchers can contrast the gene expression profiles of parasites and hosts. In addition to traditional approaches, leishmaniasis research has increasingly employed newly developed next-generation genomics technologies like NGS, RNA-seq and Cos-Seq. Sandra et al. conducted blood transcriptomics to uncover the pathways and mechanisms involved in infection conditions. The RNA-seq analysis was performed on both VL patients and healthy individuals without infections, highlighting various cellular mechanisms and differential gene expression levels. They found long non-coding RNAs (lncRNAs) such as MIR4435-2HG and other coexpressed lncRNA/mRNA pairs such as GASAL1/IFN-γ (Maruyama et al. 2022).

5.5 NOVEL VACCINE CANDIDATES AND DRUG TARGETS

Although drug therapies exist for VL, current challenges include the emergence of resistance and a chronic phase that remains unresponsive to treatment. The intense transmission of anthroponotic *L. donovani* in India appears to be characterised by resistance. Given the increasing prevalence of drug resistance, it is vital to identify targets for developing innovative anti-leishmanial drug therapies, focussing on recognising and describing proteins directly or indirectly related to resistance. Both proteomics and genomics screening have contributed to finding therapeutic targets and understanding established and novel antimicrobial agent resistance mechanisms (Table 5.2).

McNicoll et al. used an integrated approach combining proteomics and DNA microarray analysis to study the differential stages in *Leishmania*. In each developmental stage of *L. infantum*, 2200

TABLE 5.2
Novel Targets, Their Roles, and Active Drug Molecules

Leishmania Species	Target	Role of the Target Protein	Drug Molecules and Their Nature	References
Leishmania species	Enzymes involved in polyamine biosynthesis	Polyamines are involved in the growth and evasion of parasites and deregulate the peroxidation of lipid molecules, thus providing a better environment for their survival in the host cells.	Alpha-difluoromethylornithine: polyamines/enzyme inhibitors	(Heby et al. 2007)
L. donovani	Topoisomerase II	Provides genetic stability and promotes transient breakage of DNA strands in the parasites.	Etoposide and novobiocin: induce cell disruption and cellular damage in parasites	(Rosypal et al. 2010)
L. major	Metacaspases	Induces programmed cell death of the parasite.	Miltefosine: induces *Leishmania* apoptosis	(Kosec et al. 2005; Basmaciyan et al. 2019)
L. donovani, *L. mexicana*	Cyclin-dependent kinases (CDK)	CDK-related kinase 3 (CRK3) codes for the cdc2 protein kinase, mimicking the role of eukaryotic histone H1.	Olomoucine, purvalanol, butyrolactone: these compounds compete with ATP at the catalytic site, weakening the parasite by presenting it to the macrophages	(Grant et al. 2004)
L. major	Aspartic peptidases, signal peptide peptidase	Involved in the autophagy process and cleavage of the transmembrane domain.	K11777: cysteine protease inhibitor	(Yang et al. 2012)
L. chagasi, *L. braziliensis*	Fe-superoxide dismutase (FeSOD)	Guards against the toxicity of superoxide radicals for the parasites' glycosomal enzymes.	Thiazole derivatives: inhibitors of LbSOD	(Brito et al. 2019)
L. infantum	Enzymes of thiol metabolism, trypanothione ($T(SH)_2$)	Helps in reducing Fe and NO into a harmless dinitrosyl iron complex, preventing the parasite from nitric oxide toxicity	Plumbagin: modulates redox thiol metabolism of the parasite	(Sharma et al. 2012)

protein spots were visualised, with nearly 12.4% of proteins upregulated in the amastigote stage and 6.1% upregulated in the promastigote stage of its lifecycle. Potential drug targets identified in this study included F1 ATP synthetase beta-subunit, peroxiredoxins and cyclophilin D (McNicoll et al. 2006).

Bhattacharjee et al. used an in silico approach to find a polyvalent multi-subunit peptide vaccine that is immunogenic against five *Leishmania* species. The study also used IL-12 as a natural adjuvant to boost the immunogenicity of the vaccine candidates without adverse effects on humans. All proteins were checked for the presence of cytotoxic T-lymphocyte and helper T-lymphocyte epitopes (CTL and HTL). Various software and online platforms, such as PSIPRED version 4.0 server, I-TASSER server, ERRAT server, ElliPro server and the JCat tool, were used for secondary and tertiary structure prediction as well as further validation of the resultant peptide (Bhattacharjee et al. 2023).

Kushwaha and Capalash analysed histidyl-tRNA synthetase (LdHisRS) from *L. donovani* for its capacity to modulate immunity and predicted epitopes using immunoinformatic methods. This enzyme incorporates histidine into proteins during translation; thus, a recombinant LdHisRS protein (rLdHisRS) was expressed in BL-21 cells of *E. coli*. Immunomodulatory validation studies were conducted using J774A.1 murine macrophages and BALB/c mice, revealing increased levels of stimulatory and inflammatory cytokines. They identified 30 CTLs and 20 HTLs, and analysed 18 B-cell epitopes from *L. donovani* that could serve as potential multiepitope vaccines (Kushwaha and Capalash 2023).

Structure-based drug design is a potential method for identifying drug candidates against the glutathione synthetase of *L. donovani*. Homology modelling was conducted to construct the target structure, allowing drug candidates to bind specifically to the target pathogen protein and not to the proteins present in humans. Molecular docking and molecular dynamics simulations indicated that Olysio (simeprevir) and telithromycin exhibited very low binding affinities of −89.21 kcal/mol and −45.34 kcal/mol, respectively. A novel therapeutic drug candidate was developed for synthesis and treatment (Sarma et al. 2023).

Since most of the parasite's survival mechanisms depend on proteases such as metalloproteases, cysteine proteases and other serine proteases, protease inhibitors can be employed as viable treatment options for leishmaniasis. These proteases contribute to the parasite's survival in the host system by facilitating glycolysis, sterol biosynthesis, folate biosynthesis and pathogen proliferation. Targeting metacaspases, which are present in the cytosol of the parasite and help in segregating the kinetoplast from the nucleus, could prevent genetic elements from dividing, thereby inhibiting parasite proliferation within the host cells (Kosec et al. 2006).

5.6 CHALLENGES AND FUTURE DIRECTIONS

The challenges encountered in drug discovery for VL encompass a multitude of aspects that impede progress in this field. One of the foremost obstacles is funding, which poses a considerable hindrance to developing effective treatments for VL. The scarcity of financial resources allocated to research in this area restricts the ability to conduct extensive studies, impairs the exploration of innovative approaches and delays the advancement of potential therapeutic options.

Furthermore, regulatory issues exacerbate the challenges faced in drug discovery for VL. Stringent regulations imposed by governing bodies necessitate rigorous testing and approval processes, prolonging timelines and increasing costs associated with bringing new drugs to market. Additionally, accessing endemic regions where VL is prevalent presents considerable challenges due to inadequate infrastructure, healthcare resources and support systems required to conduct clinical trials and implement effective treatment strategies.

To overcome these challenges and pave the way for future advancements, several potential research directions can be outlined. One promising avenue is developing combination therapies that utilise multiple drugs with different mechanisms of action to enhance treatment efficacy and overcome drug resistance. Another future direction involves leveraging new technologies such as genomics and proteomics to identify novel drug targets and develop more precise and targeted therapies. By employing these innovative approaches, researchers may overcome the challenges in drug discovery for VL and improve patient outcomes.

5.7 CONCLUSION

The biology of VL has advanced significantly in recent years. This understanding has facilitated the development of computational drug discovery methods as well as in vitro and in vivo assays capable of finding substances with a realistic chance of curing the disease in humans. Still, further emphasis is needed on understanding the mechanisms associated with complex parasites.

Recent knowledge gathered about the disease and the parasite has increased, illuminating pathways toward drug design and development. Virtual screening of lead compounds through molecular docking and molecular dynamics approaches has significantly impacted discovering new potential drug candidates. Biomarker discovery and exploration of novel pharmacological targets and proteins through omics studies encourage progress toward effective treatment of leishmaniasis. There is also a need for effective diagnosis of asymptomatic patients alongside deriving chemical entities against this condition. Thus, better resources and bioinformatics tools can be utilised to systematically cure the disease and reduce its relapse episodes.

ACKNOWLEDGEMENT

The authors are thankful to the ICMR-Rajendra Memorial Research Institute of Medical Sciences, Patna, for the necessary support during this study.

ABBREVIATIONS

FDA: the US Food and Drug Administration; CRK12: cdc2-related kinase 12; VL: visceral leishmaniasis; CL: cutaneous leishmaniasis; EC_{50}: half maximal effective concentration; IC_{50}: half maximal inhibitory concentration; SSG-R: sodium stibogluconate-resistant; SSG-S: sodium stibogluconate-sensitive; Th1: type I T-helper cells; GROMACS: Groningen Machine for Chemical Simulations; GOLD: Genetic Optimisation for Ligand Docking; SDS PAGE: sodium dodecyl-sulphate polyacrylamide gel electrophoresis; IFN-γ: interferon-gamma.

REFERENCES

Alrajhi, A. A., E. A. Ibrahim, E. B. De Vol, M. Khairat, R. M. Faris and J. H. Maguire (2002). "Fluconazole for the treatment of cutaneous leishmaniasis caused by Leishmania major." *New England Journal of Medicine* **346**(12): 891–895.

Amos, B., C. Aurrecoechea, M. Barba, A. Barreto, E. Y. Basenko, R. Belnap, A. S. Blevins, U. Böhme, J. Brestelli and B. P. Brunk (2022). "VEuPathDB: the eukaryotic pathogen, vector and host bioinformatics resource center." *Nucleic Acids Research* **50**(D1): D898–D911.

Andersen, E. M., M. Cruz-Saldarriaga, A. Llanos-Cuentas, M. Luz-Cjuno, J. Echevarria, C. Miranda-Verastegui, O. Colina, J. D. Berman (2005). "Comparison of meglumine antimoniate and pentamidine for Peruvian cutaneous leishmaniasis." *The American Journal of Tropical Medicine and Hygiene* **72**(2): 133–137.

Ashburn, T. T. and K. B. Thor (2004). "Drug repositioning: identifying and developing new uses for existing drugs." *Nature Reviews Drug Discovery* **3**(8): 673–683.

Authorities, G. D. R. (2008). *Impavido 10/50 mg Kapseln—Fachinformation*, German Federal Institute for Drugs and Medical Devices Bonn, Germany.

Balana-Fouce, R., M. Y. P. Pertejo, B. Dominguez-Asenjo, C. Gutierrez-Corbo and R. M. Reguera (2019). "Walking a tightrope: drug discovery in visceral leishmaniasis." Drug Discovery Today **24**(5): 1209–1216.

Basmaciyan, L., N. Azas, M. Casanova (2019). "A potential acetyltransferase involved in Leishmania major metacaspase-dependent cell death." *Parasites & Vectors* **12**(1): 1–8.
Beach, D. H., L. J. Goad, G. G. Holz Jr. (1988). "Effects of antimycotic azoles on growth and sterol biosynthesis of Leishmania promastigotes." Molecular and Biochemical Parasitology **31**(2): 149–162.
Berman, J. (2003). "Current treatment approaches to leishmaniasis." Current Opinion in Infectious Diseases **16**(5): 397–401.
Bhattacharjee, M., M. Banerjee and A. Mukherjee (2023). "In silico designing of a novel polyvalent multi-subunit peptide vaccine leveraging cross-immunity against human visceral and cutaneous leishmaniasis: an immunoinformatics-based approach." Journal of Molecular Modeling **29**(4): 99.
Brito, C. C. B., H. V. C. d. Silva, D. J. Brondani, A. R. d. Faria, R. M. Ximenes, I. M. d. Silva, J. F. d. Albuquerque, M. S. Castilho (2019). "Synthesis and biological evaluation of thiazole derivatives as Lb SOD inhibitors." Journal of Enzyme Inhibition and Medicinal Chemistry **34**(1): 333–342.
Brown, D. G. (2016). "A Medicinal Chemistry Perspective on the Hit-to-Lead Phase in the Current Era of Drug Discovery." Annual Reports in Medicinal Chemistry **49:** 329–366.
Chulay, J. D., H. C. Spencer, M. Mugambi (1985). "Electrocardiographic changes during treatment of leishmaniasis with pentavalent antimony (sodium stibogluconate)." American Journal of Tropical Medicine and Hygiene **34**(4): 702–709.
Comley, J. (2015). "Phenotypic drug discovery." Drug Discovery World **16**: 25.
Croft, S., J. Hogg, W. Gutteridge, A. Hudson and A. Randall (1992). "The activity of hydroxynaphthoquinones against *Leishmania donovani*." Journal of Antimicrobial Chemotherapy **30**(6): 827–832.
Das, V., A. Ranjan, A. Sinha, N. Verma, C. Lal, A. Gupta, N. Siddiqui and S. Kar (2001). "A randomized clinical trial of low dosage combination of pentamidine and allopurinol in the treatment of antimony unresponsive cases of visceral leishmaniasis." Transactions of the Royal Society of Tropical Medicine **49**: 609–613.
Dinesh, N., N. Soumya and S. Singh (2015). "Antileishmanial effect of mevastatin is due to interference with sterol metabolism." Parasitology Research **114**: 3873–3883.
Dogra, J. and V. Saxena (1996). "Itraconazole and leishmaniasis: a randomised double-blind trial in cutaneous disease." International Journal of Dermatology **26**(12): 1413–1415.
Dorlo, T. P., M. Balasegaram, J. H. Beijnen and P. J. de Vries (2012). "Miltefosine: a review of its pharmacology and therapeutic efficacy in the treatment of leishmaniasis." Journal of Antimicrobial Chemotherapy **67**(11): 2576–2597.
Esfandiarpour, I. and A. Alavi (2002). "Evaluating the efficacy of allopurinol and meglumine antimoniate (Glucantime) in the treatment of cutaneous leishmaniasis." International Journal of Dermatology **41**(8): 521–524.
García-Davis, S., A. López-Arencibia, C. J. Bethencourt-Estrella, D. San Nicolás-Hernández, E. Viveros-Valdez, A. R. Díaz-Marrero, J. J. Fernández, J. Lorenzo-Morales and J. E. Piñero (2023). "Laurequinone, a Lead Compound against Leishmania." Molecules **21**(6): 333.
Ginsberg, A. M., M. W. Laurenzi, D. J. Rouse, K. D. Whitney, M. K. Spigelman (2009). "Safety, tolerability, and pharmacokinetics of PA-824 in healthy subjects." Antimicrobial Agents and Chemotherapy **53**(9): 3720–3725.
Gopinath, V. S., J. Pinjari, R. T. Dere, A. Verma, P. Vishwakarma, R. Shivahare, M. Moger, P. S. K. Goud, V. Ramanathan and P. J. Bose (2013). "Design, synthesis and biological evaluation of 2-substituted quinolines as potential antileishmanial agents." European Journal of Medicinal Chemistry **69**: 527–536.
Grant, K. M., M. H. Dunion, V. Yardley, A.-L. Skaltsounis, D. Marko, G. Eisenbrand, S. L. Croft, L. Meijer, J. C. Mottram (2004). "Inhibitors of Leishmania mexicana CRK3 cyclin-dependent kinase: chemical library screen and antileishmanial activity." Antimicrobial Agents and Chemotherapy **48**(8): 3033–3042.
Guo, J., R. Kufer, D. Li, S. Wohlrab, M. Greenwood-Goodwin and F. Yang (2023). "Technical advancement and practical considerations of LC-MS/MS-based methods for host cell protein identification and quantitation to support process development." *MAbs,* **15**(5): 2217467.
Gupta, D., P. K. Singh, P. K. Yadav, T. Narender, U. K. Patil, S. K. Jain and M. K. Chourasia (2023). "Emerging strategies and challenges of molecular therapeutics in antileishmanial drug development." International Immunopharmacology **115**: 109649.
Gupta, S. K., B. S. Sisodia, S. Sinha, K. Hajela, S. Naik, A. K. Shasany and A. Dube (2007). "Proteomic approach for identification and characterization of novel immunostimulatory proteins from soluble antigens of Leishmania donovani promastigotes." Proteomics **7**(5): 816–823.
Haase, R., T. Wieder, C. C. Geilen and W. Reutter (1991). "The phospholipid analogue hexadecylphosphocholine inhibits phosphatidylcholine biosynthesis in Madin—Darby canine kidney cells." FEBS Letters **288**(1–2): 129–132.

Handman, E., L. Kedzierski, A. D. Uboldi and J. W. Goding (2008). "Fishing for anti-leishmania drugs: principles and problems." *Trends in Parasitology 625*: 48–60.

Heby, O., L. Persson and M. Rentala (2007). "Targeting the polyamine biosynthetic enzymes: a promising approach to therapy of African sleeping sickness, Chagas' disease, and leishmaniasis." Amino Acids **33**: 359–366.

Jamil, K. M., R. Haque, R. Rahman, M. A. Faiz, A. T. M. R. H. Bhuiyan, A. Kumar, S. M. Hassan, H. Kelly, P. Dhalaria and S. Kochhar (2015). "Effectiveness study of paromomycin IM injection (PMIM) for the treatment of visceral leishmaniasis (VL) in Bangladesh." PLoS Neglected Tropical Diseases **9**(10): e0004118.

Katsuno, K., J. N. Burrows, K. Duncan, R. H. Van Huijsduijnen, T. Kaneko, K. Kita, C. E. Mowbray, D. Schmatz, P. Warner and B. Slingsby (2015). "Hit and lead criteria in drug discovery for infectious diseases of the developing world." Nature Reviews Drug Discovery **14**(11): 751–758.

Kinnamon, K. E., E. A. Steck, P. S. Loizeaux, W. L. Hanson, W. L. Chapman Jr. and V. B. Waits (1978). "The antileishmanial activity of lepidines." American Journal of Tropical Medicine and Hygiene **27**(4): 751–757.

Kosec, G., V. E. Alvarez, F. Agüero, D. Sánchez, M. Dolinar, B. Turk, V. Turk, J. J. Cazzulo and J. Mottram (2006). "Metacaspases of Trypanosoma cruzi: possible candidates for programmed cell death mediators." Molecular and Biochemical Parasitology **145**(1): 18–28.

Krobthong, S., Y. Yingchutrakul, P. Samutrtai, A. Hitakarun, S. Siripattanapipong, S. Leelayoova, M. Mungthin and K. Choowongkomon (2022). "Utilizing quantitative proteomics to identify species-specific protein therapeutic targets for the treatment of Leishmaniasis." ACS Omega **7**(15): 12580–12588.

Kushwaha, V. and N. Capalash (2023). "Evaluation of immunomodulatory potential of recombinant histidyl-tRNA synthetase (rLdHisRS) protein of Leishmania donovani as a vaccine candidate against visceral leishmaniasis." Acta Tropica **241**: 106867.

Lamotte, S., G. F. Späth, N. Rachidi and E. Prina (2017). "The enemy within: targeting host–parasite interaction for antileishmanial drug discovery." PLoS Neglected Tropical Diseases **11**(6): e0005480.

Lin, T. T., T. Zhang, R. B. Kitata, T. Liu, R. D. Smith, W. J. Qian and T. Shi (2023). "Mass spectrometry-based targeted proteomics for analysis of protein mutations." Molecular & Cellular Proteomics **42**(2): 796821.

Loiseau, P., S. Cojean and J. F. Schrével (2011). "Sitamaquine as a putative antileishmanial drug candidate: from the mechanism of action to the risk of drug resistance." Parasitology **18**(2): 115.

Maarouf, M., F. Lawrence, S. Brown and M. Robert-Gero (1997). "Biochemical alterations in paromomycin-treated Leishmania donovani promastigotes." Parasitology Research **83**: 198-202.

Maltezou, H. C. (2010). "Drug resistance in visceral leishmaniasis." BioMed Research International **2010**: 187342.

Maruyama, S. R., C. A. Fuzo, A. E. R. Oliveira, L. A. Rogerio, N. T. Takamiya, G. Pessenda, E. V. de Melo, A. M. da Silva, A. R. Jesus and V. Carregaro (2022). "Insight into the long noncoding RNA and mRNA coexpression profile in the human blood transcriptome upon Leishmania infantum infection." Frontiers in Immunology **13**: 784463.

Matos, A., A. Viçosa, M.-I. Ré, E. Ricci-Júnior, C. J. Holandino (2020). "A review of current treatments strategies based on paromomycin for leishmaniasis." Journal of Drug Delivery Science and Technology **57**: 101664.

McNicoll, F., J. Drummelsmith, M. Müller, É. Madore, N. Boilard, M. Ouellette and B. Papadopoulou (2006). "A combined proteomic and transcriptomic approach to the study of stage differentiation in Leishmania infantum." Journal of Proteomics **6**(12): 3567–3581.

Melcon-Fernandez, E., G. Galli, C. García-Estrada, R. Balaña-Fouce, R. M. Reguera and Y. Pérez-Pertejo (2023). "Miltefosine and nifuratel combination: a promising therapy for the treatment of Leishmania donovani visceral leishmaniasis." International Journal of Molecular Sciences **24**(2): 1635.

Mukherjee, S., B. Mukherjee, R. Mukhopadhyay, K. Naskar, S. Sundar, J. C. Dujardin, A. K. Das and S. J. P. Roy (2012). "Imipramine is an orally active drug against both antimony sensitive and resistant Leishmania donovani clinical isolates in experimental infection." PLoS Neglected Tropical Diseases **6**(12): e1987.

Nakayama, H., P. M. Loiseau, C. Bories, S. Torres de Ortiz, A. Schinini, E. Serna, A. Rojas de Arias, M. A. Fakhfakh, X. Franck, B. J. A. (2005). "Efficacy of orally administered 2-substituted quinolines in experimental murine cutaneous and visceral leishmaniases." Antimicrobial Agents and Chemotherapy **49**(12): 4950–4956.

Pacheco-Fernandez, T., H. Markle, C. Verma, R. Huston, S. Gannavaram, H. L. Nakhasi and A. R. Satoskar (2023). "Field-deployable treatments for leishmaniasis: intrinsic challenges, recent developments and next steps." Research in Tropical Medicine **14**: 61–85.

Palit, P., A. Hazra, A. Maity, R. Vijayan, P. Manoharan, S. Banerjee, N. B. Mondal, N. Ghoshal, N. J. A. (2012). "Discovery of safe and orally effective 4-aminoquinaldine analogues as apoptotic inducers with activity against experimental visceral leishmaniasis." Antimicrobial Agents and Chemotherapy **56**(1): 432–445.

Patel, H., D. Bhatt, S. Shakhreliya, N. L. Lam, R. Maurya and V. Singh (2021). "An introduction and applications of bioinformatics." In Singh, V., Kumar, A. (eds), *Advances in Bioinformatics*, 1–14. Springer, Singapore.

Patterson, S., S. Wyllie, L. Stojanovski, M. R. Perry, F. R. Simeons, S. Norval, M. Osuna-Cabello, M. De Rycker, K. D. Read, and A. H. Fairlamb (2013). "The R enantiomer of the antitubercular drug PA-824 as a potential oral treatment for visceral leishmaniasis." Antimicrobial Agents and Chemotherapy **57**(10): 4699–4706.

Pinheiro, A. C. and M. V. N. de Souza (2022). "Current leishmaniasis drug discovery." *Research Journal of Medicine and Health Sciences* **13**: 1029–1043.

Pinto, E. G., T. A. da Costa-Silva and A. G. Tempone (2014). "Histamine H1-receptor antagonists against Leishmania (L.) infantum: an in vitro and in vivo evaluation using phosphatidylserine-liposomes." Acta Tropica **137**: 206–210.

Rani, G. F., O. Preham, H. Ashwin, N. Brown, I. S. Hitchcock and P. M. Kaye (2021). "Dissecting pathways to thrombocytopenia in a mouse model of visceral leishmaniasis." Blood Advances **5**(6): 1627–1637.

Rashidi, S., K. Kalantar, D. Rostamzadeh and G. Hatam (2018). "The importance of checking Leishmania promastigotes viability in the proteomics analysis of secretions/sekresyonlarin proteomiks analizinde Leishmania promastigotlarin canliliklarinin kontrol edilmesinin onemi." Turkish Journal of Parasitology **42**(4): 245–249.

Ritmeijer, K., H. Veeken, Y. Melaku, G. Leal, R. Amsalu, J. Seaman, and R. Davidson Hygiene (2001). "Ethiopian visceral leishmaniasis: generic and proprietary sodium stibogluconate are equivalent; HIV co-infected patients have a poor outcome." Transactions of the Royal Society of Tropical Medicine and Hygiene **95**(6): 668–672.

Rosypal, A. C., S. Tripp, S. Lewis, J. Francis, M. K. Stoskopf, R. S. Larsen and D. S. Lindsay (2010). "Survey of antibodies to Trypanosoma cruzi and Leishmania spp. in gray and red fox populations from North Carolina and Virginia." Journal of Parasitology **96**(6): 1230–1231.

Salloum, T., S. Tokajian, R. P. Hirt and D. Biology (2021). "Advances in understanding Leishmania pathobiology: what does rna-seq tell us?" Frontiers in Cellular and Infection Microbiology **9**: 702240.

Saporito, L., G. M. Giammanco, S. De Grazia and C. Colomba (2013). "Visceral leishmaniasis: host–parasite interactions and clinical presentation in the immunocompetent and in the immunocompromised host." International Journal of Infectious Diseases **17**(8): e572–e576.

Sarma, M., S. Borkotoky and V. K. J. Dubey (2023). "Structure-based drug designing against Leishmania donovani using docking and molecular dynamics simulation studies: exploring glutathione synthetase as a drug target." *Journal of Biomolecular Structure and Dynamics* **42**: 7628–7636.

Sereno, D., P. Holzmuller, I. Mangot, G. Cuny, A. Ouaissi, and J.-L. Lemesre (2001). "Antimonial-mediated DNA fragmentation in Leishmania infantum amastigotes." Antimicrobial Agents and Chemotherapy **45**(7): 2064–2069.

Sharma, N., A. K. Shukla, M. Das and V. K. Dubey (2012). "Evaluation of plumbagin and its derivative as potential modulators of redox thiol metabolism of Leishmania parasite." Parasitology Research **110**: 341–348.

Sindermann, H. and J. Engel (2006). "Development of miltefosine as an oral treatment for leishmaniasis." Transactions of the Royal Society of Tropical Medicine and Hygiene **100**: S17–S20.

Singh, N., M. Kumar and R. K. Singh (2012). "Leishmaniasis: current status of available drugs and new potential drug targets." Asian Pacific Journal of Tropical Medicine **5**(6): 485–497.

Singh, U. K., R. Prasad, B. P. Jaiswal, P. K. Singh and C. Thakur (2010). "Amphotericin B therapy in children with visceral leishmaniasis: daily vs. alternate day, a randomized trial." Journal of Tropical Pediatrics **56**(5): 321-324.

Smirlis, D., F. Dingli, P. Pescher, E. Prina, D. Loew, N. Rachidi and G. F. Späth (2020). "SILAC-based quantitative proteomics reveals pleiotropic, phenotypic modulation in primary murine macrophages infected with the protozoan pathogen Leishmania donovani." Journal of Proteomics **213**: 103617.

Smith, A. C., V. Yardley, J. Rhodes and S. L. Croft (2000). "Activity of the novel immunomodulatory compound tucaresol against experimental visceral leishmaniasis." Antimicrobial Agents and Chemotherapy **44**(6): 1494–1498.

Soto, J. and P. J. Soto (2006). "Miltefosine: oral treatment of leishmaniasis." Expert Review of Anti-infective Therapy **4**(2): 177–185.

Sudhandiran, G. and C. Shaha (2003). "Antimonial-induced increase in intracellular Ca2+ through non-selective cation channels in the host and the parasite is responsible for apoptosis of intracellular Leishmania donovani amastigotes." Journal of Biological Chemistry **278**(27): 25120–25132.
Sundar, S., L. B. Gupta, V. Rastogi, G. Agrawal and H. W. Murray (2000). "Short-course, cost-effective treatment with amphotericin B-fat emulsion cures visceral leishmaniasis." Transactions of the Royal Society of Tropical Medicine and Hygiene **94**(2): 200–204.
Sundar, S., T. Jha, C. Thakur, J. Engel, H. Sindermann, C. Fischer, K. Junge, A. Bryceson and J. Berman (2002). "Oral miltefosine for Indian visceral leishmaniasis." New England Journal of Medicine **347**(22): 1739–1746.
Sundar, S., T. Jha, C. P. Thakur, P. K. Sinha and S. K. Bhattacharya (2007). "Injectable paromomycin for visceral leishmaniasis in India." New England Journal of Medicine **356**(25): 2571–2581.
Sundar, S., D. N. Lockwood, G. Agrawal, M. Rai, M. Makharia and H. W. Murray (2001). "Treatment of Indian visceral leishmaniasis with single or daily infusions of low dose liposomal amphotericin B: randomised trialCommentary: cost and resistance remain issues." BMJ **323**(7310): 419-422.
Teles, H. R., M. Valli, L. L. Ferreira and A. D. Andricopulo (2022). "Molecular modeling, virtual screening, and molecular dynamics for Leishmania infantum methionyl-tRNA synthetase." Journal of Physical Chemistry B **126**(51): 10834–10843.
Thomas, M., S. Brand, M. De Rycker, F. Zuccotto, I. Lukac, P. G. Dodd, E.-J. Ko, S. Manthri, K. McGonagle and M. J. Osuna-Cabello (2021). "Scaffold-hopping strategy on a series of proteasome inhibitors led to a preclinical candidate for the treatment of visceral leishmaniasis." Journal of Medicinal Chemistry **64**(9): 5905–5930.
Veeken, H., K. Ritmeijer, J. Seaman and R. Davidson (2000). "A randomized comparison of branded sodium stibogluconate and generic sodium stibogluconate for the treatment of visceral leishmaniasis under field conditions in Sudan." Tropical Medicine and International Health **5**(5): 312–317.
Verma, N. K. and C. S. Dey (2004). "Possible mechanism of miltefosine-mediated death of Leishmania donovani." Antimicrobial Agents and Chemotherapy **48**(8): 3010–3015.
Wasunna, M. K., J. R. Rashid, J. Mbui, G. Kirigi, D. Kinoti, H. Lodenyo, J. M. Felton, A. J. Sabin and J. Horton (2005). "A phase II dose-increasing study of sitamaquine for the treatment of visceral leishmaniasis in Kenya." American Journal of Tropical Medicine and Hygiene **73**(5): 871–876.
West, G. M., C. L. Tucker, T. Xu, S. K. Park, X. Han, J. R. Yates III and M. C. Fitzgerald (2010). "Quantitative proteomics approach for identifying protein–drug interactions in complex mixtures using protein stability measurements." Proceedings of the National Academy of Sciences **107**(20): 9078–9082.
World Health Organization, Leishmaniasis fact sheets. (https://www.who.int/news-room/fact-sheets/detail/leishmaniasis).
Wyllie, S., S. Brand, M. Thomas, M. De Rycker, C.-w. Chung, I. Pena, R. P. Bingham, J. A. Bueren-Calabuig, J. Cantizani and D. Cebrian (2019). "Preclinical candidate for the treatment of visceral leishmaniasis that acts through proteasome inhibition." Proceedings of the National Academy of Sciences **116**(19): 9318–9323.
Wyllie, S., M. L. Cunningham and A. H. Fairlamb (2004). "Dual action of antimonial drugs on thiol redox metabolism in the human pathogen Leishmania donovani." Journal of Biological Chemistry **279**(38): 39925–39932.
Wyllie, S., S. Patterson, L. Stojanovski, F. R. Simeons, S. Norval, R. Kime, K. D. Read and A. H. Fairlamb (2012). "The anti-trypanosome drug fexinidazole shows potential for treating visceral leishmaniasis." Science Translational Medicine **4**(119): 119re111–119re111.
Wyllie, S., M. Thomas, S. Patterson, S. Crouch, M. De Rycker, R. Lowe, S. Gresham, M. D. Urbaniak, T. D. Otto and L. Stojanovski (2018). "Cyclin-dependent kinase 12 is a drug target for visceral leishmaniasis." Nature **560**(7717): 192–197.
Yang, P.-Y., M. Wang, C. Y. He and S. Q. Yao (2012). "Proteomic profiling and potential cellular target identification of K11777, a clinical cysteine protease inhibitor, in Trypanosoma brucei." Chemistry and Chemical Biology **48**(6): 835–837.

6 Major Challenges in Visceral Leishmaniasis Treatment

*Rishikesh Kumar and Krishna Pandey**
Department of Clinical Medicine, Rajendra Memorial Research Institute of Medical Sciences, Patna, Bihar, India
*Corresponding Author: Dr. Krishna Pandey, Department of Clinical Medicine, Rajendra Memorial Research Institute of Medical Sciences, Patna, Bihar, India

6.1 INTRODUCTION

Visceral leishmaniasis (VL), also known as kala-azar, is a neglected tropical disease caused by the *Leishmania donovani* complex. It is transmitted through the bites of infected female sandflies belonging to the genera *Phlebotomus* and *Lutzomyia* in the Old World and New World, respectively. VL is highly prevalent in 65 countries, with an estimated 50,000 to 100,000 new cases each year, primarily in East Africa, South Asia, and Latin America (Alvar et al., 2012; WHO, 2021). The disease is characterized by fever, weight loss, hepatosplenomegaly, and anemia, and if left untreated, it can be fatal (WHO, 2021).

The treatment of VL has evolved over the years. Antimonial preparations such as Pentostam (sodium stibogluconate) and Glucantime (meglumine antimoniate) have been the first-line drugs for many years. However, the emergence of resistance and the need for long and painful parenteral administration have prompted the search for alternative treatments (Sundar et al., 2000). Amphotericin B, particularly in its liposomal form (L-AmB), has become the preferred choice due to its high efficacy and lower toxicity compared to antimonials (Sundar et al., 2011). Miltefosine, the only oral drug available for the treatment of clinical VL, is extensively utilized in endemic regions, especially in India and its subcontinent (Sundar et al., 2002). Paromomycin, another aminoglycoside, has shown promise in clinical trials and is now being used in combination therapies (Sundar et al., 2007).

Despite these advances, several challenges remain in the treatment and control of VL. These include drug resistance, particularly in regions in the Indian subcontinent where antimonials have been extensively used (Sundar et al., 2000; Rijal et al., 2007). There are also challenges related to the availability and affordability of drugs, especially L-AmB, which is expensive and often inaccessible in resource-limited settings (Sundar et al., 2011). The long treatment duration of some drugs, such as miltefosine, can also be a barrier to compliance, leading to the emergence of resistance or unresponsiveness (Sundar et al., 2002). Additionally, the lack of effective vaccines and the convoluted nature of the parasite's life cycle and transmission dynamics pose further challenges to controlling and eliminating VL (Alvar et al., 2012). This chapter aims to review the current treatment approaches for VL, encompassing the use of antimonials, amphotericin B, miltefosine, and paromomycin, while addressing the complexities and obstacles associated with these therapeutics. It will also explore emerging therapies and strategies for overcoming challenges to improve the treatment and control of VL (Sundar et al., 2004).

DOI: 10.1201/9781032643120-6

6.2 CURRENT TREATMENT APPROACHES FOR VL

VL, predominantly triggered by the protozoan parasite *L. donovani*, can be life-threatening if not treated properly. It is widespread in tropical and subtropical regions, disproportionately impacting marginalized populations with limited access to healthcare (Alvar et al., 2012). Effective therapy is critical for reducing health risks linked to VL. This section provides an overview of the current treatment approaches for VL, including their efficacy, limitations, and challenges.

6.2.1 Pentavalent Antimonials

For decades, antimonial preparations, such as sodium antimony gluconate, sodium stibogluconate, and meglumine antimoniate, have been the primary therapeutics for VL treatment. They function by impeding parasite growth through the disruption of glycolysis and protein synthesis (Sundar et al., 2000). However, their use is associated with several drawbacks, including high toxicity, long treatment duration (up to 28 days), and the appearance of drug-unresponsive strains (Rijal et al., 2007). Despite these limitations, antimonials are still used in some regions where they remain effective.

6.2.2 Amphotericin B

Amphotericin B, a polyene antibiotic, has been a longstanding option for VL treatment, demonstrating high efficacy. It operates by attaching to ergosterol in the parasite's cell membrane, causing membrane damage and subsequent cell demise (Sundar et al., 2011). However, its use is limited by significant toxicity and the need for intravenous administration. AmBisome, or L-AmB, a liposomal formulation of amphotericin B, has been developed to mitigate these drawbacks, offering improved safety profiles and potentially easier administration. Although AmBisome presents advantages in terms of tolerability and administration, its cost may restrict accessibility, particularly in resource-limited settings. Nonetheless, both amphotericin B and AmBisome play crucial roles in managing VL by providing potent therapeutic options against this life-threatening disease. The recommended dosage of AmBisome for VL treatment is 3–5 mg/kg of body weight given intravenously over a period of 5–10 days. With cure rates over 95%, AmBisome is often the primary treatment choice for VL in numerous regions (Sundar et al., 2010).

6.2.3 Miltefosine

Miltefosine is a phospholipid derivative with hydrophilic and lipophilic properties, making it amphiphilic in nature. This unique structure enables miltefosine to interact with both hydrophilic and hydrophobic components of cell membranes. This interaction disrupts the integrity and function of *Leishmania* parasite cell membranes, leading to their death (Dorlo et al., 2012). As the first oral medication approved for VL treatment, miltefosine has revolutionized therapeutic approaches against this disease. Its oral administration offers convenience over intravenous therapies like amphotericin B, particularly in resource-limited settings. While it has shown high efficacy in clinical trials, challenges such as prolonged treatment duration, potential side effects, and the risk of developing resistance exist (Nagle et al., 2014). Nonetheless, miltefosine remains a crucial component in the treatment arsenal against VL, offering a valuable oral option for patients and healthcare providers (Sundar et al., 2002).

6.2.4 Paromomycin

Paromomycin is an aminoglycoside antibiotic with a chemical structure consisting of a cyclohexane ring attached to amino sugars. It belongs to the class of aminocyclitol glycosides and exhibits broad-spectrum antimicrobial activity. To target *Leishmania* parasites, it disrupts their protein synthesis machinery by binding to the 30S ribosomal subunit, thereby interfering with mRNA reading as well as the initiation and elongation phases of protein synthesis (Carter et al., 2000). This disruption leads to the production of faulty proteins within the parasite, ultimately causing its death. Paromomycin's ability to impede protein synthesis is pivotal to its effectiveness in treating VL (Hobbie et al., 2011). The recommended dose of paromomycin for VL typically involves intramuscular administration at a standard regimen of 15 mg/kg/day divided into three doses for 21 days. It is often used in combination with other drugs, like miltefosine or antimonials, to enhance efficacy and reduce the risk of resistance (Sundar et al., 2007). However, dosing may vary based on factors such as patient weight, infection severity, and individual response to treatment. Healthcare providers must carefully assess each patient's condition and adjust dosages accordingly to ensure optimal therapeutic outcomes while minimizing potential adverse effects. Paromomycin has demonstrated good efficacy and safety profiles, making it a valuable addition to the treatment armamentarium for VL.

6.2.5 Combination Therapies

Combination therapies have been explored to improve treatment outcomes and reduce the risk of drug resistance in VL. One of the most widely studied combinations involves L-AmB and miltefosine; this pairing has shown high efficacy and safety in clinical trials (Chappuis et al., 2007). Other combinations, such as L-AmB with paromomycin, have also been investigated with promising results (Sundar et al., 2007). Combination therapies offer the advantage of targeting different pathways in the parasite's life cycle, reducing the likelihood of treatment failure and the emergence of resistance.

6.3 CHALLENGES IN VL TREATMENT

VL, caused by the parasite *Leishmania donovani*, remains a significant public health challenge, especially in resource-limited regions. Despite treatment advances, challenges like limited healthcare access, drug resistance, and co-infections hinder effective management. Improving access to care, addressing drug resistance, and enhancing diagnostic capabilities are crucial for controlling VL. This section discusses the key challenges in VL treatment, including drug resistance, the availability and affordability of drugs, treatment duration and compliance, and the lack of effective vaccines.

6.3.1 Drug Resistance

Drug resistance is a major concern in the treatment of VL, particularly in regions where antimonials have been extensively used. Resistance to antimonials has been reported in India, Nepal, and other parts of the world, leading to treatment failure and the need for alternative therapies . Resistance mechanisms include reduced drug uptake, increased drug efflux, and alterations in drug targets (Croft et al., 2006a). Emerging resistance to other drugs, such as miltefosine and amphotericin B, has also been reported, highlighting the need for continuous monitoring and surveillance (Ahmed et al., 2023; Sundar et al., 2010). Drug resistance not only compromises treatment outcomes but also limits the effectiveness of available therapies, necessitating the development of new drugs and treatment strategies.

6.3.2 Availability and Affordability of Drugs

The availability and affordability of drugs for VL treatment pose significant challenges, particularly in resource-limited settings. L-AmB, the most effective treatment for VL, is often expensive and inaccessible in many endemic regions (Singh et al., 2024). This limits its use and hinders efforts to control the disease. Similarly, miltefosine, the only oral drug for VL, is relatively costly and may not be affordable for many patients. Moreover, the availability of quality-assured drugs is a concern, as substandard or counterfeit drugs may lead to treatment failure and the development of drug resistance (Chowdhury et al., 2017).

6.3.3 Treatment Duration and Compliance

The long treatment duration required for some VL drugs, such as miltefosine and paromomycin, poses challenges for patient compliance. Miltefosine, for example, requires a 28-day treatment course, which can be difficult for patients to complete, leading to treatment failure and the development of resistance (Sundar et al., 2002). Similarly, the 21-day treatment course for paromomycin may be challenging for patients, particularly in resource-limited settings where access to healthcare is limited (Sundar et al., 2007).

6.3.4 Lack of Effective Vaccines

The lack of effective vaccines against VL is a major obstacle in disease control efforts. Despite significant research efforts, no licensed vaccine is currently available for VL (Duthie et al., 2012). Vaccine development has been hampered by the complex nature of the parasite's life cycle and the lack of suitable animal models for testing. Moreover, the development of a vaccine for VL faces financial and logistical challenges, including the need for large-scale clinical trials and the requirement for long-term funding commitments. A successful vaccine would not only prevent disease but also contribute to reducing transmission and achieving sustainable control and elimination of VL.

VL treatment faces several challenges, including drug resistance, the availability and affordability of drugs, treatment duration and compliance, and the lack of effective vaccines. Attacking these obstacles demands a comprehensive strategy involving the development of novel drugs and treatment protocols, enhancing access to quality-assured medications, and ongoing surveillance to curb drug resistance (Ahmed et al., 2019). Collaborative efforts between governments, pharmaceutical companies, research institutions, and non-governmental organizations are essential to overcoming these challenges and achieving sustainable control and elimination of VL.

6.4 EMERGING THERAPIES AND STRATEGIES IN VL

Emerging therapies and strategies in VL treatment hold promise for improving patient outcomes and disease control efforts. Efforts to address the challenges of drug resistance, limited treatment options, and treatment-related toxicity have led to the exploration of novel therapeutic approaches. These include developing combination therapies aimed at enhancing efficacy and reducing the risk of resistance. Additionally, advances in drug delivery systems, such as liposomal formulations, potentially improve drug efficacy and lower toxicity. Furthermore, repurposing existing drugs and exploring immunotherapeutic approaches shows promise for enhancing treatment outcomes. Ongoing research into new antileishmanial therapeutics and vaccines offers hope for more effective and accessible treatment options. Overall, this quest for innovative therapies and strategies demonstrates a continued dedication to combating VL while enhancing affected individuals' well-being.

6.4.1 Novel Drug Candidates

Several novel drug candidates are under investigation for the treatment of VL. These include fexinidazole, sitamaquin, and imidazole-based compounds. Additionally, researchers are exploring natural products, such as plant-derived compounds, for their antileishmanial properties. These drug candidates offer potential alternatives to existing therapies, aiming to address challenges such as drug resistance and treatment toxicity. Continued research into these novel compounds may lead to more effective and safer treatment options for VL in the future.

- **Fexinidazole:** Fexinidazole is an oral nitroimidazole drug that has shown promising results in preclinical studies for the treatment of VL. It is currently being evaluated in clinical trials for safety and efficacy (Mowbray et al., 2020).
- **Sitamaquine:** Sitamaquine is an antimalarial drug that has shown activity against *Leishmania* parasites in preclinical studies. Clinical trials are ongoing to evaluate its efficacy and safety for treating VL (Sundar et al., 2014).
- **Imidazole-based compounds:** Imidazole-based compounds, such as WR279396 and WR279396B, have shown activity against *Leishmania* parasites in preclinical studies. These compounds are being evaluated for their potential as new drugs for VL treatment (Croft et al., 2006b).

6.4.2 Immunotherapy

Immunotherapy is an emerging approach aimed at boosting the host's immune response against *Leishmania*. Several immunotherapeutic strategies are being investigated, including:

- **Monoclonal antibodies:** Monoclonal antibodies targeting specific *Leishmania* antigens have shown promise in preclinical studies for the treatment of VL. These antibodies can enhance the host's immune response and promote parasite clearance (Gomes et al., 2012).
- **Immunomodulators:** Immunomodulatory drugs, such as imiquimod and toll-like receptor agonists, can stimulate the host's immune response and enhance parasite clearance. Clinical trials are ongoing to evaluate these drugs' efficacy for VL treatment (Ghalib et al., 2010).

6.5 VECTOR CONTROL AND ENVIRONMENTAL MANAGEMENT

Vector control and environmental management strategies are essential for reducing the transmission of *Leishmania* parasites and preventing the spread of VL. These strategies include:

- **Insecticide-treated bed nets:** Insecticide-treated bed nets can reduce the density of sandfly vectors and prevent their bites, thereby reducing the transmission of *Leishmania* parasites (Mukhopadhyay et al., 2012).
- **Environmental management:** Environmental modifications, such as improving housing conditions and reducing vegetation density around dwellings, can decrease sandfly breeding sites and limit human contact with vectors (Quinnell et al., 2019).

The emerging therapies and strategies show promise for improving the treatment and control of VL. Novel drug candidates offer potential new treatment options, while immunotherapy and vector control measures can enhance the host immune responses and reduce *Leishmania* transmission risk. Continued research and development in these areas remain essential for overcoming the challenges posed by VL and achieving sustainable control and elimination of the disease.

6.6 ADDRESSING CHALLENGES AND IMPROVING TREATMENT OF VL

Addressing challenges and improving the treatment of VL requires a multifaceted approach that enhances access to healthcare facilities ensures the availability of quality-assured medications in endemic regions. Research into new drugs and treatment strategies, such as combination therapies and innovative drug delivery systems, aims to overcome issues like drug resistance and treatment toxicity. Moreover, investing in diagnostic tools and surveillance systems is essential for early detection and prompt treatment. Collaboration among healthcare providers, researchers, policymakers, and affected communities is crucial for implementing effective control measures and advancing VL management. Collectively, these initiatives aim to alleviate the burden of VL, ultimately leading to improved patient outcomes and public health. This section discusses strategies to address these challenges and enhance VL treatment.

6.6.1 Access to Affordable and Quality-Assured Drugs

One major challenge in VL treatment is the availability and affordability of drugs, particularly in resource-limited settings where the disease is endemic. L-AmB, the most effective treatment for VL, is often prohibitively expensive and inaccessible to many patients (Singh et al., 2023). To tackle this issue, efforts should focus on negotiating lower prices with pharmaceutical companies and ensuring the availability of generic versions of essential drugs. Additionally, quality assurance mechanisms should be implemented to prevent the distribution of substandard or counterfeit drugs, which can lead to treatment failure and the development of drug resistance (Chowdhury et al., 2017).

6.6.2 Surveillance and Monitoring

Monitoring drug resistance is crucial for effective VL treatment. Regular surveillance of drug efficacy and resistance patterns can help identify emerging resistance early and guide treatment strategies. Molecular surveillance approaches, including polymerase chain reaction and sequencing, can identify genetic mutations linked to drug unresponsiveness or resistance (Coelho et al., 2020). Surveillance should also encompass assessments of treatment outcomes and adverse reactions to guarantee the safety and effectiveness of treatment regimens.

6.6.3 Strengthening Health Systems

Strengthening health systems is essential for improving VL treatment and control. This includes training healthcare workers to diagnose and treat VL effectively, ensuring the availability of essential drugs and supplies, and improving access to healthcare services in remote and underserved areas. Integrated vector management strategies should also be implemented to reduce the transmission of *Leishmania* parasites and prevent the spread of VL (WHO, 2014). These strategies include the use of insecticide-treated bed nets, indoor residual spraying, and environmental modifications to reduce sandfly breeding sites.

6.6.4 Development of New Treatment Regimens

Advancing new treatment regimens is crucial for combating drug resistance and enhancing treatment outcomes in VL. Combination therapies that employ two or more drugs with distinct mechanisms of action have proven effective in reducing the risk of drug resistance while improving cure rates (Singh et al., 2022). Novel drug candidates, such as fexinidazole and sitamaquine, are currently being investigated for their potential in VL treatment (Mowbray et al., 2020; Sundar et al., 2014). These new approaches should be rigorously tested in clinical trials to ensure their safety and efficacy before being adopted for routine use.

6.6.5 Community Engagement and Education

Community engagement and education are key components of successful VL control programs. Educating communities about the causes, symptoms, and prevention of VL can help reduce the stigma associated with the disease while promoting early diagnosis and treatment-seeking behavior. Community health workers can play a crucial role in raising awareness about VL and encouraging individuals to seek timely treatment. Additionally, community-based surveillance programs can help monitor disease trends and identify outbreaks early, allowing for prompt intervention (WHO, 2014).

In conclusion, addressing the challenges associated with VL treatment requires a multifaceted approach that includes improving access to affordable and quality-assured drugs, strengthening health systems, developing new treatment regimens, and engaging communities in disease control efforts. By implementing these strategies, it is possible to improve the treatment and control of VL and reduce the burden of this devastating disease on affected populations.

6.7 CONCLUSION AND FUTURE DIRECTIONS

VL, caused by the protozoan parasite *Leishmania donovani*, continues to pose significant challenges to public health, particularly in resource-limited regions of the world. Despite several available treatment options, including pentavalent antimonials, liposomal amphotericin B, miltefosine, and paromomycin, the management and control of VL are complicated by factors such as drug resistance, treatment duration and compliance, availability and affordability of drugs, as well as the lack of effective vaccines.

Managing these obstacles demands a holistic approach that encompasses developing new drugs, combination therapies, immunotherapeutic strategies, and vector control measures. The development of new drugs is crucial for overcoming drug resistance and improving treatment outcomes in VL. Novel drug candidates with improved efficacy, safety, and affordability are being explored to target different aspects of the parasite's biology and pathogenesis.

Combination therapies utilizing two or more drugs with different mechanisms of action have shown promise in reducing the risk of drug resistance and improving cure rates. Immunotherapeutic strategies, including monoclonal antibodies and immunomodulatory drugs, offer the potential for enhancing the host immune responses against *Leishmania* parasites, further improving treatment outcomes.

Vector control and environmental management are essential components of VL control programs aimed at reducing the transmission of *Leishmania* parasites and preventing the spread of the disease. Strategies such as insecticide-treated bed nets, indoor residual spraying, and environmental modifications can help reduce sandfly vector density and limit their contact with humans.

Community engagement and education are also critical for improving disease awareness and promoting early diagnosis and treatment-seeking behavior. By involving communities in disease control efforts and increasing awareness about VL and its preventive measures, the burden of the disease can be significantly reduced.

In conclusion, addressing the challenges associated with VL treatment requires a multifaceted approach that includes developing of new drugs, combination therapies, immunotherapeutic strategies, and vector control measures. Community engagement and education are also crucial for improving disease awareness and promoting early diagnosis and treatment. By implementing these strategies, it is possible to improve VL management and alleviate the burden of this devastating disease on affected populations.

ACKNOWLEDGMENT

The authors sincerely thank the Indian Council of Medical Research-Rajendra Memorial Research Institute of Medical Sciences (ICMR-RMRIMS), Patna, for their essential support throughout this study.

REFERENCES

Ahmed, G., Jamal, F., Tiwari, R. K., Singh, V., Rai, S. N., Chaturvedi, S. K., & Vamanu, E. (2023). Arsenic exposure to mouse visceral leishmaniasis model through their drinking water linked to the disease exacerbation via modulation in host protective immunity: a preclinical study. *Scientific Reports, 13*(1), 21461.

Ahmed, G., Thakur, A. K., Pushpanjali, S., Chaturvedi, S. K., Shivam, P., & Narayan, S. (2019). Modulation of the immune response and infection pattern to *Leishmania donovani* in visceral leishmaniasis due to arsenic exposure: an in vitro study. *PLoS One, 14*(2), e0210737.

Alvar, J., Vélez, I. D., Bern, C., Herrero, M., Desjeux, P., Cano, J., & Jannin, J. (2012). Leishmaniasis worldwide and global estimates of its incidence. *PLoS One, 7*(5), e35671.

Carter, A. P., Clemons, W. M., Brodersen, D. E., Morgan-Warren, R. J., Wimberly, B. T., & Ramakrishnan, V. (2000). Functional insights from the structure of the 30S ribosomal subunit and its interactions with antibiotics. *Nature, 407*(6802), 340–348.

Chappuis, F., Sundar, S., Hailu, A. et al. (2007). Visceral leishmaniasis: what are the needs for diagnosis, treatment and control?. *Nat Rev Microbiol*, 5, 873–882.

Chowdhury, R., Mondal, D., Chowdhury, V., Faria, S., Alvar, J., Nabi, S. G., ..., & Akter, S. (2017). How far are we from visceral leishmaniasis elimination in Bangladesh? An assessment of epidemiological surveillance data. *PLoS Neglected Tropical Diseases, 11*(9), e0005896.

Coelho, A. C., Boisvert, S., Mukherjee, A., Leprohon, P., Corbeil, J., & Ouellette, M. (2020). Multiple mutations in heterogeneous miltefosine-resistant Leishmania major population as determined by whole genome sequencing. *PLoS Neglected Tropical Diseases, 14*(1), e0007943.

Croft, S. L., Seifert, K., & Yardley, V. (2006a). Current scenario of drug development for leishmaniasis. *Indian Journal of Medical Research, 123*(3), 399–410.

Croft, S. L., Sundar, S., & Fairlamb, A. H. (2006b). Drug resistance in leishmaniasis. *Clinical Microbiology Reviews, 19*(1), 111–126.

Dorlo, T. P., Balasegaram, M., Beijnen, J. H., & de Vries, P. J. (2012b). Miltefosine: a review of its pharmacology and therapeutic efficacy in the treatment of leishmaniasis. *Journal of Antimicrobial Chemotherapy, 67*(11), 2576–2597.

Duthie, M. S., Reed, S. G., & Notkins, A. L. (2012). A historical perspective on the development of vaccines to prevent leishmaniasis. *Human Vaccines & Immunotherapeutics, 8*(11), 1768–1773.

Ghalib, H. W., Whittle, J. A., Kubin, M., Hashim, F. A., El-Hassan, A. M., & Grabstein, K. H. (2010). IL-12 enhances Th1-type responses in human *Leishmania donovani* infections. *Journal of Immunology, 154*(9), 4623–4629.

Gomes, C. M., Cesetti, M. V., de Paula, C. D., Laurenti, M. D., da Matta, V. L. R., & Saldanha, L. C. (2012). Monoclonal antibodies against a *Leishmania infantum* chagasi antigen as a tool for in vitro diagnosis of canine visceral leishmaniasis. *Parasitology Research, 110*(5), 2133–2141.

Hobbie, S. N., Kaiser, M., Schmidt, S., Shcherbakov, D., Janusic, T., Brun, R., & Böttger, E. C. (2011). Genetic reconstruction of protozoan rRNA decoding sites provides a rationale for paromomycin activity against Leishmania and Trypanosoma. *PLoS Neglected Tropical Diseases, 5*(5), e1161.

Mowbray, C. E., Braimah, Z., Matthews, H., Gedeon, J., Hill, J., Amewu, R. K., ..., & O'Neill, P. M. (2020). The discovery of fexinidazole for the treatment of human African trypanosomiasis. *Bioorganic & Medicinal Chemistry, 28*(12), 115553.

Mukhopadhyay, A. K., Hati, A. K., Chakraborty, A. K., Saxena, N., Goswami, A., Sarkar, A., & Bhattacharya, S. K. (2012). The presence of deltamethrin resistant Phlebotomus argentipes in Bihar (India) and its impact on the kala-azar control programme. *Memórias do Instituto Oswaldo Cruz, 107*(8), 1029–1032.

Nagle, A. S., Khare, S., Kumar, A. B., Supek, F., Buchynskyy, A., Mathison, C. J., & Molteni, V. (2014). Recent developments in drug discovery for leishmaniasis and human African trypanosomiasis. *Chemical Reviews, 114*(22), 11305–11347.

Quinnell, R. J., Carson, C., Reithinger, R., & Garcez, L. M. (2019). Vector control in leishmaniasis. *Current Tropical Medicine Reports, 6*(3), 128–141.

Rijal, S., Yardley, V., Chappuis, F., Decuypere, S., Khanal, B., Singh, R., ... & Dujardin, J. C. (2007). Antimonial treatment of visceral leishmaniasis: are current in vitro susceptibility assays adequate for prognosis of in vivo therapy outcome? *Microbes and Infection, 9*(4), 529–535.

Singh, V., Singh, J., Singh, N., Verma, M. K., Praveena, S. M., & Mishra, V. (2023). Sustainable removal of Pb (II) from aqueous medium by using chitosan functionalized Citrus limetta peels biomass. *Bioremediation Journal, 28*(3), 245–265.

Singh, V., Singh, J., Singh, N., Rai, S. N., Verma, M. K., Verma, M., Singh, V., Chivate, M. S., Bilal, M., & Mishra, V. (2022). Simultaneous removal of ternary heavy metal ions by a newly isolated Microbacterium paraoxydans strain VSVM IIT(BHU) from coal washery effluent. *BioMetals, 36*(4), 829–845.

Singh, V., Singh, N., Rai, S. N., Chaturvedi, V. K., Singh, S. K., Kumar, A., Vamanu, E., & Mishra, V. (2024). Chitosan functionalized recyclable and eco-friendly nanoadsorbent for Pb(II) adsorption from water. *Nanocomposites, 10*(1), 214–226.

Sundar, S., Chakravarty, J., & Agarwal, D. (2010). Visceral leishmaniasis elimination program in India, Bangladesh, and Nepal: reshaping the case finding/case management strategy. *PLoS Neglected Tropical Diseases, 4*(10), e852.

Sundar, S., Jha, T. K., Thakur, C. P., Engel, J., Sindermann, H., Fischer, C., & Junge, K. (2002). Oral miltefosine for Indian visceral leishmaniasis. *New England Journal of Medicine, 347*(22), 1739–1746.

Sundar, S., Jha, T. K., Thakur, C. P., Sinha, P. K., & Bhattacharya, S. K. (2000). Injectable paromomycin for Visceral leishmaniasis in India. *New England Journal of Medicine, 343*(20), 1475–1479.

Sundar, S., Jha, T. K., Thakur, C. P., Sinha, P. K., Bhattacharya, S. K., & Rai, M. (2007). Injectable paromomycin for Visceral leishmaniasis in India. *New England Journal of Medicine, 356*(25), 2571–2581.

Sundar, S., Mehta, H., Suresh, A. V., Singh, S. P., Rai, M., & Murray, H. W. (2004). Amphotericin B treatment for Indian visceral leishmaniasis: conventional versus lipid formulations. *Clinical Infectious Diseases, 38*(3), 377–383.

Sundar, S., Singh, A., Rai, M., Prajapati, V. K., Singh, A. K., Ostyn, B., ..., & Rijal, S. (2014). Efficacy of miltefosine in the treatment of visceral leishmaniasis in India after a decade of use. *Clinical Infectious Diseases, 59*(2), 451–457.

Sundar, S., Sinha, P. K., Rai, M., Verma, D. K., Nawin, K., Alam, S., ..., Modabber, F. (2011). Comparison of short-course multidrug treatment with standard therapy for visceral leishmaniasis in India: an open-label, non-inferiority, randomised controlled trial. *Lancet, 377*(9764), 477–486.

World Health Organization. (2014). Control of the leishmaniases: report of a meeting of the WHO Expert Committee on the Control of Leishmaniases, Geneva, 22–26 March 2010.

World Health Organization. (2021). Leishmaniasis. Retrieved from https://www.who.int/news-room/fact-sheets/detail/leishmaniasis.

7 Post-kala-azar Dermal Leishmaniasis Development from Visceral Leishmaniasis

*Ghufran Ahmed, Veer Singh, Dayakar Alti, Sonali Vedika, Ashish Kumar**

Department of Biochemistry, ICMR-Rajendra Memorial Research Institute of Medical Sciences, Patna, Bihar, India
*Corresponding Author: Dr. Ashish Kumar, Scientist - D, Department of Biochemistry, ICMR-Rajendra Memorial Research Institute of Medical Sciences, Patna, Bihar, India

7.1 INTRODUCTION

Visceral leishmaniasis (VL), also called kala-azar, is a deadly disease caused by protozoan parasites of the *Leishmania donovani* complex and disseminated by infected female phlebotomine sandflies. The disease is highly endemic in tropical and subtropical regions, affecting economically disadvantaged residents. Individuals in these communities often keep cattle for their livelihoods, and the presence of cattle manure creates favorable conditions for sandfly breeding, which is responsible for transmitting the disease (Ahmed et al., 2023). Following the injection of leishmanial parasites through the bite of an infected sandfly, individuals may develop VL characterized by intermittent or remittent fever, enlargement of the liver and spleen, pancytopenia, and anemia. Hence, timely and effective therapeutics are essential for containing this infection; presently, only a few antileishmanial medications are available, including antimonial preparations, amphotericin-B, AmBisome, miltefosine, and paromomycin (Singh et al., 2021a; Singh et al. 2021b). The decreasing effectiveness of these drugs and regional variations in response contribute to differing treatment guidelines for leishmaniasis across various territories (Singh et al., 2021c). Furthermore, the emergence of co-infections, such as HIV and tuberculosis with VL, and of drug-resistant *Leishmania* parasites significantly lowers the usefulness of current therapeutic modalities. For example, the efficacy of sodium stibogluconate (SSG) has declined by over 60% in the Indian subcontinent, prompting clinicians in this region to discontinue its use. Similarly, the effectiveness of amphotericin-B (AmB), its liposomal formulation (AmBisome), and miltefosine has been gradually declining (Ahmed et al., 2023; Panda et al., 2019; Darshani et al., 2018; Chaudhuri et al., 2022).

Additionally, post-kala-azar dermal leishmaniasis (PKDL) has become an increasingly prevalent and growing concern in VL endemic areas, mainly in Southeast Asia (India, Nepal, and Bangladesh) and East Africa (Sudan, Ethiopia, and Kenya) over the past few decades (Kumar et al., 2021). Individuals with PKDL dermal lesions act as reservoirs for *L. donovani*, facilitating parasite transmission to sandflies during blood meals. A study estimated that PKDL may account for up to 25% of VL cases in the Bangladeshi population (Swain et al., 2023). Additionally, the inadequate healthcare-seeking tendencies among affected individuals necessitate proactive surveillance to deal with concealed cases. The emergence of parasite resistance to current therapeutic regimens presents another hurdle in managing VL. Thus, active surveillance can engage individuals who are hesitant

DOI: 10.1201/9781032643120-7

to seek healthcare and encourage them to pursue treatment options (Ghosh et al., 2021a). PKDL is recognized as a significant challenge to VL elimination programs in South Asia; therefore, public health initiatives addressing VL and PKDL, such as strategies for active case surveillance and identification, diagnosis, and treatment, should be prioritized (Zijlstra et al., 2017). Moreover, the etiopathogenesis and contributing factors leading to the development of PKDL from VL remain unknown. In this chapter, we explore the geographical distribution, clinical manifestations, immunological insights, and parasitological perspectives of PKDL development to enrich understanding and pave the way for future interventions to manage and control the disease.

7.2 EPIDEMIOLOGY OF PKDL

PKDL, a dermal complication, is recognized as a distinct clinical entity in a significant proportion of patients several months to years after completing therapeutic regimens recommended by the World Health Organization (WHO), depending on geographical areas (World Health Organization, 2022). Over a century ago, in 1922, Dr. U.N. Brahmachari identified Leishman-Donovan bodies in a skin lesion smear from Indian patients, termed dermal leishmanoid, and subsequently confirmed this through animal studies (Ghosh et al., 2021b; Gupta 1927; Brahmachari et al., 1922). It was later designated as post-kala-azar dermal leishmaniasis because it arises following VL. Predicting who will develop PKDL is a challenging task. Factors such as inadequate treatment adherence, youthful age (5–17 years), malnutrition, VL-HIV co-infection, HIV treatment history, and prolonged exposure to environmental arsenic may all contribute to the development of PKDL (Joshi et al., 2023). The exact magnitude of the PKDL burden remains uncertain; however, information gleaned from diverse studies conducted in various geographical areas contributes to a better understanding of the disease's scope. Since 2014, the WHO has actively enhanced surveillance of PKDL, which is recognized as a potential *Leishmania* parasite reservoir during inter-endemic periods (Kumar et al., 2023). Between 2014 and 2022, three WHO regions documented more than 10,094 PKDL cases: the African region accounted for 5%, the Eastern Mediterranean for 6%, and Southeast Asia for 89% of the global burden; fewer cases were reported in the Western Pacific and the Americas (Figure 7.1) (Ruiz-Postigo et al., 2021; 2022). Recently, more than 90% of documented PKDL cases originated from Southeast Asia, particularly from India, Bangladesh, and Nepal. In India specifically, PKDL is primarily reported from Bihar and Jharkhand, collectively contributing over 80% of India's reported cases between 2014 and 2023 (Singh et al., 2020) (Figure 7.1).

7.3 CLINICAL PRESENTATION OF PKDL

Researchers have explored the spectrum of clinical indices associated with PKDL since its recognition as a distinct clinical entity. It has minimal mortality compared to VL, and a significant

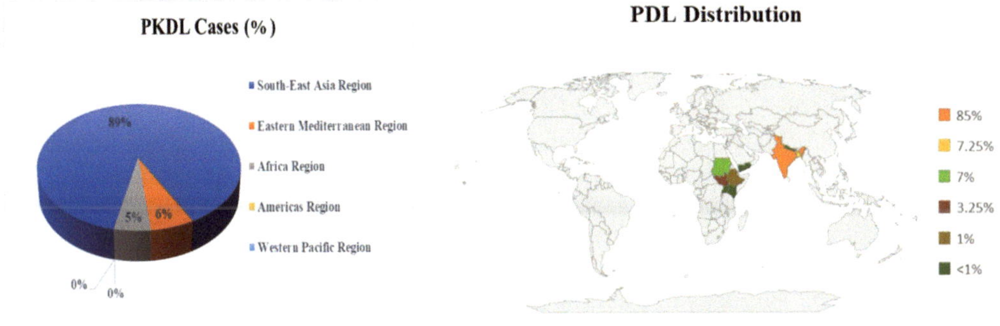

FIGURE 7.1 Distribution of PKDL worldwide (Sources: WHO and NVBDCP, 2014–2023)

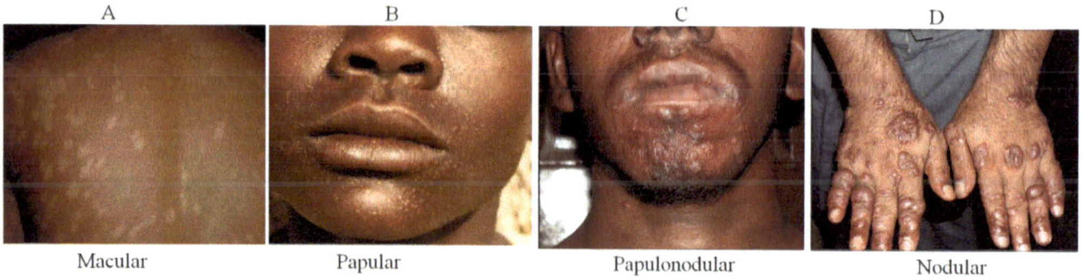

FIGURE 7.2 Clinical presentations of PKDL, ranging from subtle skin alterations (Image A: Macular) to prominent lesions (Image D: Nodular). This depiction provides insights for researchers and clinicians to competently comprehend the diagnostic challenges and treatment complexities of PKDL (Zijlstra et al. 2019).

proportion of VL patients develop PKDL after completing WHO-recommended therapeutic regimens. Due to its occurrence following a VL episode, it was designated as post-kala-azar dermal leishmaniasis by Acton and Napier in 1927 to distinguish it from cutaneous leishmaniasis, also known as oriental sore (Singh et al., 2022a; Singh et al., 2024). The clinical presentation of PKDL is mainly characterized by hypopigmented rashes or skin lesions that can be macular, papular, or nodular. These lesions can develop anywhere on the body but are primarily located on the face, torso, legs, arms, and genital area (Kumar et al., 2021; Mondal et al., 2014; Singh et al., 2023a). In South Asia, approximately 10–20% of seemingly cured VL patients develop PKDL within 2–7 years; while in East Africa, this figure is significantly higher at 50–60% within 3–6 months, often occurring alongside VL (Roy et al., 2023; Sengupta et al., 2019) (Figure 7.2).

7.4 DIAGNOSIS OF PKDL

PKDL presents in distinct clinical forms: macular, papular, nodular, and mixed (papulonodular). The hypopigmented macular form is often characterized by discolored flat skin and typically has a low parasite count. The papular variant manifests as small, raised lesions on the skin, often accompanied by hypopigmentation. In contrast, the nodular form exhibits larger nodules that may ulcerate. Additionally, the mixed form combines elements of both papular and nodular presentations (Kumar et al., 2021; Ramesh et al., 2008). These diverse clinical manifestations pose a challenge in diagnosis because they resemble other dermatological conditions such as vitiligo, leprosy, and psoriasis. Diagnosis depends on detecting *Leishmania* parasites in skin tissue samples, commonly obtained through methods such as slit skin smears, punch biopsies, or skin scrapings. The primary method for PKDL diagnosis continues to be the use of a microscope to identify amastigote parasites in skin slit smears. However, the sensitivity of this method varies across lesion types; accuracy ranges from 67–100% in nodular PKDL skin lesions to 7–33% in macular skin lesions (Ghosh et al., 2022).

Serological tests like rK39 are crucial in diagnosing PKDL as they detect *Leishmania*-specific antibodies in the patient's blood serum. They facilitate early detection of PKDL with high sensitivity and specificity (80–90%), even before the onset of visible skin lesions, and can provide a diagnosis within minutes. This early detection enables healthcare providers to promptly initiate treatment, preventing further disease progression and reducing the risk of transmission within the community (Adams et al., 2013). Moreover, it requires minimal specialized equipment to perform in remote and resource-limited areas where more sophisticated laboratory facilities may not be available. Overall, it offers a convenient, reliable, and rapid diagnostic procedure for PKDL, contributing to timely diagnosis, patient management, and disease control in endemic regions.

Quantitative polymerase chain reaction (qPCR) is a valuable tool for PKDL diagnosis due to its high sensitivity and specificity in detecting *Leishmania* DNA in skin tissue samples. The sensitivity of qPCR for diagnosing PKDL has been reported to range from 75–100% in various studies, depending on factors such as the target gene sequence used and the geographic location of the study

TABLE 7.1
Diagnostic Modalities for PKDL

Methods	Advantages	Disadvantages
Microscopy	The gold standard method in resource-limited settings	Invasive; low sensitivity; time-consuming; requires skilled personnel
Serological methods (rK39)	Quick; non-intrusive; improved case identification	Limited sensitivity and specificity; unable to discriminate between present and prior infections
PCR	Higher sensitivity; ability to quantify parasite number	Requiring trained personnel and specialized laboratory facilities
LAMP	Rapid; highly sensitive; suitable for field tests	Moderate sensitivity and specificity; susceptible to contamination
RPA	Rapid and sensitive detection	Requires strict temperature control; potential for false positives due to primer dimerization

(Ghosh et al., 2018). Different target sequences, including kinetoplast DNA (kDNA), ribosomal RNA, mini-exon-derived RNA (med RNA), β-tubulin, and GP63, can be employed for this application. Additionally, its ability to quantify parasite load in skin tissue samples further enhances its utility in monitoring disease progression and treatment efficacy (Salotra et al., 2001). Overall, qPCR demonstrates promise as an effective and reliable diagnostic tool for PKDL, contributing to improved patient care and disease management strategies.

Traditional diagnostic methods for PKDL, such as microscopy and serological tests, have limitations in terms of sensitivity and specificity. The loop-mediated isothermal amplification (LAMP) assay is a molecular method that offers several advantages over conventional techniques, particularly in field settings, due to its higher sensitivity and tolerance to inhibitors in clinical samples (Verma et al., 2013). The isothermal nature of LAMP reactions enables DNA amplification at a constant temperature, eliminating the need for sophisticated thermal cycling equipment and reducing the time required for diagnosis. The assay provides rapid results (typically within 15–60 minutes) and clear visual differentiation of positive and negative results, making it suitable for point-of-care testing in endemic regions. It has demonstrated an accuracy of approximately 96% in PKDL diagnosis (Dixit et al., 2021). Nonetheless, the risk of false-positive results due to cross-contamination exists, and its application for diagnosing *Leishmania* infections is still limited.

Recombinase polymerase amplification (RPA) presents a notable advancement in diagnosing PKDL, offering a new approach to detecting *Leishmania* parasites and managing the disease. RPA offers high sensitivity and specificity (over 90%) in clinical samples acquired from PKDL lesions, facilitating rapid and accurate identification of *Leishmania* parasites (Tan et al., 2022). The method amplifies specific sequences in the *Leishmania* genome, allowing the detection of low levels of *Leishmania* DNA. RPA is simple and requires minimal equipment and expertise, making it suitable for use in remote or resource-limited areas where PKDL is common (Roy et al., 2023; Ghosh et al., 2021a). Furthermore, continued research and implementation efforts are essential to optimize and expand the use of RPA in routine PKDL diagnosis, ultimately improving patient outcomes and disease management strategies (Table 7.1).

7.5 TREATMENT STRATEGIES FOR PKDL

Managing PKDL requires a strategic approach that considers the patient's health status, the specific *Leishmania* species, local drug resistance patterns, and available treatment options. Antileishmanial medications are the cornerstone of PKDL treatment. Historically, SSG served as the mainstay, but increasing resistance has prompted the exploration of alternative drugs. Clinicians administered

SSG through daily injections of 20 mg/kg of body weight for 20 days each month over six months (Datta et al., 2021). However, SSG has several limitations, including potential toxicity, prolonged hospitalization, and the need for daily painful injections. Additionally, rising antimony resistance, particularly in regions like India, has led to the discontinuation of SSG for treating both VL and PKDL (Ramakrishna et al., 2023).

Miltefosine, a phospholipid derivative initially developed as an anti-cancer agent, was later repurposed for leishmaniasis, including PKDL. It is the only oral medication available and was first licensed in India in 2002 for the treatment of VL. Miltefosine made its debut for PKDL treatment in 2006. Currently, in the Indian subcontinent, miltefosine is recommended as the primary treatment option for PKDL as a 12-week course, with doses adjusted based on the patient's age and weight (Sundar et al., 2023). It is administered orally after meals at specific dosages: 2.5 mg/kg/day for patients under 12 years old; 50 mg/day for those aged 12 years and older and weighing more than 25 kg; and 100 mg/day (divided doses) for patients in the same age group and weight range (Datta et al., 2021). Major side effects are related to the gastrointestinal tract and may include vomiting, abdominal discomfort, loss of appetite, and ocular complications. Recent reports suggest a significant (~15%) increase in relapse rates with miltefosine monotherapy.

A promising alternative treatment for PKDL is amphotericin-B, especially its liposomal formulation, AmBisome, which has shown high efficacy rates. While conventional amphotericin-B exhibits high effectiveness against PKDL, its use is limited by its toxicity and prolonged infusion periods. Studies have shown high cure rates for PKDL with AmBisome. In Sudan, a regimen involving six infusions of a 5 mg/kg dose over three weeks resulted in a 96% cure rate for macular cases. AmBisome at a dose of 2.5 mg/kg per day for 20 days achieved an 83% cure rate in SSG-unresponsive PKDL cases, leading to the regression of papular and macular lesions without adverse events (Rabi Das et al., 2017). In Bangladesh, a 15-day short-course regimen of AmBisome at 15 mg/kg, administered in five biweekly infusions of 3 mg/kg, improved lesions in 89% of PKDL cases without serious adverse events (Younis et al., 2023). While AmBisome has demonstrated effectiveness in treating various types of lesions in Indian PKDL patients, the treatment response may vary depending on the type of lesion. However, some studies have raised concerns about potential adverse effects like hypokalemia-induced rhabdomyolysis, necessitating careful monitoring during treatment.

Paromomycin, an aminoglycoside, has proven effective against *Leishmania* and bacterial infections. Its effectiveness in visceral and cutaneous leishmaniasis treatment is well-established. Several studies highlight its potential in treating PKDL (Pokharel et al., 2021). In one study, patients received intramuscular paromomycin at a daily dose of 11 mg/kg for 45 days; however, poor patient compliance was observed (Sundar et al., 2014). Another study compared a combination of paromomycin and miltefosine given for two 10-day courses with the same combination given for three 10-day courses. The final cure rate was reported at 83%, with a relapse rate of 17% upon follow-up after 12 months (Musa et al., 2023). Although paromomycin is naturally safe, infrequent damaging effects include pain at the injection site, reversible ototoxicity, nephrotoxicity, hypokalemia, and minor liver function abnormalities.

The current treatment regimen designates miltefosine as the preferred first-line drug for PKDL treatment in the Indian subcontinent due to its status as the only oral medication with high efficacy. However, recent studies showing a drop in miltefosine's effectiveness – along with a significant (15%) rise in relapses – have prompted the exploration of alternative approaches, such as combination regimens or repurposing of the drugs (Sundar et al., 2015). The underlying principle of combination therapy is to boost effectiveness by using medications that are individually effective in treating the infection when used alone. Both liposomal amphotericin-B (LAmB) and miltefosine therapy individually exhibit high effectiveness in treating PKDL; their synergistic effect, when combined, is expected to improve the cure rate and decrease the relapse rate. The combination regimen of LAmB and miltefosine proved effective when administered at half the recommended dose of either drug when used individually. One study aimed to evaluate the safety and efficacy

TABLE 7.2
Therapeutic Approaches for PKDL

Therapeutic Modality	Dose	Duration	Cure Rate	Toxicity	Location
Miltefosine	100–150 mg/day	12 weeks	75–80%	Gastrointestinal disturbances; renal toxicities; teratogenic effects	Parts of India, Bangladesh, Sudan
Amphotericin-B (AmB)	3–4 mg/kg/day	4–6 weeks	90–95%	Infusion-related reactions; nephrotoxicity	Global
Liposomal amphotericin-B (LAmB)	2.5–10 mg/kg/day	4 days	90–95%	Chills and rigors	Global
Sodium stibogluconate (SSG)	20 mg/kg/day	30–60 days	50–85%	Cardiotoxicity; pancreatitis; hepatotoxicity	East Africa, Indian subcontinent
Paromomycin	11 mg/kg/day	45–60 days	78–85%	Injection site reactions; ototoxicity; nephrotoxicity	East Africa, Indian subcontinent
Combination therapy (MF + LAmB)	MF (100 mg/kg) + LAmB (5 mg/kg)	45 days	95–100%	Vomiting and nausea	Indian subcontinent, East Africa

of this combination therapy. Three intravenous injections of LAmB (5 mg/kg body weight) were administered, followed by daily miltefosine pills (50 mg/kg body weight), starting after the first LAmB dose and continuing until the end of the 45-day course. Treatment with the combination resulted in a notable decrease in parasite burden during the mid-treatment phase (Ramesh et al., 2020). Following the treatment, the parasite load decreased to a negligible level during the follow-up period, resulting in a 100% cure rate with no reports of relapse at the 18-month follow-up. The combination therapy of LAmB and miltefosine spans a concise duration of just 45 days, as opposed to the lengthy 90-day regimen required for miltefosine monotherapy. This shorter duration fosters enhanced patient adherence without any instances of dropout or relapse reported. The therapy is immensely effective with minimal side effects. Moreover, opting for combination treatment over monotherapy is expected to ease both the duration of hospitalization and the risk of resistance or relapse (Table 7.2).

7.6 FACTORS IN PKDL DEVELOPMENT

The onset of PKDL involves multiple factors, such as the host immune system, genetic predispositions, insufficient treatment, failure to adhere to drug regimens, and exposure to environmental elements like UV-B radiation and arsenic. Treatment for PKDL typically consists of a course of antileishmanial drugs, including amphotericin-B, miltefosine, and paromomycin, or combinations of these therapeutics (Singh et al., 2016, Datta et al., 2021).

7.6.1 Host Immune Dysregulation

Numerous studies have investigated the immune system's role in PKDL, revealing the dysregulated immune responses associated with the condition. Understanding the interconnected issues within the immune system is crucial for developing improved strategies to treat and manage PKDL (Chatterjee et al., 2020; Singh et al., 2022b).

A notable shift occurs in macrophage polarization toward an alternatively activated M2 phenotype in PKDL lesions. This shift is characterized by decreased production of reactive oxygen and nitrogen radicals, which indicates impaired antimicrobial activity against *Leishmania* parasites. Molecular analyses have demonstrated enhanced mRNA expression of nuclear receptor PPAR-γ and ARG1, key markers associated with M2 polarization, further supporting the predominance of M2 macrophages in PKDL lesions. Additionally, the increased presence of CD68+ macrophages in dermal lesions underscores the role of M2 polarization in PKDL pathogenesis. This dysregulated immune response, characterized by the predominance of M2 macrophages, contributes to disease chronicity and the persistence of *Leishmania* parasites within the host (Singh et al., 2022c; Singh et al., 2023b).

Another study examined gene expression changes in cytokines, chemokine receptors, and apoptotic molecules during active PKDL. Molecules like CD40, CD40L, and interferon regulatory factor 1 (IRF1) exhibited modulation, indicating their potential involvement in PKDL progression. Increased IL-17 expression suggests tissue inflammation, while TNF-α and nitric oxide production implicate their roles in PKDL pathogenesis. Investigating the reciprocal regulation of regulatory T cells (Tregs) and Th17 cells could provide deeper insights for targeted interventions (Haribhai et al., 2016). Repolarizing alternatively activated macrophages to the classically activated M1 phenotype using antileishmanial drugs or immune stimulators presents a promising intervention for combating PKDL and achieving better treatment outcomes (Haribhai et al., 2016; Mukhopadhyay et al., 2014).

7.6.2 Host Genetic Factors

Genetic factors play a pivotal role in the complex pathogenesis of leishmaniasis, specifically in PKDL. The precise mechanisms of PKDL pathogenesis remain elusive; however, evidence suggests that host genetic factors are crucial in determining susceptibility and the clinical course of disease severity (Gedda et al., 2020). Research into PKDL pathogenesis has explored the influence of host genetic factors, particularly within the human leukocyte antigen (HLA) gene complex. One significant finding is the association between polymorphism in the interferon-gamma receptor (IFN-γR) and PKDL, which affects cytokine balance and responsiveness to IFN-γ, potentially facilitating parasite persistence and growth (Salih et al., 2014, 2007). However, no correlation has been established between IFN-γ or IL-10 promoter polymorphism and disease susceptibility (Farouk et al., 2010; Salih et al., 2007). Interestingly, lower expression of IFN-γR1 in Indian PKDL patients indicates a compromised response to IFN-γ, which may facilitate parasite growth. Moreover, studies in Sudan have revealed uniformly reduced expression of IFN-γ and its receptor in the dermal tissue of PKDL patients, shedding light on mechanisms of parasite persistence (Zijlstra et al., 2016). Additionally, elevated levels of plasma C-reactive protein (CRP) in Sudanese PKDL patients prompt inquiries into genetic polymorphisms in CRP genes that may contribute to PKDL susceptibility. Variations in genes encoding proteins involved in innate immunity, such as FCN-2, MBL2, and SLC11A1, have also been linked to susceptibility to cutaneous and visceral leishmaniasis, suggesting their potential role in PKDL pathogenesis (Braliou et al., 2019). Furthermore, studies in India have identified genetic polymorphisms in parasites causing both VL and PKDL, underscoring the importance of comparative gene expression profiling to elucidate differences contributing to disease manifestations. Polymorphisms in genes encoding interleukins (IL-4, IL-10), tumor necrosis factor-alpha (TNF-α), IFN-γ, and toll-like receptors have been investigated for their role in modulating immune responses and disease outcomes in PKDL (Bharati et al., 2022). In addition to immune-related genes, genetic factors related to drug metabolism and host-pathogen interactions may also contribute to PKDL pathogenesis. Variations in genes encoding drug-metabolizing enzymes or drug transporters could influence the efficacy and toxicity of antileishmanial drugs used in PKDL treatment. Epigenetic investigations in relevant genes offer promising avenues for uncovering the underlying mechanisms driving PKDL development (Yasmin et al., 2022; Volpedo et al., 2021). These findings underscore the importance of genetic research in understanding PKDL pathogenesis. Further studies are warranted to elucidate the genetic determinants underlying PKDL susceptibility, with the potential for epigenetic investigations to provide deeper insights into the mechanisms driving PKDL development (Mondal et al. 2014).

7.6.3 Inadequate Treatment

Suboptimal treatment results may occur when the chosen drug therapy fails to eliminate the *Leishmania* parasites from the host. This deficiency may stem from multiple factors, including drug resistance or inadequate drug concentrations reaching the target tissues. Antimonial resistance, in particular, has been a concern in regions where VL is endemic, potentially leading to treatment failure and increasing the risk of PKDL development (Wijnant et al., 2022). Epidemiological and clinical data from different areas, such as East Africa (specifically Sudan) and the Indian subcontinent, unfailingly indicate a high risk of PKDL development following sodium antimony gluconate (SAG) treatment. SAG resistance has emerged in the Indian subcontinent, with approximately 73% of PKDL patients developing the condition after SAG treatment. The remaining 27% of PKDL cases in this region were associated with other drugs such as AmB, miltefosine, or paromomycin. Additionally, in Sudan, Bangladesh, and Nepal, all observed PKDL cases had previously undergone SAG treatment for VL (Kumar et al., 2023).

Research indicates that employing alternative treatments such as paromomycin, miltefosine, or a combination of miltefosine and AmB in VL patients lowers the risk of PKDL development (Kumar et al., 2009). Adverse effects associated with antileishmanial drugs can sometimes lead to treatment interruptions or modifications, compromising therapy effectiveness. For instance, severe adverse reactions to antimonials may necessitate treatment discontinuation, leaving the infection incompletely treated and predisposing patients to PKDL (WHO, 2022; Rijal et al., 2013). Moreover, the immune status of the host significantly influences the effectiveness of treatment. Immunocompromised individuals face a higher risk of treatment failure and subsequent development of PKDL (Sundar et al., 2012). Overall, addressing issues related to inadequate treatment, such as optimizing treatment protocols, monitoring drug efficacy, and ensuring timely diagnosis and initiation of treatment, is essential for reducing the incidence of PKDL following VL treatment.

7.6.4 Environmental Factors

PKDL lesions often manifest on sun-exposed areas, reflecting clothing habits, and can vary from macules to nodules. The pathogenesis of PKDL involves abnormal immune responses induced by ultraviolet (UV) light, specifically UVB radiation, which suppresses the Th1 immune response necessary for parasite elimination (Mukhopadhyay et al., 2014). The energy of UVB photons penetrates the skin, where photoreceptors absorb it. This absorption leads to mutagenic effects and immune suppression, including the upregulation of cytokine genes such as TNF-α, IL-6, IL-8, IL-10, and IL-12. TNF-α, primarily produced by keratinocytes, appears to be a key mediator of UVB-induced immunosuppression (Ismail et al., 2006). Furthermore, exposure to UV light alters the morphology and function of Langerhans cells. This damage includes the antigen-presenting epidermal Langerhans cells, resulting in a decrease in their numbers and a downregulation of the expression of HLA-DR and B7-1. Ultimately, these effects impact both alloantigen and hypersensitivity responses (Gedda et al., 2020). These changes lead to a shift toward a Th2 immune response characterized by the production of IL-10 and IL-4. In PKDL lesions, this Th2 response antagonizes the Th1 response required for parasite clearance, thus contributing to chronic inflammation.

Additionally, arsenic exposure plays a role in the progression of PKDL, especially in regions with widespread arsenic contamination. Arsenic exposure impairs the immune system, rendering individuals more susceptible to *Leishmania* parasites. Moreover, arsenic can directly affect skin integrity and the immune system, creating favorable conditions for PKDL lesions to develop (Ahmed et al., 2019; Das et al., 2016). Furthermore, arsenic exposure may interact with other environmental factors, such as UV light, further increasing the risk of PKDL development. Therefore, arsenic exposure is considered a significant environmental factor in the pathogenesis of PKDL, particularly in areas with widespread arsenic contamination.

7.7 CHALLENGES IN PKDL MANAGEMENT

PKDL is a complex disease that is difficult to diagnose and manage; it is often misdiagnosed as other skin diseases. Additionally, there is limited access to drugs for managing it, and existing treatments are often ineffective or have serious side effects (Pal et al., 2023). Asymptomatic PKDL cases can harbor a substantial parasite load, making them a probable source of new infections for vector sandflies. This is especially concerning during VL inter-epidemic periods, as PKDL can keep the parasite alive in the community even when VL cases are low (Le Rutte et al., 2019). The lack of a definitive diagnostic test exacerbates the situation. While laboratory approaches such as slit skin smears and molecular techniques exist, they are frequently unavailable in resource-constrained regions where PKDL is frequent. Healthcare personnel who are unfamiliar with PKDL may misdiagnose it as another skin disorder, resulting in inadequate management.

Managing PKDL is critical for breaking the transmission loop and ensuring ongoing VL eradication. Insufficient knowledge about PKDL among healthcare professionals and the general public may lead to inadequate reporting of cases. Systematic epidemiological investigations are required to fill this information imbalance and inform evidence-based management strategies (Adams et al., 2013). Treating PKDL is complicated; the treatment plan used is determined by criteria such as clinical presentation, parasite burden, and drug availability. The same antileishmanial drugs used to treat VL are also used for PKDL; however, results can vary. Some PKDL cases respond well to treatment, while others fail or relapse. The longer treatment period necessary for PKDL, as well as potential side effects from certain medications, can impair patient compliance and further complicate management efforts (Sundar et al., 2014).

The challenges faced in managing PKDL are numerous and involve clinical variability, therapeutic complexities, disease transmission dynamics, insufficient data, health system integration issues, and research shortages. Addressing these difficulties necessitates a comprehensive and interdisciplinary approach that includes advancements in diagnostics, treatment options, healthcare system integration, awareness campaigns, and joint research activities (Haribhai et al., 2016; Mondal et al., 2014). By successfully managing PKDL, we can reduce its impact on public health and contribute to the long-term management and eventual eradication of VL.

7.8 FUTURE PROSPECTS AND RESEARCH PRIORITIES FOR CONTROLLING PKDL PROGRESSION

The Kala-azar Elimination Program (KAEP) is shifting from the attack phase to the consolidation phase. The objective is to pinpoint new low-endemic areas via active case detection, strengthen vector control measures, and tackle potential sources of infection, with a particular focus on PKDL (Zijlstra et al., 2017, 2018). PKDL is recognized as a challenge in the elimination effort, and the KAEP aims to develop solutions for identifying cases, diagnosis, and treatment. Incorporating PKDL management into broader health goals ensures that efforts continue beyond the initial VL eradication targets, preventing recurrence.

Point-of-care tests that can detect PKDL cases efficiently and readily, especially in resource-constrained settings, are critically important. Strengthening healthcare provider capacity through training and education can enhance PKDL diagnosis and treatment at the primary care level (Bengtson et al., 2020; Cruz et al., 2019). Furthermore, research efforts should be targeted toward increasing the sensitivity and specificity of diagnostic procedures while ensuring their usability by healthcare practitioners with varying levels of experience. Understanding the factors that contribute to treatment failure and recurrence in PKDL instances can influence the development of personalized therapeutics. Examining other pharmaceutical options and combination therapies may help conquer drug resistance and treatment complexities. Molecules with superior efficacy, shorter treatment durations, and fewer adverse effects can lead to better patient outcomes.

Additionally, promoting awareness about the impact of PKDL among the general population can encourage early healthcare-seeking behavior and minimize stigma (Basher et al., 2015). Engaging communities in disease surveillance, reporting, and treatment adherence can foster a more hands-on approach to PKDL control. The horizon of PKDL management shows bright potential, spanning enhanced diagnostics, advancements in drug development, deeper immunological understanding, refined vector control, integration into public health measures, and other promising avenues.

Research priorities should focus on creating precise interventions, cultivating interdisciplinary cooperation, and promoting worldwide awareness and attention (Sengupta et al., 2019; Garapati et al., 2018). By tackling these forthcoming opportunities and research emphases, we can lay a foundation for successful PKDL management, lessening its impact on public health and propelling us closer to the overarching objective of eradicating VL.

7.9 CONCLUSION

The development of PKDL from VL presents a multifaceted challenge encompassing diagnosis, treatment, pathogenesis, and management. The transition from VL to PKDL underscores the complexity of *Leishmania* parasite dynamics within the human host. Diagnosis remains problematic due to the lack of definitive tests, leading to frequent misdiagnosis and delayed treatment initiation. Treatment options are limited, with existing drugs often ineffective or associated with significant side effects, exacerbating the burden on affected individuals.

Contributing factors in PKDL pathogenesis, including immune dysregulation, genetic factors, environmental factors, asymptomatic cases, and parasite persistence, further complicate disease management and transmission dynamics. Addressing these challenges necessitates a multifaceted approach that integrates advancements in diagnostics, treatment strategies, and understanding of disease pathogenesis. Future directions should focus on the development of accurate diagnostic tools, novel therapeutic interventions, and a deeper comprehension of host-parasite interactions. Additionally, enhancing healthcare provider capacity, promoting community engagement, and integrating PKDL management into broader health initiatives are essential for effective disease control.

ACKNOWLEDGMENT

The authors are thankful to the ICMR-Rajendra Memorial Research Institute of Medical Sciences, Patna, for its necessary support during this study.

REFERENCES

Adams ER, Versteeg I, Leeflang MM. Systematic review into diagnostics for post-kala-azar dermal leishmaniasis (PKDL). *Journal of Tropical Medicine.* 2013 Oct;2013:150746.

Ahmed G, Jamal F, Tiwari RK, Singh V, Rai SN, Chaturvedi SK, Pandey K, Singh SK, Kumar A, Narayan S, Vamanu E. Arsenic exposure to mouse visceral leishmaniasis model through their drinking water linked to the disease exacerbation via modulation in host protective immunity: a preclinical study. *Scientific Reports.* 2023 Dec 5;13(1):21461.

Ahmed G, Thakur AK, Pushpanjali S, Chaturvedi SK, Shivam P, Jamal F, Singh MK, Bimal S, Singh SK, Choudhary SK. Modulation of the immune response and infection pattern to *Leishmania donovani* in visceral leishmaniasis due to arsenic exposure: An in vitro study. *PLoS One.* 2019 Feb 5;14(2):e0210737.

Basher A, Nath P, Nabi SG, Selim S, Rahman MF, Sutradhar SR, Faiz A, Bhuiyan MR, Ahmed BN, Rahman R. A study on health seeking behaviors of patients of post-kala-azar dermal leishmaniasis. *BioMed Research International*. 2015 Oct;2015(1):314543.

Bengtson M, Bharadwaj M, Ten Bosch A, Nyakundi H, Matoke-Muhia D, Dekker C, Diehl JC. Matching development of point-of-care diagnostic tests to the local context: a case study of visceral leishmaniasis in Kenya and Uganda. *Global Health: Science and Practice*. 2020 Sep 30;8(3):549–65.

Bharati K. Human genetic polymorphism and Leishmaniasis. *Infection, Genetics and Evolution*. 2022 Mar 1;98:105203.

Brahmachari UN. A new form of cutaneous leishmaniasis—dermal leishmanoid. *The Indian Medical Gazette*. 1922 Apr;57(4):125.

Braliou GG, Kontou PI, Boleti H, Bagos PG. Susceptibility to leishmaniasis is affected by host SLC11A1 gene polymorphisms: a systematic review and meta-analysis. *Parasitology Research*. 2019 Aug 1;118:2329–42.

Chatterjee M, Sengupta R, Mukhopadhyay D, Mukherjee S, Dighal A, Moulik S, Sengupta S. Immune responses in post kala-azar dermal leishmaniasis. *Indian Journal of Dermatology*. 2020 Nov 1;65(6):452–60.

Chaudhuri, R., Prasanth, T., & Dash, J. (2022). Expanding the Toolbox of Target Directed Bio-Orthogonal Synthesis: In Situ Direct Macrocyclization by DNA Templates. *Angewandte Chemie*, *135*(7), e202215245.

Cruz I, Albertini A, Barbeitas M, Arana B, Picado A, Ruiz-Postigo JA, Ndung'u JM. Target product profile for a point-of-care diagnostic test for dermal leishmaniases. *Parasite Epidemiology and Control*. 2019 May 1;5:e00103.

Darshani, P., Gumpu, M. B., Thumpati, P., Rayappan, J. B. B., Ravichandiran, V., Pazhani, G. P., Veerapandian, M. (2018). Chemically synthesized butein and butin: Optical, structure and electrochemical redox functionality at electrode interface. *Journal of Photochemistry and Photobiology B Biology*, *182*, 122–129.

Das S, Mandal R, Rabidas VN, Verma N, Pandey K, Ghosh AK, Kesari S, Kumar A, Purkait B, Lal CS, Das P. Chronic arsenic exposure and risk of post kala-azar dermal leishmaniasis development in India: a retrospective cohort study. *PLoS Neglected Tropical Diseases*. 2016 Oct 24;10(10):e0005060.

Datta A, Podder I, Das A, Sil A, Das NK. Therapeutic modalities in post kala-azar dermal leishmaniasis: A systematic review of the effectiveness and safety of the treatment options. *Indian Journal of Dermatology*. 2021 Jan;66(1):34.

Dixit KK, Ramesh V, Gupta R, Negi NS, Singh R, Salotra P. Real-time fluorimetry loop-mediated isothermal amplification for diagnosis of leishmaniasis and as a tool for assessment of cure for post–kala-azar dermal leishmaniasis. *The American Journal of Tropical Medicine and Hygiene*. 2021 Jun;104(6):2097.

Farouk S, Salih MA, Musa AM, Blackwell JM, Miller EN, Khalil EA, Elhassan AM, Ibrahim ME, Mohamed HS. Interleukin 10 gene polymorphisms and development of post kala-azar dermal leishmaniasis in a selected Sudanese population. *Public Health Genomics*. 2010 Aug 1;13(6):362–7.

Garapati P, Pal B, Siddiqui NA, Bimal S, Das P, Murti K, Pandey K. Knowledge, stigma, health seeking behaviour and its determinants among patients with post kalaazar dermal leishmaniasis, Bihar, India. *PloS One*. 2018 Sep 7;13(9):e0203407.

Gedda MR, Singh B, Kumar D, Singh AK, Madhukar P, Upadhyay S, Singh OP, Sundar S. Post kala-azar dermal leishmaniasis: a threat to elimination program. *PLoS Neglected Tropical Diseases*. 2020 Jul 2;14(7):e0008221.

Ghosh P, Chowdhury R, Maruf S, Picado A, Hossain F, Owen SI, Nath R, Baker J, Hasnain MG, Shomik MS, Ghosh D. Gauging the skin resident Leishmania parasites through a loop mediated isothermal amplification (LAMP) assay in post-kala-azar dermal leishmaniasis. *Scientific Reports*. 2022 Oct 27;12(1):18069.

Ghosh P, Hasnain MG, Hossain F, Khan MA, Chowdhury R, Faisal K, Mural MA, Baker J, Nath R, Ghosh D, Maruf S. Evaluation of real-time PCR for diagnosis of post-kala-azar dermal leishmaniasis in endemic foci of Bangladesh. *InOpen Forum Infectious Diseases*. 2018 Oct;5(10):ofy234.

Ghosh P, Roy P, Chaudhuri SJ, Das NK. Epidemiology of post-kala-azar dermal leishmaniasis. *Indian Journal of Dermatology*. 2021b Jan 1;66(1):12–23.

Ghosh P, Sharma A, Bhattarai NR, Abhishek K, Nisansala T, Kumar A, Böhlken-Fascher S, Chowdhury R, Khan MA, Faisal K, Hossain F. A multi-country, single-blinded, phase 2 study to evaluate a point-of-need system for rapid detection of leishmaniasis and its implementation in endemic settings. *Microorganisms*. 2021a Mar 12;9(3):588.

Gupta BD. A Note on the Parasite of "Dermal Leishmanoid". *The Indian Medical Gazette*. 1927 Jan;62(1):11.

Haribhai D, Ziegelbauer J, Jia S, Upchurch K, Yan K, Schmitt EG, Salzman NH, Simpson P, Hessner MJ, Chatila TA, Williams CB. Alternatively Activated Macrophages Boost Induced Regulatory T and Th17 Cell Responses during Immunotherapy for Colitis. *The Journal of Immunology*. 2016;196(8):3305–3317.

Ismail A, Khalil EA, Musa AM, Hassan IE, Ibrahim ME, Theander TG, Hassan AE. The pathogenesis of post kala-azar dermal leishmaniasis from the field to the molecule: does ultraviolet light (UVB) radiation play a role?. *Medical Hypotheses.* 2006 Jan 1;66(5):993–9.

Joshi AB, Banjara MR, Das ML, Ghale P, Pant KR, Parajuli N, Pyakurel UR, Dahal GR, Das CL, Kroeger A, Aseffa A. Prevalence of post kala-azar dermal leishmaniasis (PKDL) and treatment seeking behavior of PKDL patients in Nepal. *PLoS Neglected Tropical Diseases.* 2023 Feb 9;17(2):e0011138.

Kumar A, Singh VK, Tiwari R, Madhukar P, Gautam V, Sundar S, Kumar R. Post kala-azar dermal leishmaniasis in the Indian sub-continent: challenges and strategies for elimination. *Frontiers in Immunology.* 2023 Aug 11;14:1236952.

Kumar D, Ramesh V, Verma S, Ramam M, Salotra P. Post-kala-azar dermal leishmaniasis (PKDL) developing after treatment of visceral leishmaniasis with amphotericin B and miltefosine. *Annals of Tropical Medicine & Parasitology.* 2009 Dec 1;103(8):727–30.

Kumar P, Chatterjee M, Das NK. Post kala-azar dermal leishmaniasis: clinical features and differential diagnosis. *Indian Journal of Dermatology.* 2021 Jan;66(1):24.

Le Rutte EA, Zijlstra EE, de Vlas SJ. Post-kala-azar dermal leishmaniasis as a reservoir for visceral leishmaniasis transmission. *Trends in Parasitology.* 2019 Aug 1;35(8):590–2.

Mondal D, Hamano S, Hasnain G, Satoskar AR. Challenges for management of post kala-azar dermal leishmaniasis and future directions. *Research and Reports in Tropical Medicine.* 2014 Nov 14:105–11.

Mukhopadhyay D, Dalton JE, Kaye PM, Chatterjee M. Post kala-azar dermal leishmaniasis: an unresolved mystery. *Trends in Parasitology.* 2014 Feb 1;30(2):65–74.

Musa AM, Mbui J, Mohammed R, Olobo J, Ritmeijer K, Alcoba G, Muthoni Ouattara G, Egondi T, Nakanwagi P, Omollo T, Wasunna M. Paromomycin and miltefosine combination as an alternative to treat patients with visceral leishmaniasis in eastern Africa: a randomized, controlled, multicountry trial. *Clinical Infectious Diseases.* 2023 Feb 1;76(3):e1177–85.

NVBDCP (2014–2023): Kala-Azar Situation in India. National Center for Vector Borne Diseases Control (NCVBDC). Available at: https://ncvbdc.mohfw.gov.in/index4.php?lang=1&level=0&linkid=467&lid=3750.

Pal B, Kumari S, Kaur M, Wadhwa P, Murti K, Kumar R, Pandey K, Siddiqui NA, Dhingra S, Padmakar S. Barriers to the effective management and prevention of post kala-azar dermal leishmaniasis (PKDL) in the Indian subcontinent. *Medical Journal Armed Forces India.* 2023 May 15;79(5):500–505.

Panda D, Saha P, Chaudhuri R, Prasanth T, Ravichandiran V, Dash, J. A Competitive Pull-Down Assay Using G-quadruplex DNA Linked Magnetic Nanoparticles To Determine Specificity of G-quadruplex Ligands. *Analytical Chemistry.* 2019;91(12):7705–7711.

Pokharel P, Ghimire R, Lamichhane P. Efficacy and safety of paromomycin for visceral leishmaniasis: a systematic review. *Journal of Tropical Medicine.* 2021 Jul 26;2021(1):8629039.

Rabi Das VN, Siddiqui NA, Pal B, Lal CS, Verma N, Kumar A, Verma RB, Kumar D, Das P, Pandey K. To evaluate efficacy and safety of amphotericin B in two different doses in the treatment of post kala-azar dermal leishmaniasis (PKDL). *PLoS One.* 2017 Mar 29;12(3):e0174497.

Ramakrishna, K., Nalla, L. V., Naresh, D., Venkateswarlu, K., Viswanadh, M. K., Nalluri, B. N., Chakravarthy, G., Duguluri, S., Singh, P., Rai, S. N., Kumar, A., Singh, V., & Singh, S. K. (2023). WNT-β Catenin Signaling as a Potential Therapeutic Target for Neurodegenerative Diseases: Current Status and Future Perspective. *Diseases*, *11*(3), 89.

Ramesh V, Dixit KK, Sharma N, Singh R, Salotra P. Assessing the efficacy and safety of liposomal amphotericin B and miltefosine in combination for treatment of post kala-azar dermal leishmaniasis. *The Journal of Infectious Diseases.* 2020 Feb 3;221(4):608–17.

Ramesh V, Ramam M, Singh R, Salotra P. Hypopigmented post-kala-azar dermal leishmaniasis. *International Journal of Dermatology.* 2008 Apr;47(4):414–6.

Rijal S, Ostyn B, Uranw S, Rai K, Bhattarai NR, Dorlo TP, Beijnen JH, Vanaerschot M, Decuypere S, Dhakal SS, Das ML. Increasing failure of miltefosine in the treatment of Kala-azar in Nepal and the potential role of parasite drug resistance, reinfection, or noncompliance. *Clinical Infectious Diseases.* 2013 Jun 1;56(11):1530–8.

Roy M, Ceruti A, Kobialka RM, Roy S, Sarkar D, Wahed AA, Chatterjee M. Evaluation of Recombinase Polymerase Amplification assay for monitoring parasite load in patients with kala-azar and post kala-azar dermal leishmaniasis. *PLOS Neglected Tropical Diseases.* 2023 Apr 19;17(4):e0011231.

Ruiz-Postigo JA, Jain S, Madjou S, Agua JF, Maia-Elkhoury AN, Valadas S, Warusavithana S, Osman M, Yajima A, Lin Z, Beshah A. Global leishmaniasis surveillance, 2022: assessing trends over the past 10 years/Surveillance mondiale de la leishmaniose, 2022: evaluation des tendances des 10 dernieres annees. *Weekly Epidemiological Record.* 2023 Oct 6;98(40):471–88.

Ruiz-Postigo JA, Jain S, Madjou S, Maia-Elkhoury AN, Valadas S, Warusavithana S, Osman M, Yajima A, Lin Z, Beshah A, Kim S. Global leishmaniasis surveillance: 2021, assessing the impact of the COVID-19 pandemic.

Salih MA, Fakiola M, Abdelraheem MH, Younis BM, Musa AM, ElHassan AM, Blackwell JM, Ibrahim ME, Mohamed HS. Insights into the possible role of IFNG and IFNGR1 in Kala-azar and Post Kala-azar Dermal Leishmaniasis in Sudanese patients. *BMC Infectious Diseases*. 2014 Dec;14:1–9.

Salih MA, Ibrahim ME, Blackwell JM, Miller EN, Khalil EA, ElHassan AM, Musa AM, Mohamed HS. IFNG and IFNGR1 gene polymorphisms and susceptibility to post-kala-azar dermal leishmaniasis in Sudan. *Genes & Immunity*. 2007 Jan;8(1):75–8.

Salotra P, Sreenivas G, Pogue GP, Lee N, Nakhasi HL, Ramesh V, Negi NS. Development of a species-specific PCR assay for detection of *Leishmania donovani* in clinical samples from patients with kala-azar and post-kala-azar dermal leishmaniasis. *Journal of Clinical Microbiology*. 2001 Mar 1;39(3):849–54.

Sengupta R, Chaudhuri SJ, Moulik S, Ghosh MK, Saha B, Das NK, Chatterjee M. Active surveillance identified a neglected burden of macular cases of post kala-azar dermal leishmaniasis in West Bengal. *PLoS Neglected Tropical Diseases*. 2019 Mar 11;13(3):e0007249.

Singh N, Singh V, Rai SN, Mishra V, Vamanu E, Singh MP. Deciphering the gut microbiome in neurodegenerative diseases and metagenomic approaches for characterization of gut microbes. *Biomedicine & Pharmacotherapy*. 2022c;156:113958.

Singh N, Singh V, Rai SN, Vamanu E, Singh MP. Metagenomic Analysis of Garden Soil-Derived Microbial Consortia and Unveiling Their Metabolic Potential in Mitigating Toxic Hexavalent Chromium. *Life*. 2022;12(12):2094.

Singh O, Singh B, Chakravarty J, Sundar S. Current challenges in treatment options for visceral leishmaniasis in India: a public health perspective. *Infectious Diseases of Poverty*. 2016 Apr 1;5(02):1–5.

Singh V, Abhishek K, Rai SN, Singh SK, Vamanu E, Kumar A. Source of Cr(VI) in the aquatic ecosystem, its genotoxic effects and microbial removal from contaminated water. *Green Chemistry Letters and Reviews*. 2023;16(1):2267079.

Singh V, Ahmed G, Vedika S, Kumar P, Chaturvedi SK, Rai SN, Vamanu E, Kumar A. Toxic heavy metal ions contamination in water and their sustainable reduction by eco-friendly methods: isotherms, thermodynamics and kinetics study. *Scientific Reports*. 2024;14(1):7595.

Singh V, Jain SK, Mishra V. A Recent Update on SARS-CoV-2 Transmission and its Variants: Transmission, Pathogenic Mechanism, and Treatment. *Coronaviruses*. 2023;4(1):1–13.

Singh V, Mishra V. Microbial removal of Cr (VI) by a new bacterial strain isolated from the site contaminated with coal mine effluents. *Journal of Environmental Chemical Engineering*. 2021a;9(5):10627.

Singh V, Singh J, Mishra V. Sorption kinetics of an eco-friendly and sustainable Cr (VI) ion scavenger in a batch reactor. *Journal of Environmental Chemical Engineering*. 2021b;9(2):105125.

Singh V, Singh J, Mishra V. Development of a cost-effective, recyclable and viable metal ion doped adsorbent for simultaneous adsorption and reduction of toxic Cr (VI) ions. *Journal of Environmental Chemical Engineering*. 2021c;9(2):105124.

Singh V, Singh MP, Mishra V. Bioremediation of toxic metal ions from coal washery effluent. *Desalination and Water Treatment*. 2020;197:300–318.

Singh V, Singh N, Verma M, Kamal R, Tiwari R, Chivate MS, Rai SN, Kumar A, Singh A, Singh MP, Vamanu E, Mishra V. Hexavalent-Chromium-Induced Oxidative Stress and the Protective Role of Antioxidants against Cellular Toxicity. *Antioxidants*. 2022b;11(12):2375.

Sundar S, Singh A, Chakravarty J, Rai M. Efficacy and safety of miltefosine in treatment of post-kala-azar dermal leishmaniasis. *The Scientific World Journal*. 2015 Oct;2015(1):414378.

Sundar S, Singh A, Rai M, Prajapati VK, Singh AK, Ostyn B, Boelaert M, Dujardin JC, Chakravarty J. Efficacy of miltefosine in the treatment of visceral leishmaniasis in India after a decade of use. *Clinical Infectious Diseases*. 2012 Aug 15;55(4):543–50.

Sundar S, Singh A, Singh OP. Strategies to overcome antileishmanial drugs unresponsiveness. *Journal of Tropical Medicine*. 2014 Oct;2014(1):646932.

Sundar S, Singh J, Dinkar A, Agrawal N. Safety and effectiveness of miltefosine in Post–Kala-Azar dermal leishmaniasis: an observational study. *InOpen Forum Infectious Diseases*. 2023 May;10(5):ofad231.

Swain BR, Jena SR, Beriha SK, Mahanta CS, Jena BB, Prasanth T, Samanta L, Satapathy R, Dash BP. Synthesis and anticancer properties of dendritic glycoconjugates containing multiple o-carborane clusters. *New Journal of Chemistry*. 2023;47(21):10296–308.

Tan M, Liao C, Liang L, Yi X, Zhou Z, Wei G. Recent advances in recombinase polymerase amplification: principle, advantages, disadvantages and applications. *Frontiers in Cellular and Infection Microbiology*. 2022 Nov 28;12:1744.

Verma S, Avishek K, Sharma V, Negi NS, Ramesh V, Salotra P. Application of loop-mediated isothermal amplification assay for the sensitive and rapid diagnosis of visceral leishmaniasis and post-kala-azar dermal leishmaniasis. *Diagnostic Microbiology and Infectious Disease.* 2013 Apr 1;75(4):390–5.

Volpedo G, Pacheco-Fernandez T, Holcomb EA, Cipriano N, Cox B, Satoskar AR. Mechanisms of immunopathogenesis in cutaneous leishmaniasis and post kala-azar dermal leishmaniasis (PKDL). *Frontiers in Cellular and Infection Microbiology.* 2021 Jun 8;11:512.

Wijnant GJ, Dumetz F, Dirkx L, Bulté D, Cuypers B, Van Bocxlaer K, Hendrickx S. Tackling drug resistance and other causes of treatment failure in leishmaniasis. *Frontiers in Tropical Diseases.* 2022 May 12;3:837460.

World Health Organization. Report of the meeting of programme managers and the Regional Technical Advisory Group for the kala-azar elimination programme: virtual meeting 18–20 April 2022. World Health Organization. Regional Office for South-East Asia; 2022.

Yasmin H, Adhikary A, Al-Ahdal MN, Roy S, Kishore U. Host–pathogen interaction in leishmaniasis: immune response and vaccination strategies. *Immuno.* 2022 Mar 9;2(1):218–54.

Younis BM, Mudawi Musa A, Monnerat S, Abdelrahim Saeed M, Awad Gasim Khalil E, Elbashir Ahmed A, Ahmed Ali M, Noureldin A, Muthoni Ouattara G, Nyakaya GM, Teshome S. Safety and efficacy of paromomycin/miltefosine/liposomal amphotericin B combinations for the treatment of post-kala-azar dermal leishmaniasis in Sudan: a phase II, open label, randomized, parallel arm study. *PLOS Neglected Tropical Diseases.* 2023 Nov 21;17(11):e0011780.

Zijlstra EE. The immunology of post-kala-azar dermal leishmaniasis (PKDL). *Parasites & Vectors.* 2016 Dec;9(1):1–9.

Zijlstra EE. Biomarkers in Post-kala-azar Dermal Leishmaniasis. *Frontiers in Cellular and Infection Microbiology.* 2019;9:1-10.

Zijlstra EE, Alves F, Rijal S, Arana B, Alvar J. Post-kala-azar dermal leishmaniasis in the Indian subcontinent: a threat to the South-East Asia Region Kala-azar elimination programme. *PLoS Neglected Tropical Diseases.* 2017 Nov 16;11(11):e0005877.

Zijlstra EE, Kumar A, Sharma A, Rijal S, Mondal D, Routray S. Report of the fifth post-kala-azar dermal leishmaniasis consortium meeting, Colombo, Sri Lanka, 14–16 May 2018.

8 Symptoms and Diagnosis of Visceral Leishmaniasis and Post-kala-azar Dermal Leishmaniasis

Priyanka Yadav[1]*, *Kusum Yadav*[1], *Gyanendra Tripathi*[2], *and Pratibha Tiwari*[3]
[1]Department of Biochemistry, University of Lucknow, Uttar Pradesh, India
[2]Department of Bioengineering, Integral University, Lucknow, Uttar Pradesh, India
[3]Department of Biotechnology, IIMT College, Greater Noida, Uttar Pradesh, India
*Corresponding Author: Priyanka Yadav, Department of Biochemistry, University of Lucknow, Uttar Pradesh, India, Email: priyankayadav3294@gmail.com

8.1 INTRODUCTION

According to studies, post-kala-azar dermal leishmaniasis (PKDL), which can appear during or after treatment, is a cutaneous complication of visceral leishmaniasis (VL). It has been commonly observed in both Sudan and India, but it has only occasionally been documented in South America. Since both disorders can present clinically with hypopigmented macular lesions, and histological examinations can reveal neural involvement, it may be difficult to distinguish between the two. In the nations where both diseases are endemic, research on the coinfection of leprosy and VL has been conducted. The coinfection of leprosy and VL, which may develop into PKDL, along with discussions of the clinical and pathological aspects in patients, as well as reviews of the relevant literature, have also been reported (1, 2).

The *Leishmania donovani* complex – specifically *L. donovani* sensu stricto in East Africa and the Indian subcontinent, and *Leishmania infantum* in North Africa, Europe, and Latin America – is responsible for the systemic disease known as VL, which is found to be fatal if left untreated (3). There are two forms of VL: zoonotic VL and anthroponotic VL. These forms differ in their transmission from human to human and from animal to human, respectively. In zoonotic VL, the parasite is stored in animals, primarily dogs, with humans serving as sporadic hosts. Zoonotic VL is found in areas where *L. infantum* is transmitted; conversely, anthroponotic VL occurs in locations where *L. donovani* prevails. PKDL, an after-treatment consequence of VL characterised by a macular, maculopapular, or nodular rash, is more frequently seen in Sudan than in other East African nations and the Indian subcontinent. In regions where *L. infantum* is endemic, it can also occur in immunocompromised people. In Sudan, the time between treated VL and PKDL ranges from 0 to 6 months, whereas in India, it is 3 months to 9 months. Because the nodular lesions of PKDL cases contain a large number of parasites, these cases are thought to serve as a reservoir for anthroponotic VL between epidemic cycles (4).

 DOI: 10.1201/9781032643120-8

8.2 VISCERAL LEISHMANIASIS: SYMPTOMS

VL has several symptoms; some important symptoms are illustrated in .

As a protozoan infection, VL typically presents as a prolonged fever lasting for weeks or months, primarily due to neutropenia, a condition characterised by a drastic decrease in white blood cell count. It may appear as a constitutional syndrome or manifest as a long-standing fever accompanied by other symptoms such as weight loss, fatigue and weakness, splenomegaly (enlarged spleen), hepatomegaly (enlarged liver), anaemia (low red blood cell count), leukopaenia (low white blood cell count), loss of appetite, swollen lymph nodes, abdominal discomfort, vomiting (children), diarrhoea (children), night sweat, systemic illness (VL), and scaly skin.

VL is a systemic illness brought on by the *L. donovani* complex that has a high morbidity and fatality rate in underdeveloped countries. Patients with VL may exhibit signs and indications of a systemic infection or may be asymptomatic. In patients with typical symptoms, the positive predictive value of a clinical diagnosis is typically strong; however, signs and symptoms are often ambiguous and can be confused with those of other co-endemic diseases. The heterogeneity of the species of *Leishmania* species prevalent in various endemic regions – as well as the frequent tendency of HIV coinfections to result in unusual presentations – contributes to this diagnostic challenge. Despite these challenges, some parasitological techniques are still regarded as the most reliable methods for diagnosing VL due to their specificity.

The introduction of molecular and serological assays has further improved the diagnostic methods of VL. The direct agglutination test (DAT) and the rK39 antigen-based immunochromatographic test are the major suitable approaches for serological diagnosis of VL. Recombinant antigens have enhanced the performance of serodiagnostic testing. Although their use is usually difficult and impractical, molecular methods' outstanding sensitivity and specificity, as well as their ability to investigate diverse chemicals, have increased their relevance. Quantitative polymerase chain reaction (qPCR) was found to be superior to conventional polymerase chain reaction (PCR) in distinguishing between active and asymptomatic VL infections, such as for the identification of VL-HIV coinfection.

This chapter focuses on the most recent advancements in VL diagnostics and summarises of the currently used diagnostic approaches (5). Up to 350 million people worldwide are at high risk of developing leishmaniasis – a vector-borne disease – of which VL remains the most severe from among its multiple clinical presentations. Accurate identification is essential since the clinical signs of VL might resemble those of several other widespread diseases; this is particularly critical given that treatment can carry high toxicity.

The most effective methods for diagnosing VL continue to be invasive and hazardous approaches that entail demonstrating parasites in stained preparations from splenic and bone marrow aspirates. A significant portion of asymptomatic people test positive for rK39 using the DAT, immunoblotting, and enzyme-linked immunosorbent assay (ELISA) serological tests; this complicates relapse identification as positive results can persist for months or years post-cure.

PCR remains one of the most widely used molecular techniques for diagnosing and differentiating between species. This chapter compares the numerous VL diagnostic tools now on the market, reviewing their relative advantages and disadvantages alongside recent advancements in the field. A rapid, painless, precise, and affordable marker for active VL applicable in field settings is needed for improving diagnosis (6). Epidemiological studies have investigated clinical trends and risk factors for VL in an endemic region of Brazil; findings indicate that 3.1% of children under 15 years had the condition, averaging 4.3 cases per 1,000 children annually. Within the endemic region, disease prevalence varied seasonally and annually; malnutrition prior to disease onset and young age (median age 3 years) were risk factors. Factors such as intestinal parasitism or migration into the area had no impact on infection progression into illness. Serological testing indicates that 7.5% of children become infected with *Leishmania* annually, with an overall disease-to-infection ratio across the entire region being 1:18.5 – and 1:6.5 for the area with the highest disease frequency.

Early detection and treatment have changed the disease's clinical trends (7). The *L. donovani* complex is the protozoan responsible for causing the devastating disease VL across both the Old and New Worlds. It is estimated that there are approximately 500,000 cases of human VL annually worldwide, with many types of clinical presentations of the disease. The primary reservoir hosts for the Mediterranean variant of VL are domestic dogs (*Canis familiaris*). Cases have been reported in both human and animal reservoir hosts across various Iranian locations; notably, VL has recently become a serious endemic disease in northwest Iran. Recent research findings have revealed that leptomonad infections occur naturally in the sandfly (*Phlebotomus*) species *Phlebotomus major* and *Phlebotomus keshishiani* found in southern and central Iran, as well as *Phlebotomus kandelakii* and *Phlebotomus perfiliewi* in northwestern regions.

The DAT was developed based on the methodology established by Allain and Kagan; it is applicable for both sero-diagnosis and sero-epidemiological studies of VL. A modified version of the DAT has been utilised in Iran as a quick, accurate, reasonably priced method for diagnosing VL in humans as well as canine reservoir hosts. As domestic dogs have been identified as the major significant reservoir hosts for this protozoan, the study focused on ascertaining dogs' infection rates as well as their function in transmitting the infection to humans (8).

Leishmania completes its life cycle in two hosts: humans and sandflies. The adult female sandfly is a bloodsucker that feeds primarily at night on sleeping victims; when a sandfly bites an infected person, *Leishmania* is ingested along with the blood of its prey. The protozoan exists initially in one of its two smaller forms as a spherical, non-motile amastigote measuring only 3–7 micrometres in diameter before transforming rapidly into promastigotes – longer motile forms – within the sandfly's stomach. The promastigotes have a single flagellum, allowing mobility; they are spindle-shaped – approximately three times larger than amastigotes – and migrate towards the proximal end of the gut, where they prepare for a regurgitative transmission while living extracellularly and reproducing asexually in the alimentary canal.

Upon biting a human host again during feeding sessions, promastigotes are expelled from the sandfly's proboscis at the biting site, where they enter their new host's body and infiltrate macrophages (9). Within these cells, they revert back into amastigote form – their smaller variant – and proliferate within the phagolysosome, the most hostile area inside macrophage cells, by suppressing and evading this organelle's natural defence mechanism. Recent conjecture suggests that amastigotes may also exit macrophages by inducing exocytosis responses from these cells. After repeated replication cycles, they ultimately destroy their host cell through sheer mass pressure. Protozoans from the daughter cells subsequently migrate towards new cells or circulate through the blood, seeking new hosts. This process facilitates the infection's advance, eventually affecting the liver and spleen as well as the mononuclear phagocyte system of the host. Sandflies then consume the liberated amastigotes in peripheral tissues to begin a new cycle (10).

8.2.1 Diagnosis

Clinical presentation: The symptoms mentioned above, along with the patient's history and geographical location, can provide diagnostic clues.

Microscopic examination: The identification of *Leishmania* amastigotes in tissue samples (such as spleen, bone marrow, or lymph node aspirates) is the gold standard for diagnosis (Figure 8.1).

Serological tests: Medical professionals may examine blood samples for indicators of *Leishmania* infection via small needle pricks or venous draws from the arms; however, blood tests often lack efficacy in identifying live infections since asymptomatic individuals may yield positive results despite not currently exhibiting any illness symptoms. Various serological tests, like rK39 rapid diagnostic tests and DATs, can detect antibodies against *Leishmania* in blood samples; these tests provide the diagnostic methods used most often. The indirect fluorescent antibody test, along with ELISA and Western blot methods, are commonly employed throughout Africa, Asia, Europe, and Latin America, but each of these methods calls for equipment that has not been optimised

FIGURE 8.1 Microscopic examination of *Leishmania*

for field use, and standardised protocols, antigens, and other reagents are frequently insufficient. The Montenegro test, or leishmanin skin test, has similar limitations – it might not even serve as a species-specific test for *Leishmania*. It typically turns negative during active kala-azar and turns positive within two months post successful treatment (Figure 8.2) (11).

Polymerase chain reaction: PCR tests can detect *Leishmania* DNA in tissue samples and provide species-specific identification (12). Both direct confirmation (parasitic) and indirect confirmation (immunologic) should be used to make a clinical diagnosis of leishmaniasis, given the frequent difficulty in diagnosing it and the complexities arising from limited resources available across many countries involved. Practitioners should remain vigilant regarding patient history, epidemiology, clinical symptoms, and physical exam results indicating that leishmaniasis may be behind their health problems. For instance, cutaneous leishmaniasis should be considered if an individual exhibits painless, non-purulent skin ulcer on their face or extremities following recent travel into endemic

FIGURE 8.2 Serological tests

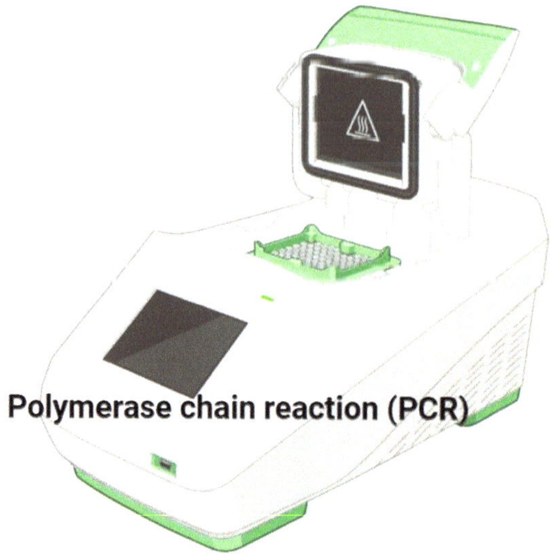

FIGURE 8.3 PCR machine

areas. Contrarily, VL (along with HIV, TB, and cancer) may be indicated in a patient from a region where leishmaniasis is common, such as Africa, by persistent fevers, tiredness, weight loss, anaemia, leucopenia, and hepatosplenomegaly. Given that leishmaniasis therapy can be dangerous and vary widely depending on the identification of the species, a clinical diagnosis of leishmaniasis should ideally be backed by evidence of the amastigotes in the tissue and molecular tests like that of PCR (Figure 8.3) (13).

8.3 POST-KALA-AZAR DERMAL LEISHMANIASIS

Although "kala-azar" and "visceral leishmaniasis" are sometimes used interchangeably, kala-azar – meaning "black fever" in Hindi – typically refers specifically to severe or advanced cases. In its most severe forms, VL usually leads to death if left untreated, either directly from the illness or indirectly from its complications, such as secondary (myco-) bacterial infection or haemorrhage. Following (or during) treatment for VL, some patients may develop PKDL, a syndrome marked by skin lesions, such as papules, erythematous or hypopigmented macules, patches, and nodules, which are usually first noticed on the face. Chronic PKDL patients may act as significant reservoir hosts for infection (14–17).

A patient who has recovered from VL and is otherwise healthy may develop PKDL, which is defined by a macular, maculopapular, and nodular rash. Depending on its severity, the rash typically begins around the mouth and then spreads to other parts of the body. PKDL is mainly observed in Sudan and India, where it occurs in 50% and 5–10% of cases after the treatment of VL, respectively. Consequently, it is mostly limited to regions that harbour the causative parasite, *Leishmania donovani*. In India, PKDL typically develops 2–3 years for PKDL following VL treatment, but in Sudan, it can occur 0–6 months later. As a parasite reservoir, PKDL likely has a considerable impact on the interepidemic stages of VL. A growing body of research shows that the aetiology of PKDL is primarily immunologically mediated; elevated levels of interleukin-10 in the peripheral blood of VL patients can predict the emergence of PKDL. During VL, peripheral blood mononuclear cells (PBMCs) do not produce interferon. After VL therapy, PBMCs begin to produce interferon, coinciding with the emergence of PKDL lesions due to skin inflammation caused by interferon-producing cells responding to persistent parasites in the skin.

While parasites can occasionally be detected microscopically in smears, most diagnoses are made clinically. In more than 80% of instances, parasites can be found using PCR and monoclonal antibodies. Leishmanin skin testing and serological assays have limited utility. In Sudan, the majority of cases may self-cure, but severe and persistent cases require treatment. Treatment is always necessary for Indian PKDL. Sodium stibogluconate is administered for two months at a dose of 20 mg/kg in both India and Sudan. Liposomal amphotericin B appears to be effective, while newer substances like miltefosine, which can be used topically or orally, show great promise. While research has produced a wealth of new knowledge on the pathophysiology and treatment of PKDL, a number of problems, particularly those pertaining to control, remain unresolved and require immediate attention (18).

Following VL treatment, PKDL typically manifests as a macular, papular, or nodular rash on the face, upper arms, and trunk. It occurs predominantly in East Africa, especially Sudan, and in the Indian subcontinent, where it is documented in 5–10% of kala-azar patients. Although rare, it has also been documented in Brazil among patients with HIV coinfection caused by *L. infantum*. While it can occur earlier, PKDL typically manifests anywhere from 6 months to over a year after kala-azar has supposedly been treated. Individuals with PKDL may serve as potential sources of *Leishmania* infection (19).

8.3.1 Symptoms

- **Macular, papular, or nodular skin lesions:** These can appear on the patient's face, arms, or other parts of the body.
- **Skin discolouration:** Hyperpigmented (dark) or hypopigmented (light) patches can appear on the skin. Lesions may be flat or raised, and can be single or multiple.

8.3.2 Diagnosis

- **Slit-skin smear:** Scraping the edge of a skin lesion and examining it under a microscope can reveal the presence of *Leishmania* amastigotes (Figure 8.4) (20, 21).
- **PCR:** Polymerase chain reaction tests can detect *Leishmania* DNA in blood and skin samples and aid in diagnosis (22). It is essential that both diagnosis and treatment for VL and PKDL are conducted by healthcare professionals experienced in managing these diseases.

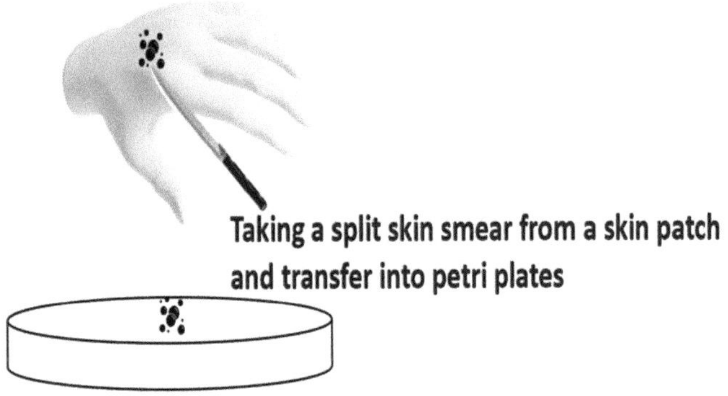

FIGURE 8.4 Slit-skin smear test

8.4 CONCLUSION

In conclusion, understanding the symptoms and diagnosis of VL and PKDL is crucial for the effective management and control of these diseases. VL, also known as kala-azar, is a severe systemic infection caused by protozoan parasites of the *Leishmania* genus. The disease affects the liver, spleen, and bone marrow, leading to a wide range of symptoms, such as prolonged fever, weight loss, fatigue, enlarged liver and spleen, anaemia, and immune system suppression. Early diagnosis of VL is challenging due to non-specific symptoms and limited access to healthcare facilities in endemic areas; however, prompt and accurate diagnosis is essential for timely treatment and the prevention of complications.

Various diagnostic methods are available for VL, including molecular techniques like PCR, serological tests such as rK39 immunochromatographic tests, and parasite detection through microscopy or culture. Serological tests are commonly applied due to their simplicity and relatively high sensitivity and specificity. Molecular techniques are more sensitive and specific but require specialised laboratory facilities. Microscopy and culture methods are time-consuming and may lack sensitivity in certain cases.

PKDL is a complication that may occur after an apparent cure from VL. Characterised by the appearance of skin lesions, which can be macular, papular, nodular, or mixed and may be accompanied by hypopigmentation or hyperpigmentation, PKDL serves as an important reservoir for the parasite, contributing to transmission cycles. Timely diagnosis and treatment of PKDL are vital for interrupting the transmission of the disease.

The diagnosis of PKDL primarily relies on clinical evaluation based on characteristic skin lesions and a previous history of VL; laboratory confirmation through slit-skin smears or skin biopsies can provide additional evidence, while PCR-based techniques can facilitate molecular diagnosis.

In both VL and PKDL cases, early diagnosis and treatment are crucial for preventing complications while reducing transmission rates and improving patient outcomes. Effective control strategies should focus on strengthening healthcare systems in endemic areas, ensuring access to diagnostic facilities, and training healthcare professionals on accurate diagnosis and treatment protocols. Additionally, ongoing research efforts should continue exploring new diagnostic methods that are sensitive, specific, affordable, and suitable for resource-limited settings.

Overall, a comprehensive understanding of the symptoms and diagnostic approaches for VL and PKDL is essential for early detection, proper management, and effective control of these neglected tropical diseases.

ACKNOWLEDGEMENT

The authors are thankful to the University of Lucknow for their necessary support during this study.

REFERENCES

1. Zijlstra, E.E., Musa, A.M., Khalil, E.A.G., El Hassan, I.M. and El-Hassan, A.M., 2003. Post-kala-azar dermal leishmaniasis. *The Lancet Infectious Diseases*, *3*(2), pp. 87–98.
2. Díaz, J.M., De las Cuevas León, D., González, A.M. and Escuin, R.R., 2018. Fever of unknown origin: Don't forget kala-azar even in Europe. *European Journal of Case Reports in Internal Medicine*, *5*(2), 1–3.
3. Lukeš, J., Mauricio, I.L., Schönian, G., Dujardin, J.C., Soteriadou, K., Dedet, J.P., Kuhls, K., Tintaya, K.W.Q., Jirků, M., Chocholová, E. and Haralambous, C., 2007. Evolutionary and geographical history of the *Leishmania donovani* complex with a revision of current taxonomy. *Proceedings of the National Academy of Sciences*, *104*(22), pp. 9375–9380.
4. Addy, M. and Nandy, A., 1992. Ten years of kala-azar in West Bengal, Part I. Did post-kala-azar dermal leishmaniasis initiate the outbreak in 24-Parganas?. *Bulletin of the World Health Organization*, *70*(3), p. 341.

5. Sakkas, H., Gartzonika, C. and Levidiotou, S., 2016. Laboratory diagnosis of human visceral leishmaniasis. *Journal of Vector Borne Diseases*, *53*(1), p. 8.
6. Srivastava, P., Dayama, A., Mehrotra, S. and Sundar, S., 2011. Diagnosis of visceral leishmaniasis. *Transactions of the Royal Society of Tropical Medicine and Hygiene*, *105*(1), pp. 1–6.
7. Badaro, R., Jones, T.C., Lorenco, R., Cerf, B.J., Sampaio, D., Carvalho, E.M., Rocha, H., Teixeira, R. and Johnson Jr, W.D., 1986. A prospective study of visceral leishmaniasis in an endemic area of Brazil. *Journal of Infectious Diseases*, *154*(4), pp. 639–649.
8. Mohebali, M., Edrissian, G.H., Shirzadi, M.R., Akhoundi, B., Hajjaran, H., Zarei, Z., Molaei, S., Sharifi, I., Mamishi, S., Mahmoudvand, H. and Torabi, V., 2011. An observational study on the current distribution of visceral leishmaniasis in different geographical zones of Iran and implication to health policy. *Travel Medicine and Infectious Disease*, *9*(2), pp. 67–74.
9. Sacks, D.L., 2001. Leishmania–sand fly interactions controlling species-specific vector competence: Microreview. *Cellular Microbiology*, *3*(4), pp. 189–196.
10. Lodge, R. and Descoteaux, A., 2008. Leishmania invasion and phagosome biogenesis. *Molecular Mechanisms of Parasite Invasion: Subcellular Biochemistry*, 47, pp. 174–181.
11. Ready, P.D., 2014. Epidemiology of visceral leishmaniasis. *Clinical Epidemiology*, 6, pp. 147–154.
12. de Bruijn, M.H. and Barker, D.C., 1992. Diagnosis of new world leishmaniasis: specific detection of species of the Leishmania braziliensis complex by amplification of kinetoplast DNA. *Acta Tropica*, *52*(1), pp. 45–58.
13. Mann, S., Frasca, K., Scherrer, S., Henao-Martínez, A.F., Newman, S., Ramanan, P. and Suarez, J.A., 2021. A review of leishmaniasis: current knowledge and future directions. *Current Tropical Medicine Reports*, *8*, pp. 121–132.
14. Zijlstra, E.E. and El-Hassan, A.M., 2001. Leishmaniasis in Sudan. 4. Post kala-azar dermal leishmaniasis. *Transactions of the Royal Society of Tropical Medicine and Hygiene*, *95*(Supplement_1), pp. S59–S76.
15. Kumar, P., Chatterjee, M. and Das, N.K., 2021. Post kala-azar dermal leishmaniasis: Clinical features and differential diagnosis. *Indian Journal of Dermatology*, *66*(1), p. 24.
16. Singh, S., Sharma, U. and Mishra, J., 2011. Post-kala-azar dermal leishmaniasis: recent developments. *International Journal of Dermatology*, *50*(9), pp. 1099–1108.
17. Pandey, K., Das, V.N.R., Singh, D., Das, S., Lal, C.S., Verma, N., Bimal, S., Topno, R.K., Siddiqui, N.A., Verma, R.B. and Sinha, P.K., 2012. Post-kala-azar dermal leishmaniasis in a patient treated with injectable paromomycin for visceral leishmaniasis in India. *Journal of Clinical Microbiology*, *50*(4), pp. 1478–1479.
18. Zijlstra, E.E., Musa, A.M., Khalil, E.A.G., El Hassan, I.M. and El-Hassan, A.M., 2003. Post-kala-azar dermal leishmaniasis. *The Lancet Infectious Diseases*, *3*(2), pp. 87–98.
19. Leta, S., Dao, T.H.T., Mesele, F. and Alemayehu, G., 2014. Visceral leishmaniasis in Ethiopia: an evolving disease. *PLoS Neglected Tropical Diseases*, *8*(9), p. e3131.
20. Wijesinghe, H., Gunathilaka, N., Semege, S., Pathirana, N., Manamperi, N., de Silva, C. and Fernando, D., 2020. Histopathology of cutaneous leishmaniasis caused by *Leishmania donovani* in Sri Lanka. *BioMed Research International*, *2020*, pp. 1–8..
21. Tareen, A., Afaq, S. and ul Haque, A., 2014. Comparative study of the diagnosis of cutaneous leishmaniasis by slit skin smear and skin biopsy for histopathology. *Journal of Rawalpindi Medical College (JRMC)*, *18*(1), pp. 83–86.
22. Roura, X., Fondevila, D., Sánchez, A. and Ferrer, L., 1999. Detection of Leishmania infection in paraffin-embedded skin biopsies of dogs using polymerase chain reaction. *Journal of Veterinary Diagnostic Investigation*, *11*(4), pp. 385–387.

9 Molecular Mechanism of PKDL Development

*Shobha Kumari[1], Pradeep Das[2], and Ashish Kumar[1]**
[1]Department of Biochemistry, ICMR Rajendra Memorial Research Institute of Medical Sciences, Patna, Bihar, India
[2]ICMR-National Institute of Cholera and Enteric Diseases, Kolkata, West Bengal, India
*Corresponding Author: Dr. Ashish Kumar, Department of Biochemistry, ICMR-Rajendra Memorial Research Institute of Medical Sciences, Patna, Bihar, India

9.1 INTRODUCTION

Post-kala-azar dermal leishmaniasis (PKDL) is a dermatological disorder that might develop after individuals have been infected with visceral leishmaniasis (VL). As we know, the causative agent behind VL is the protozoan parasite *Leishmania donovani*, which primarily infiltrates internal organs like the spleen, liver, and bone marrow. Occasionally, the parasites migrate from these visceral organs to peripheral skin cells, resulting in the manifestation of PKDL. This chronic skin disorder is predominantly found in the Indian subcontinent and East Africa, including regions like Kenya and Sudan. In both areas, VL is attributed to *L. donovani* and is infrequently associated with *L. chagasi* and *L. infantum* (El Hassan et al. 1992; Thakur and Kumar 1992). In South Asia, PKDL is regarded as the exclusive reservoir for the parasite, given that VL transmission is anthroponotic (Thakur and Kumar 1992; Zijlstra et al. 2003; Rahman et al. 2010). In the African context, PKDL may arise during therapy or shortly after treatment and could spontaneously regress. Conversely, PKDL observed in the Indian subcontinent tends to be more persistent, emerging around six months to a year after the kala-azar cure or even decades later. There may be an absence of any recorded kala-azar history, indicating the possibility of a subclinical infection (Desjeux et al. 2013). The clinical presentation of PKDL often encompasses hypopigmented macules, erythematous plaques (raised, solid, flat-topped red lesions), and papulonodular lesions. These lesions frequently cluster around the central facial area before expanding to affect the entire body in a generalized distribution (Kumar et al. 2021).

9.2 HOW PKDL DEVELOPS?

The development of PKDL is intricately linked to prior experiences with VL, which is considered a severe form of leishmaniasis. In VL, the parasite invades and multiplies within vital internal organs like the spleen, liver, and bone marrow, leading to life-threatening complications. Individuals who have previously battled VL are particularly at risk of developing PKDL, a skin manifestation that can emerge after an apparent recovery from the disease's visceral form.

However, PKDL's path to development isn't solely paved by a history of VL. Intriguingly, it can also arise in individuals who have never been diagnosed with VL or displayed its characteristic symptoms. This raises the possibility of subclinical or asymptomatic *Leishmania* infections, where the parasite resides quietly in the body without causing noticeable illness until PKDL surfaces.

 DOI: 10.1201/9781032643120-9

Therefore, we can assert that the development of PKDL is not universal among individuals who have had VL or have been exposed to *Leishmania* parasites. The factors determining whether an individual will develop PKDL are complex and not fully understood. Genetic factors, immune responses, and interactions between the parasite and host all play a role in this process. Thus, while certain populations are more prone to PKDL due to their exposure to *Leishmania*, the specific individuals who will develop it cannot be definitively predicted.

9.3 WHY NODULAR, MACULAR, AND PAPULAR FORMS AND SKIN LESIONS DEVELOP IN PKDL CASES?

In PKDL, the clinical presentation can vary, and different forms of skin lesions can be observed, including nodular, macular, and papular forms. Patients can be classified into two distinct subgroups based on the features of their lesions: polymorphic and monomorphic PKDL. Polymorphic PKDL is characterized by the emergence of papules and/or nodules alongside hypopigmented macules. In contrast, monomorphic PKDL patients typically exhibit only one type of lesion – predominantly papulonodules (Zijlstra and Alvar 2012; Kaushal et al. 2016). While the precise molecular mechanisms underlying these specific lesion types remain unclear, several potential explanations exist.

Polymorphic PKDL lesions exhibit a densely infiltrated dermis with elevated macrophages and B cells but fewer dendritic cells (DCs) and Langerhans cells (diagrammatically represented in Figure 9.1) (Mukherjee et al. 2015, 2019; Sengupta et al. 2019). This fosters a parasite-friendly environment marked by increased interleukin-10 (IL-10) and interleukin-12p40 expression (Mukherjee et al. 2015) alongside diminished activity of toll-like receptors (TLRs) 2 and 4 as well as reduced reactive oxygen species (ROS) and nitric oxide (NO) production (Mukhopadhyay et al. 2015). Conversely, monomorphic PKDL lesions have sparse infiltrates, lower parasitic loads, and reduced Leishman-Donovan bodies, as well as $CD4^+$ T helper cells, DCs, and macrophage presence (Sengupta et al. 2019). Polymorphic PKDL also shows higher parasite burdens and elevated interferon-gamma (IFN-γ) and tumour necrosis factor-alpha (TNF-α) levels compared to monomorphic cases, with increased $CD3^+$ T cells and natural killer cells, likely due to a higher number of circulating parasites (Kaushal et al. 2016; Sengupta et al. 2019).

In acute PKDL cases, the cell-mediated immune response (CMI) exhibits greater strength, while chronic cases are characterized by diminished responses (Mukhopadhyay et al. 2012). Similarly, in macular cases, there is a robust CMI characterized by fewer parasites and low antibody levels

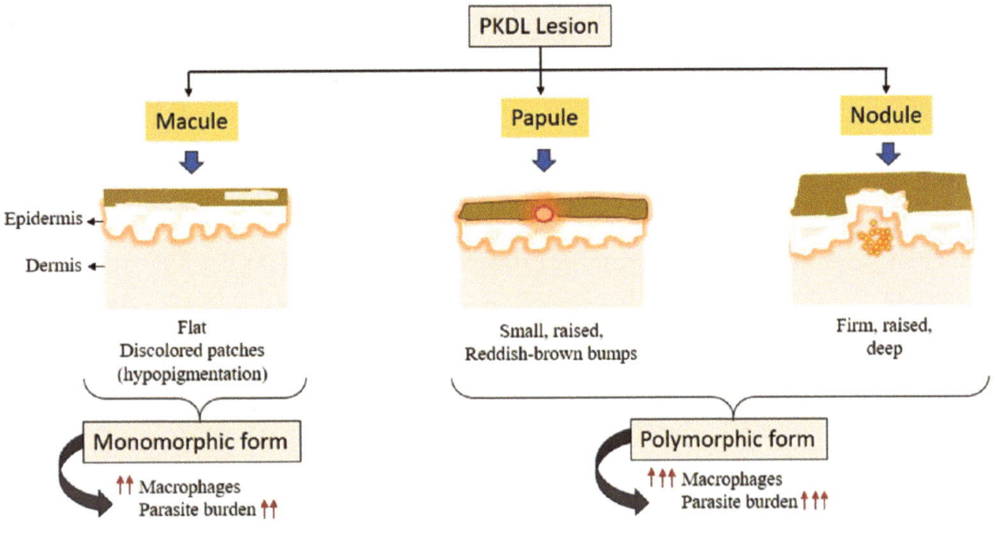

FIGURE 9.1 Different forms of PKDL lesions

(elevated only in Ig1). In contrast, polymorphic (papulonodular) forms of PKDL exhibit diminished CMI, influenced by factors such as TGF-β and IL-10, resulting in higher levels of regulatory T cell markers, increased parasite presence, and elevated antibody levels, encompassing both Ig1 and Ig3 (markers associated with IL-10) (Katara et al. 2011; Haldar et al. 1983; Saha et al. 2007; Mukhopadhyay et al. 2012) (Figure 9.1).

Overall, the development of these various lesion forms in PKDL appears to be influenced by a combination of factors, including the parasite burden, host immune response, and local tissue factors (Volpedo et al. 2021). Therefore, we can describe the diversity in PKDL forms according to the following factors.

9.3.1 Nodular Form

Nodular lesions in PKDL are characterized by raised, firm, and often ulcerated nodules on the skin. The molecular mechanisms contributing to nodular lesions are not well-defined. However, it is believed that higher parasite burdens in the skin, local immune dysregulation, and selection and activation of immune cells, such as macrophages and neutrophils, may contribute to nodule formation.

9.3.2 Macular Form

Macular lesions in PKDL are flat, discoloured patches on the skin. The molecular mechanisms underlying macular lesions are also not fully elucidated; however, limited lymphohistiocytic infiltrates accompanied by small numbers of plasma cells, primarily situated around superficial perivascular areas, have been observed in macular lesions (Kar et al. 2019; Ramesh and Ramam 2020). It is hypothesized that dysregulation of cytokines and chemokines, such as increased production of IL-10 and transforming growth factor beta (TGF-β), along with alterations in matrix metalloproteinase (MMP) activity, could contribute to the development of macular lesions. These factors may influence the recruitment and activation of immune cells and the remodelling of the extracellular matrix in the skin.

9.3.3 Papular Form

Papular lesions in PKDL appear as raised, reddish-brown bumps on the skin. The molecular mechanisms underlying papular lesions are not well characterized.

However, it is believed that in papular lesions, immune cells enter around blood vessels and near hair follicles, reaching the middle layer of the skin. The extent of these infiltrates varies depending on the type of raised skin lesion. In some cases, the infiltrates spread throughout the entire skin layer, forming nodules (Kar et al. 2019).

These explanations are based on clinical and histopathological studies. However, the exact molecular mechanisms are still not fully understood. Therefore, further research is required to elucidate the processes underlying the development of nodular, macular, and papular forms in PKDL.

9.4 PKDL DEVELOPMENT IN INDIVIDUALS WITH A HISTORY OF VL

PKDL remains enigmatic, as the exact mechanism by which the parasite migrates into skin cells is not yet fully understood. However, several hypotheses attempt to explain this phenomenon.

9.4.1 Localized Immune Response

Localized immune responses are thought to trigger the parasite's migration to skin cells. During VL, the immune system mounts a response against the parasite, and it is thought that this response may persist in some individuals even after the parasite has been cleared from the visceral organs (Zijlstra 2014, 2016). This persistent immune response could attract the parasites to the skin, where they continue to proliferate.

9.4.2 Parasite Dissemination

Another hypothesis suggests that the parasite may circulate through the bloodstream or lymphatic system, reaching skin cells from the visceral organs. *Leishmania* parasites can invade and survive within various cell types, including immune cells as well as endothelial cells that line blood and lymphatic vessels. Through these routes, they could potentially reach the skin cells and establish an infection (Lambert and Barragan 2010).

9.4.3 Altered Host Immune Response

Certain host factors or immune responses may play a role in the parasite's migration to the skin. For example, genetic variations or immune system dysregulation in some individuals may promote the development of PKDL (Volpedo et al. 2021). Further research is required to better understand the host-parasite interactions and immune mechanisms involved in PKDL.

It's crucial to emphasise that the migration of the parasite to the skin in PKDL is not a universal occurrence. Only a subset of individuals with a history of VL develop PKDL, and the factors that determine this transition are not yet fully elucidated. Further research is needed to unravel the underlying mechanisms and identify potential targets for the prevention and treatment of PKDL.

9.5 MOLECULAR MECHANISMS OF PKDL DEVELOPMENT

The exact disease mechanism underlying PKDL remains unclear. Some hypotheses suggest immunosuppression may enable the reactivation of dormant parasites from visceral tissues or lead to skin reinfection (Zijlstra et al. 2003). While our understanding of the molecular mechanisms leading to PKDL is not comprehensive, several factors have been identified as potential contributors to its pathogenesis.

9.5.1 Immune-Based Signalling Pathways in PKDL Development

Recent immunological findings in Indian PKDL patients reveal distinct immune patterns in the skin and peripheral regions. In the skin, IL-10 and FoxP3 primarily orchestrate immune regulation (Ganguly et al. 2010; Katara et al. 2011). Despite elevated IFN-γ and TNF-α levels, their corresponding receptors are downregulated (Ansari et al. 2006, 2008). The skin also exhibits increased Th17 cells and IL-17 (Katara et al. 2012). In contrast, peripheral immunity is predominantly governed by $CD8^+$ T cells, which are the primary sources of IL-10 and display anergic characteristics (Ganguly et al. 2010). This anergy is accompanied by noticeably reduced co-stimulation, as evidenced by decreased CD28 in circulating $CD8^+$ T cells and diminished CD86 in $CD14^+$ monocytes (Ganguly et al. 2010; Mukhopadhyay et al. 2011).

PKDL development involves intricate interactions among the parasite, host immune system, and various signalling pathways. While not fully understood, several key pathways have been implicated:

- **TGF-β pathway:** The cytokine TGF-β plays a crucial role in immune regulation and tissue remodelling. *Leishmania* parasites can induce TGF-β production, potentially modulating the host immune response and leading to immunosuppression and PKDL progression. Patients with treatment failure exhibit notably high TGF-β levels (Elmekki et al. 2016), consistent with previous findings in VL (Gantt et al. 2003). TGF-β, along with IL-10, likely contributes to PKDL reactivation following kala-azar. Conversely, ambisome treatment for VL reduced TGF-β and eliminated IL-10, potentially decreasing PKDL incidence (Saha et al. 2007). In Indian PKDL patients, disease severity is inversely correlated with lymphoproliferation and directly related to TGF-β, IL-10, and antibody production, underscoring the importance of TGF-β regulation in PKDL development.

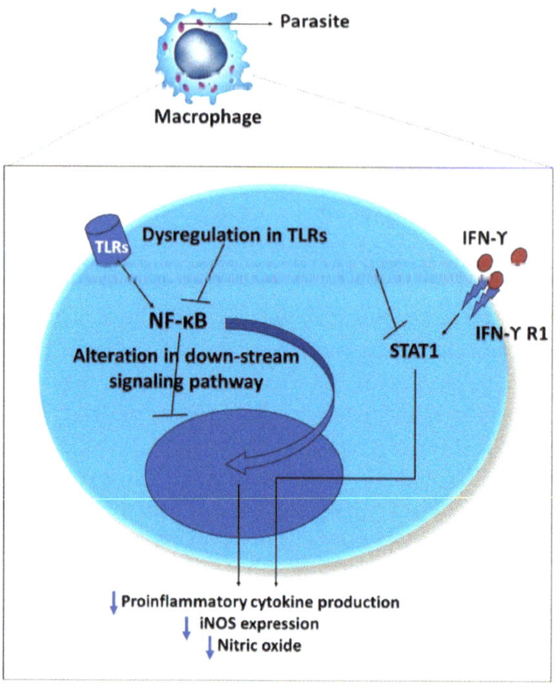

FIGURE 9.2 Diagrammatic presentation of dysfunction in the immunological pathway and alteration in downstream molecules

- **IFN-γ pathway:** IFN-γ's role in PKDL formation is complex. IFN-γ is a vital cytokine secreted by NK cells and T helper 1 cells, well-known for its function in stimulating macrophages to efficiently fight against parasites (Murray et al. 1983). In contrast to lesions in kala-azar, PKDL lesions exhibit significantly higher levels of IFN-γ mRNA (Ismail et al. 1999; Gasim et al. 2000). This suggests that IFN-γ plays a proactive role in the immune response against the parasite in PKDL (Figure 9.2).

  In the context of PKDL, the effectiveness of IFN-γ faces several challenges: Firstly, the IFN-γ R1 gene, responsible for producing the receptor crucial for IFN-γ, might not be expressed adequately. This limitation in IFN-γ receptors hampers the cellular response to IFN-γ, reducing its ability to manage the infection. This can be better understood by elaborating on the specific pathway underlying the mechanism of IFN-γ. IFN-γ plays a crucial role in activating intracellular signalling by binding to its receptor, IFN-γR. This interaction triggers STAT1 phosphorylation, prompting its translocation into the cell nucleus. Within the nucleus, STAT1 initiates the transcription of genes IFN-γ-associated, such as iNOS. Consequently, this process boosts NO production, a key mechanism for combatting intracellular pathogens. However, it's important to note that *Leishmania* parasites can disrupt the STAT1 pathway. These parasites release elongation factor 1 alpha (EF1α), which activates SHP-1, a cytosolic phosphatase (Nandan et al. 2002; Saha et al. 2010). SHP-1 dephosphorylates STAT1, hindering its movement into the nucleus and consequently impeding the expression of IFN-γ-induced genes. Additionally, PKDL lesions display elevated levels of various cytokines, including IL-10, IL-6, and TGF-β. These cytokines have counteracting effects, potentially reducing the extent of the IFN-γ-mediated immune response. IFN-γ may not be able to fully exert its antiparasitic effect when these counteracting cytokines are present. However, studies suggest that with appropriate treatment, there is a restoration of IFN-γ R1 at both the transcriptional and translational levels, accompanied by a decrease in the levels of opposing cytokines (Ansari et al. 2006). This paves the way for a more favourable environment where IFN-γ can effectively function and eliminate the parasite.

Therefore, dysregulation of the IFN-γ pathway might contribute to impaired parasite clearance and persistent PKDL development. This may occur through alterations in downstream signalling molecules or regulatory factors.

- **Nuclear factor-kappa B (NF-κB) pathway and toll-like receptor (TLR) signalling:** NF-κB is a transcription factor that controls diverse immune and inflammatory reactions. *Leishmania* parasites activate this pathway, facilitating their survival and replication within host cells. An irregularity in NF-κB signalling has been linked to skin lesion development, potentially contributing to the persistent inflammation and tissue damage (Bell et al. 2003). Similarly, TLRs are essential receptors in the immune system, recognizing specific pathogen-associated molecular patterns. This recognition is a pivotal step in the body's ability to identify and respond to infections, with TLR activation initiating innate immune responses against pathogens. Evidence suggests TLR signalling may be involved in PKDL initiation (Hosseini et al. 2015). Dysregulation of TLRs or alterations in downstream signalling pathways could contribute to parasite persistence and the sustained inflammatory response seen in PKDL. *Leishmania* parasites can inhibit TLR signalling, which is activated by the detection of pathogen-associated molecular patterns. When TLRs engage with these patterns, it triggers the translocation of NF-κB to the nucleus, initiating the transcription of proinflammatory cytokines like IL-1β, TNF, and IL-12 as a response to the presence of pathogens. However, Leishmania parasites release proteases, such as GP63, which cleave the NF-κB p65 subunit in the cytoplasm, thereby interfering with its normal function (Terrazas et al. 2016).
- **IL-10 pathway:** IL-10, an immunosuppressive cytokine, plays a crucial role in governing the immune response to *Leishmania* infection. Ganguly et al. (2008) found significantly increased IL-10 levels in PKDL patients, suggesting a direct link between IL-10 and PKDL pathogenesis. This IL-10 overexpression suppresses the activation of macrophages, which are instrumental in the host's defence mechanism against *Leishmania*, responsible for the phagocytosis and destruction of the parasites. Excessively high IL-10 levels reduce macrophages' effectiveness in eliminating the parasites. Intriguingly, PKDL patients exhibit not only elevated IL-10 levels but also increased levels of IgG3 and IgG1, two immunoglobulin classes identified as surrogate markers for IL-10 activity (Ganguly et al. 2008). This phenomenon suggests that altered IL-10 signalling could significantly contribute to the establishment and persistence of chronic *Leishmania* infections in PKDL patients.

While these signalling pathways may offer insights into how PKDL emerges, our understanding of the exact mechanisms and interactions involved in PKDL remains incomplete, necessitating additional research to unravel these signalling pathways and their roles in PKDL pathogenesis.

9.6 OTHER FACTORS THAT SUPPORT PKDL DEVELOPMENT

9.6.1 UVB Light Exposure

Research supports the hypothesis that abnormal skin immune responses triggered by exposure to ultraviolet (UV) light are the primary drivers. Studies have demonstrated that simulated solar UV radiation (comprising UVA and UVB; wavelength range 295–400 nm) provided after immunization can suppress the immune memory and the body's ability to respond to common pathogens like *Candida albicans*. Within this context, UVB light is proposed to contribute to PKDL's development. This occurs by causing changes in Langerhans cells, which result in alterations in skin immune responses (Ismail et al. 1999; Zijlstra et al. 2003; Zijlstra and El-Hassan 2001; Musa et al. 2002;

Nghiem et al. 2002). Therefore, we can say that eliminating *Leishmania* parasites from the body relies on activating parasitized macrophages through a Th1 immune response, which UVB light can suppress. The variations observed in the skin tissue of PKDL patients are consistent with the effects of UVB light, suggesting that UVB exposure is a significant contributing factor to PKDL development.

9.6.1.1 Pathogenesis Induced by UV Light

DCs are a diverse and multifaceted group of immune cells derived from bone marrow that are found throughout the body. In the skin, they are considered the most effective antigen-presenting cells. Their specialty lies in their ability to capture, process, and subsequently display antigens to T cells, which are crucial players in our immune system (Karnati and Baumgart-Vogt 2009). When it comes to patrolling for pathogens, DCs act as our immune system's detectives.

There are different types of DCs in the skin. One group is called epidermal Langerhans cells (E-LCs), and the other is made up of two subsets of migratory dermal DCs (dDCs), which can be distinguished by a marker called CD1a. What makes DCs so vital is their unique talent for scouting potential trouble spots, responding to signals from microbes, and then activating naive T cells. All of this plays a critical role in starting our body's defence against infections (Pracharoenwattana and Smith 2008; Zaba et al. 2009). So, in the context of diseases like leishmaniasis, including PKDL, DCs play a crucial role. They're like the conductors of the immune orchestra, especially in the skin. They help orchestrate our body's defence against the parasite. Now, let's discuss how UV light fits into this.

Keratinocytes are skin cells that release substances like vitamin D3 and cis-urocanic acid in the presence of UV light. This seems to affect E-LCs, which are critical for immune responses. This leads to a reduction in the number of these cells and changes in their appearance (Ismail et al. 2006). E-LCs start to exhibit changes in their behaviour and protein expression. The levels of essential immune proteins, such as MHCII, CD80, and CD86, decrease in E-LCs, which are vital for activating immune responses (Ismail et al. 2006; Ganguly et al. 2010). Simultaneously, E-LC production of IL-10 increases. IL-4, produced by another type of immune cell called Th2 cells, steps in and activates dermal dendritic cells (dDCs) to produce even more IL-10 (shown in Figure 9.3). This surge in IL-10 prompts the increased presence of regulatory T cells (Tregs). The increased presence of Tregs creates an environment in the body that is less effective in fighting off *Leishmania* parasites.

This immunosuppressive environment allows the parasites to persist and evade the immune system's defences (Mukhopadhyay et al. 2014). TGF-β, another substance released by infected macrophages, can also activate Tregs. These activated Tregs then produce more TGF-β, which further facilitates the persistence of the *Leishmania* parasites. UV light may also trigger macrophages to adopt an alternative activation state. The change in macrophage behaviour, where vitamin D3 pushes them towards an alternative activation mode, along with the increased presence of Tregs, collectively creates an immune environment that is less effective in eliminating the parasites.

In summary, exposure to UV light seems to affect various components of the immune system, including Langerhans cells, DCs, and regulatory T cells. This results in an immunosuppressive environment that allows *Leishmania* parasites to persist and potentially contribute to the pathogenesis of PKDL.

9.6.2 The Role of Iron in PKDL Development

Iron plays a multifaceted role in the development of PKDL. *L. donovani* thrives within human macrophages, which are integral to the body's immune defence. The regulation of iron within host cells significantly influences the complex interplay between the host's immune response and the pathogen's survival strategies.

Iron is essential for both pathogens and hosts. Microbes, including pathogens like *L. donovani*, depend on iron for their growth and survival (Niu et al. 2016). Iron is indispensable for mammalian

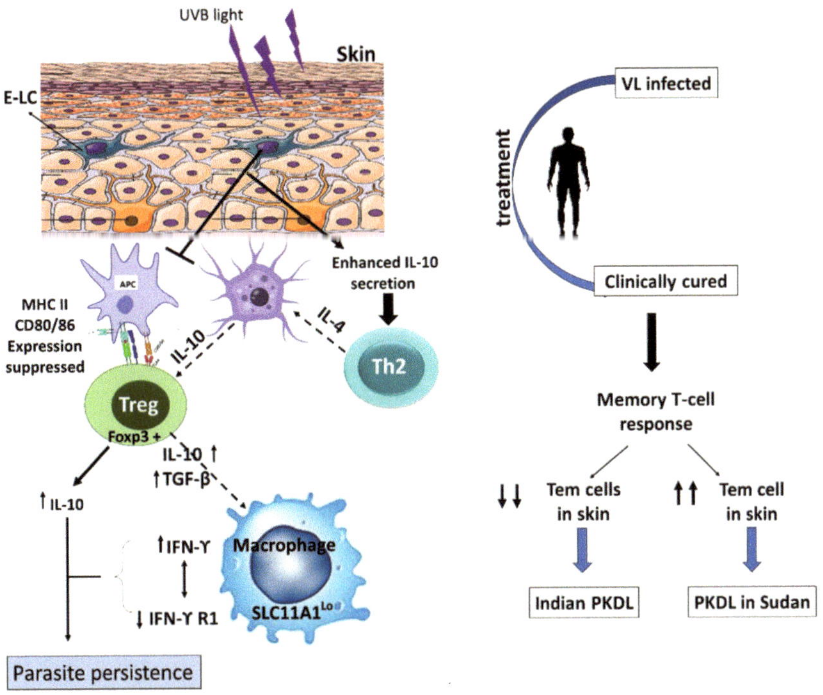

FIGURE 9.3 The overall mechanism supporting PKDL development and pathogenesis. Exposure to UVB radiation weakens the skin's immune response, allowing parasites to persist. It causes morphological alterations in E-LC cells, leading to decreased expression of MHC-II and co-stimulatory molecules; however, IL-10 secretion increases. Consequently, Th2 cells secrete IL-4, which activates dDCs and contributes to more IL-10 production. This provides an immunosuppressive environment that activates the function of Treg cells and supports parasite persistence. Similarly, activated macrophages secrete TGF-β, which can activate Treg cells and stimulate more TGF-β secretion, leading to parasite persistence. T cells secrete both IFN-γ and IL-10. People with IFN-γ receptor polymorphism may experience immunosuppression IL-10 dominance. Individuals with high IFN-γR expression are least prone to developing PKDL, whereas the lack of proper antileishmanial treatment may lead to the development of PKDL. The right side of the diagram shows a model for the failure of organ-specific T cell memory response. In some cases, PKDL is characterized by a weak immune response (in the Indian region), while self-healing patients show an enhanced immune response in the skin (in Sudan)

cells as well, as it is essential for activating the immune system and facilitating the production of reactive oxygen species, which are vital for combating infections. However, the host and pathogen's shared need for iron creates a competitive environment, necessitating the tight control of its availability to maintain the host's core biological functions while preventing the proliferation of intracellular pathogens (Silva-Gomes et al. 2013).

A comprehensive study that investigated iron dynamics within macrophages of PKDL patients revealed intriguing insights into the disease's pathogenesis. In PKDL, monocytes and macrophages, usually involved in protecting the host from infections, exhibit altered iron-handling mechanisms. This shift is characterized by an increased influx of iron into these immune cells, driven by the upregulation of iron transport molecules CD71 and CD163 (Dighal et al. 2020).

This elevated iron influx leads to the accumulation of labile iron pools and elevated ferritin levels within the infected macrophages. Ferritin, a protein responsible for storing excess iron, becomes more prominent in this context. Paradoxically, despite this increased iron availability, the study also identified an amplified expression of iron exporters such as NRAMP-1 and Fpn-1. This suggests that the host's immune response is actively trying to restrict *L. donovani*'s access to iron, creating a ferritin^{high}/ferroportin^{high} phenotype within the infected macrophages (Niu et al. 2016; Dighal et al. 2020).

Remarkably, this phenotype contrasts with the conventional ferritin^{low}/ferroportin^{high} profile seen in alternatively activated M2-type macrophages. This highlights the host's unique manipulation of iron homeostasis in response to the pathogenic challenge *L. donovani* poses.

The intricate orchestration of iron availability and regulation significantly contributes to the host-pathogen dynamic in PKDL. This delicate control of iron creates a tug-of-war between the host's immune response and the pathogen's survival mechanisms, ultimately influencing the trajectory of the infectious disease, either favouring the host's defence mechanisms or benefiting the invading pathogen.

In summary, the interplay between iron regulation, the host's immune response, and *L. donovani*'s survival strategy is a complex dynamic that significantly impacts PKDL development (Huynh and Andrews 2008; Dighal et al. 2020; Ferreira et al. 2021). The study's findings underscore the pivotal role of iron homeostasis in shaping the course of infectious diseases by influencing the delicate balance between the host and the pathogen.

9.6.3 PKDL in Relation to Treatment

Infection with the *Leishmania* pathogen has been associated with peroxisomal damage in the host (Gupta et al. 2009). Remarkably, reports have indicated that the host's peroxisomal characteristics may not fully recover following sodium antimony gluconate treatment for VL, potentially contributing to the onset of PKDL (Gupta et al. 2009; Verma et al. 2007).

Peroxisomes serve as essential organelles responsible for synthesizing ether lipids and polyunsaturated fatty acids and facilitating beta-oxidation for both bioactive and potentially harmful lipid derivatives. Additionally, they are pivotal in regulating the metabolism of reactive oxygen species (Karnati and Baumgart-Vogt 2009; Pracharoenwattana and Smith 2008; Kovacs et al. 2002). Given their involvement in various metabolic pathways crucial for host cell function, it is imperative to fully repair *Leishmania*-induced peroxisomal dysfunction during VL treatment. Moreover, disruptions in peroximal function can lead to a multitude of human diseases, including conditions like defects in bile acid biosynthesis (Ferdinandusse et al. 2009), neonatal adrenoleukodystrophy, rhizomelic chondrodysplasia punctata, Zellweger syndrome, and infantile Refsum's disease (Leclercq et al. 2008; Wanders and Waterham 2005).

In conclusion, *Leishmania* infection has been linked to peroxisomal damage in the host, and the failure to restore normal peroxisomal function following VL treatment may contribute to PKDL development.

9.7 EXPECTED OR SUGGESTED MOLECULAR PATHWAY THAT HELPS IN THE DEVELOPMENT OF PKDL

The emergence of PKDL is a multifaceted process that hinges on intricate interactions between the infectious parasite, *L. donovani*, and the host's immune system. While the specific molecular pathways underlying PKDL development are not fully elucidated, several molecular mechanisms have been proposed to contribute to its pathogenesis. Here are some expected or suggested molecular pathways that may play a role in PKDL:

- ***Leishmania*-induced dysregulation of cytokines and chemokines**: *Leishmania* parasites can influence the generation and communication of various cytokines and chemokines responsible for regulating immune responses and inflammation. When molecules such as IL-10, IFN-γ, TNF-α, and TGF-β become dysregulated, they may advance the development of PKDL. These perturbations within the cytokine and chemokine network can impair the body's immune reactions, ultimately allowing the parasite to persist within the skin.

- **Activation of matrix metalloproteinases (MMPs)**: MMPs are enzymes involved in degrading components of the extracellular matrix. It has been suggested that *Leishmania* infection can induce the production of MMPs, particularly MMP-9, which can contribute to tissue remodelling and the characteristic skin lesions observed in PKDL. Increased MMP activity may facilitate the migration of immune cells and the infiltration of inflammatory mediators into the skin (Maretti-Mira et al. 2011).
- **Immunological tolerance and Tregs**: Immunological tolerance refers to the suppression of immune responses against self-antigens or persistent pathogens. In PKDL, the prolonged presence of *Leishmania* parasites in the skin may lead to the induction of immunological tolerance. Tregs, a subset of immune cells known for their suppressive function, may play a role in this process. The expansion or activation of Tregs can inhibit effector immune responses and contribute to PKDL chronicity.
- **Failure of organ-specific memory T cell response**: The hypothesis regarding the development of PKDL due to the failure of organ-specific memory T cell response suggests that PKDL may occur when the immune system fails to generate or maintain a robust memory T cell response specifically within the skin, where the disease manifests (Mukhopadhyay et al. 2014). This hypothesis is based on observations and research findings indicating the importance of tissue-resident memory T cells, particularly effector memory (EM) T cells, in controlling *Leishmania* parasites. Typically, when individuals are infected with *L. donovani*, they develop VL. Successful treatment of VL often results in lifelong immunity due to the presence of systemic central memory (CM) and EM T cells, which are capable of recognizing and responding to *Leishmania* antigens. However, in cases where PKDL develops, it is proposed that the immune response within the skin, which is the site of PKDL lesions, may not be adequately primed or maintained. This could lead to insufficient EM T cells in the skin to control parasite persistence or reactivation in this tissue.

  The hypothesis suggests a dichotomy between self-limiting and non-self-healing PKDL cases. In self-limiting cases, as seen in some regions like Sudan, an increased EM T cell response within the skin may enable the spontaneous resolution of PKDL lesions (Ganguly et al. 2008; Mukhopadhyay et al. 2014). Conversely, in non-self-healing cases, as seen in India, a weaker or insufficient EM T cell response in the skin may allow the disease to persist.
- **Genetic and host susceptibility factors**: Genetic variations in host immune response genes can influence susceptibility to PKDL. Polymorphisms in genes responsible for cytokine production, immune regulation, and antigen presentation correlate with an elevated risk of developing PKDL. These genetic factors directly impact the host's immune response and its ability to effectively combat *Leishmania* infection, ultimately playing a role in the development and severity of PKDL.

  A model (illustrated in Figure 9.3) highlights the potential interplay among genes contributing to host susceptibility in PKDL (Mukhopadhyay et al. 2014). For instance, the SLC11A1 gene, part of the solute carrier family 11A1, plays a role in enhancing macrophage activation, particularly its antimicrobial functions, such as iron sequestration from parasites (Blackwell et al. 2001). Polymorphisms in SLC11A1 (referred to as SLC11A1lo) can increase the host's susceptibility to VL. In addition to this susceptibility, individuals with these polymorphisms are also more prone to developing PKDL, largely due to a polymorphism in the IFN-γ receptor gene (IFN-γRlo), which diminishes the functionality of the IFN-γ receptor gene (Salih et al. 2007; Mohamed et al. 2003).

  In summary, the hypothesis underscores that genetic diversity in host immune response genes is a significant factor in determining susceptibility to PKDL. Polymorphisms in these genes influence immune responses and can dictate an individual's ability to combat *Leishmania* infection, thus contributing to the likelihood and severity of PKDL development. Understanding these genetic factors is essential for unravelling the mechanisms behind PKDL and developing tailored strategies for its prevention and treatment.

All the above-described molecular pathways are based on current understanding and ongoing research. Further investigation is needed to fully elucidate the molecular mechanisms related to PKDL development. The complexity of PKDL pathogenesis suggests that multiple pathways and interactions are likely involved in its progression.

9.8 CONCLUSION

Based on several studies, information has been gathered about the complex and diverse characteristics of this particular skin disorder. However, PKDL remains a mysterious skin disease. Generally, a complex interplay of factors is thought to play a role in the development of PKDL, which emerges as a sequel to VL. However, the development of PKDL does not depend on a history of VL; it can also occur in individuals who have never had symptoms of VL, indicating the presence of a subclinical condition.

Multiple factors influence PKDL development, including genetic predisposition, immune response, and complex interactions between the host and *Leishmania* parasites. Several potential molecular pathways have been suggested to be involved in PKDL pathogenesis, such as the regulation of cytokines (including IL-10, IFN-γ, and TGF-β) and chemokines, activation of matrix metalloproteinases, changes in TLR signalling, and immunological tolerance mediated by regulatory T cells. However, more research is needed to fully understand the complex molecular structure of PKDL development. By unveiling molecular pathways, we can develop targeted treatments and interventions to manage and prevent this chronic skin disorder. Ongoing research continues to explore the complex molecular mechanisms underlying the development of PKDL.

ACKNOWLEDGEMENT

The authors are thankful to the ICMR-Rajendra Memorial Research Institute of Medical Sciences, Patna, for their necessary support during this study.

REFERENCES

Ansari, N A, G K Katara, V Ramesh, and P Salotra. 2008. "Evidence for Involvement of TNFR1 and TIMPs in Pathogenesis of Post-Kala-Azar Dermal Leishmaniasis." *Clinical and Experimental Immunology* 154 (3): 391–98. https://doi.org/10.1111/j.1365-2249.2008.03761.x.

Ansari, N A, V Ramesh, and P Salotra. 2006. "Interferon (IFN)-γ, Tumor Necrosis Factor-α, Interleukin-6, and IFN-γ Receptor 1 are the Major Immunological Determinants Associated with Post-Kala Azar Dermal Leishmaniasis." *Journal of Infectious Diseases* 194 (7): 958–65. https://doi.org/10.1086/506624.

Bell, S, K Degitz, M Quirling, N Jilg, S Page, and K Brand. 2003. "Involvement of NF-κB Signalling in Skin Physiology and Disease." *Cellular Signalling* 15 (1): 1–7.

Blackwell, J M, T Goswami, C AW Evans, D Sibthorpe, N Papo, J K White, S Searle, et al. 2001. "SLC11A1 (Formerly NRAMP1) and Disease Resistance." *Cellular Microbiology* 3 (12): 773–84. https://doi.org/10.1046/j.1462-5822.2001.00150.x.

Desjeux, P, R S Ghosh, P Dhalaria, N Strub-Wourgaft, and Ed E Zijlstra. 2013. "Report of the Post Kala-Azar Dermal Leishmaniasis (PKDL) Consortium Meeting, New Delhi, India, 27–29 June 2012." *Parasites & Vectors* 6 (1): 196.

Dighal, A, D Mukhopadhyay, R Sengupta, S Moulik, S Mukherjee, S Roy, S J Chaudhuri, N K Das, and M Chatterjee. 2020. "Iron Trafficking in Patients with Indian Post Kala-Azar Dermal Leishmaniasis." *PLoS Neglected Tropical Diseases* 14 (2): 1–18. https://doi.org/10.1371/journal.pntd.0007991.

Elmekki, M A, M M Elhassan, H A Ozbak, and M M Mukhtar. 2016. "Elevated TGF-Beta Levels in Drug-Resistant Visceral Leishmaniasis." *Annals of Saudi Medicine* 36 (1): 73–77.

Ferdinandusse, S, S Denis, P L Faust, and R J A Wanders. 2009. "Bile Acids: The Role of Peroxisomes." *Journal of Lipid Research* 50 (11): 2139–47.

Ferreira, C, J Estaquier, and R Silvestre. 2021. "Immune-Metabolic Interactions between Leishmania and Macrophage Host." *Current Opinion in Microbiology* 63: 231–37.

Ganguly, S, N K Das, M Panja, S Pal, D Modak, M Rahaman, S Mallik, et al. 2008. "Increased Levels of Interleukin-10 and IgG3 are Hallmarks of Indian Post-Kala-Azar Dermal Leishmaniasis." *Journal of Infectious Diseases* 197 (12): 1762–71.

Ganguly, S, D Mukhopadhyay, N K Das, M Chaduvula, S Sadhu, U Chatterjee, M Rahman, et al. 2010. "Enhanced Lesional Foxp3 Expression and Peripheral Anergic Lymphocytes Indicate a Role for Regulatory T Cells in Indian Post-Kala-Azar Dermal Leishmaniasis." *Journal of Investigative Dermatology* 130 (4): 1013–22. https://doi.org/10.1038/jid.2009.393.

Gantt, K R, S Schultz-Cherry, N Rodriguez, S Jeronimo, E T Nascimento, T L Goldman, T J Recker, M A Miller, and M E Wilson. 2003. "Activation of TGF β by Leishmania Chagasii Importance for Parasite Survival in Macrophages." *The Journal of Immunology* 170 (5): 2613–20.

Gasim, S, A M Elhassan, A Kharazmi, E A G Khalil, A Ismail, and T G Theander. 2000. "The Development of Post-Kala-Azar Dermal Leishmaniasis (PKDL) is Associated with Acquisition of Leishmania Reactivity by Peripheral Blood Mononuclear Cells (PBMC)." *Clinical & Experimental Immunology* 119 (3): 523–29.

Gupta, S, B Raychaudhury, and S C Datta. 2009. "Host Peroxisomal Properties Are Not Restored to Normal after Treatment of Visceral Leishmaniasis with Sodium Antimony Gluconate." *Experimental Parasitology* 123 (2): 140–45.

Haldar, J P, S Ghose, K C Saha, and AC264486 Ghose. 1983. "Cell-Mediated Immune Response in Indian Kala-Azar and Post-Kala-Azar Dermal Leishmaniasis." *Infection and Immunity* 42 (2): 702–7.

Hassan, A M El, H W Ghalib, E E Zijlstra, I A Eltoum, M Satti, M S Ali, and H M A Ali. 1992. "Post Kala-Azar Dermal Leishmaniasis in the Sudan: Clinical Features, Pathology and Treatment." *Transactions of the Royal Society of Tropical Medicine and Hygiene* 86 (3): 245–48.

Hosseini, A M, J Majidi, B Baradaran, and M Yousefi. 2015. "Toll-Like Receptors in the Pathogenesis of Autoimmune Diseases." *Advanced Pharmaceutical Bulletin* 5 (Suppl 1): 605.

Huynh, C, and N W Andrews. 2008. "Iron Acquisition within Host Cells and the Pathogenicity of Leishmania." *Cellular Microbiology* 10 (2): 293–300.

Ismail, A, A M El Hassan, K Kemp, S Gasim, A E Kadaru, T Moller, A Kharazmi, and T G Theander. 1999. "Immunopathology of post kala-azar dermal leishmaniasis (PKDL): T-cell phenotypes and cytokine profile." *J Pathol.* 189: 615–622.

Ismail, A, E A G Khalil, A M Musa, I M El Hassan, M E Ibrahim, T G Theander, and A M El Hassan. 2006. "The Pathogenesis of Post Kala-Azar Dermal Leishmaniasis from the Field to the Molecule: Does Ultraviolet Light (UVB) Radiation Play a Role?" *Medical Hypotheses* 66 (5): 993–99. https://doi.org/10.1016/j.mehy.2005.03.035.

Kar, C, P Saha, N Chakrabarti, and P Sarkar. 2019. "Histopathological study of Post Kala-Azar Dermal Leishmaniasis and Its Importance for Diagnosis in a Peripheral Tertiary Institution." *IP Indian Journal of Clinical and Experimental Dermatology*·5: 9–15 https://doi.org/10.18231/2581-4729.2019.0003.

Karnati, S, and E Baumgart-Vogt. 2009. "Peroxisomes in Airway Epithelia and Future Prospects of these Organelles for Pulmonary Cell Biology." *Histochemistry and Cell Biology* 131: 447–54.

Katara, G K, N A Ansari, A Singh, V Ramesh, and P Salotra. 2012. "Evidence for Involvement of Th17 Type Responses in Post Kala Azar Dermal Leishmaniasis (PKDL)." *PLoS Neglected Tropical Diseases* 6 (6): e1703.

Katara, G K, N A Ansari, S Verma, V Ramesh, and P Salotra. 2011. "Foxp3 and IL-10 Expression Correlates with Parasite Burden in Lesional Tissues of Post Kala Azar Dermal Leishmaniasis (PKDL) Patients." *PLoS Neglected Tropical Diseases* 5 (5): e1171.

Kaushal, H, R Bras-Gonçalves, K Avishek, D Kumar Deep, E Petitdidier, J-L Lemesre, G Papierok, S Kumar, V Ramesh, and P Salotra. 2016. "Evaluation of Cellular Immunological Responses in Mono- and Polymorphic Clinical Forms of Post-Kala-Azar Dermal Leishmaniasis in India." *Clinical & Experimental Immunology* 185 (1): 50–60.

Kovacs, W J, L M Olivier, and S K Krisans. 2002. "Central Role of Peroxisomes in Isoprenoid Biosynthesis." *Progress in Lipid Research* 41 (5): 369–91.

Kumar, P, M Chatterjee, and N Kanti Das. 2021. "Post Kala-Azar Dermal Leishmaniasis: Clinical Features and Differential Diagnosis." *Indian Journal of Dermatology* 66 (1): 24–33.

Lambert, H, and A Barragan. 2010. "Modelling Parasite Dissemination: Host Cell Subversion and Immune Evasion by Toxoplasma Gondii." *Cellular Microbiology* 12 (3): 292–300.

Leclercq, S, J Skrzypski, A Courvoisier, C Gondcaille, F Bonnetain, A André, J-M Chardigny, et al. 2008. "Effect of Dietary Polyunsaturated Fatty Acids on the Expression of Peroxisomal ABC Transporters." *Biochimie* 90 (10): 1602–7.

Maretti-Mira, A C, K M de Pinho Rodrigues, M P de Oliveira-Neto, C Pirmez, and N Craft. 2011. "MMP-9 Activity is Induced by Leishmania Braziliensis Infection and Correlates with Mucosal Leishmaniasis." *Acta Tropica* 119 (2–3): 160–64.

Mohamed, H S, M E Ibrahim, E N Miller, C S Peacock, E A G Khalil, H J Cordell, J M M Howson, A M El Hassan, R E H Bereir, and J M Blackwell. 2003. "Genetic Susceptibility to Visceral Leishmaniasis in the Sudan: Linkage and Association with IL4 and IFNGR1." *Genes & Immunity* 4 (5): 351–55.

Mukherjee, S, D Mukhopadhyay, C Braun, J N Barbhuiya, N K Das, U Chatterjee, E von Stebut, and M Chatterjee. 2015. "Decreased Presence of Langerhans Cells is a Critical Determinant for Indian Post Kala-Azar Dermal Leishmaniasis." *Experimental Dermatology* 24 (3): 232–34.

Mukherjee, S, R Sengupta, D Mukhopadhyay, C Braun, S Mitra, S Roy, N K Das, U Chatterjee, E von Stebut, and M Chatterjee. 2019. "Impaired Activation of Lesional CD8+ T-Cells is Associated with Enhanced Expression of Programmed Death-1 in Indian Post Kala-Azar Dermal Leishmaniasis." *Scientific Reports* 9 (1): 762.

Mukhopadhyay, D, J E Dalton, P M Kaye, and M Chatterjee. 2014. "Post Kala-Azar Dermal Leishmaniasis: An Unresolved Mystery." *Trends in Parasitology* 30 (2): 65–74.

Mukhopadhyay, D, N K Das, S De Sarkar, A Manna, D N Ganguly, J Nath Barbhuiya, A K. Maitra, A Hazra, and M Chatterjee. 2012. "Evaluation of Serological Markers to Monitor the Disease Status of Indian Post Kala-Azar Dermal Leishmaniasis." *Transactions of the Royal Society of Tropical Medicine and Hygiene* 106 (11): 668–76. https://doi.org/10.1016/j.trstmh.2012.07.005.

Mukhopadhyay, D, N Kanti Das, S Roy, S Kundu, J N Barbhuiya, and M Chatterjee. 2011. "Miltefosine Effectively Modulates the Cytokine Milieu in Indian Post Kala-Azar Dermal Leishmaniasis." *Journal of Infectious Diseases* 204 (9): 1427–36.

Mukhopadhyay, D, S Mukherjee, S Roy, J E Dalton, S Kundu, A Sarkar, N K Das, P M Kaye, and M Chatterjee. 2015. "M2 Polarization of Monocytes-Macrophages is a Hallmark of Indian Post Kala-Azar Dermal Leishmaniasis." *PLoS Neglected Tropical Diseases* 9 (10): e0004145.

Murray, H W, B Y Rubin, C D Rothermel, et al. 1983. "Killing of Intracellular Leishmania Donovani by Lymphokine-Stimulated Human Mononuclear Phagocytes. Evidence that Interferon-Gamma is the Activating Lymphokine." *The Journal of Clinical Investigation* 72 (4): 1506–10.

Musa, A M, E A G Khalil, M A Raheem, E E Zijlstra, M E Ibrahim, I M Elhassan, M M Mukhtar, and A M El Hassan. 2002. "The Natural History of Sudanese Post-Kala-Azar Dermal Leishmaniasis: Clinical, Immunological and Prognostic." *Ann Trop Med Parasitol* 96 (8): 765–72.

Nandan, D, T Yi, M Lopez, C Lai, and N E Reiner. 2002. "Leishmania EF-1α Activates the Src Homology 2 Domain Containing Tyrosine Phosphatase SHP-1 Leading to Macrophage Deactivation." *Journal of Biological Chemistry* 277 (51): 50190–97.

Nghiem, D X, N Kazimi, D L Mitchell, A A Vink, H N Ananthaswamy, M L Kripke, and S E Ullrich. 2002. "Mechanisms Underlying the Suppression of Established Immune Responses by Ultraviolet Radiation." *Journal of Investigative Dermatology* 119 (3): 600–8.

Niu, Q, S Li, D Chen, Q Chen, and J Chen. 2016. "Iron Acquisition in Leishmania and its Crucial Role in Infection." *Parasitology* 143 (11): 1347–57.

Pracharoenwattana, I, and S M Smith. 2008. "When is a Peroxisome Not a Peroxisome?" *Trends in Plant Science* 13 (10): 522–25.

Rahman, K M, S Islam, M W Rahman, E Kenah, C M Galive, M M Zahid, J Maguire, et al. 2010. "Increasing Incidence of Post-Kala-Azar Dermal Leishmaniasis in a Population-Based Study in Bangladesh." *Clinical Infectious Diseases* 50 (1): 73–76.

Ramesh, V, and M Ramam. 2020. "Histopathology of Post Kala-Azar Dermal Leishmaniasis." *Indian Journal of Dermatology* 65 (6): 461–64.

Saha, B, S Bhardwaj, N Srivastava, and R Sudan. 2010. "Leishmania Interferes with Host Cell Signaling to Devise a Survival Strategy." *Journal of Biomedicine and Biotechnology* 2010: 109189. https://doi.org/10.1155/2010/109189.

Saha, S, S Mondal, R Ravindran, S Bhowmick, D Modak, S Mallick, M Rahman, et al. 2007. "IL-10- and TGF-β-Mediated Susceptibility in Kala-Azar and Post-Kala-Azar Dermal Leishmaniasis: The Significance of Amphotericin B in the Control of Leishmania Donovani Infection in India." *The Journal of Immunology* 179 (8): 5592–5603.

Salih, M A, M E Ibrahim, J M Blackwell, E Nancy Miller, E A G Khalil, A M ElHassan, A M Musa, and H S Mohamed. 2007. "IFNG and IFNGR1 Gene Polymorphisms and Susceptibility to Post-Kala-Azar Dermal Leishmaniasis in Sudan." *Genes & Immunity* 8 (1): 75–78.

Sengupta, R, S Mukherjee, S Moulik, S Mitra, S Jyati Chaudhuri, N K Das, U Chatterjee, and M Chatterjee. 2019. "In-Situ Immune Profile of Polymorphic vs. Macular Indian Post Kala-Azar Dermal Leishmaniasis." *International Journal for Parasitology: Drugs and Drug Resistance* 11: 166–76.

Silva-Gomes, S, S Vale-Costa, R Appelberg, and M S Gomes. 2013. "Iron in Intracellular Infection: To Provide or to Deprive?" *Frontiers in Cellular and Infection Microbiology* 3: 96.

Terrazas, C, S Oghumu, B K Jha, G Natarajan, M Drew, E Y Denkers, A R Satoskar, and B S McGwire. 2016. "Subverting Immunity from the Inside: Strategies of Intracellular Survival-Protozoans." Michael J.H. Ratcliffe (ed.), *Encyclopedia of Immunobiology*, Academic Press, Elsevier Inc., Amsterdam, the Netherlands 4, 83–93.

Thakur, C P, and K Kumar. 1992. "Post Kala-Azar Dermal Leishmaniasis: A Neglected Aspect of Kala-Azar Control Programmes." *Annals of Tropical Medicine \& Parasitology* 86 (4): 355–59.

Verma, S K, S Ahmad, N Shirazi, A Kusum, R M Kaushik, and S P Barthwal. 2007. "Sodium Stibogluconate-Sensitive Visceral Leishmaniasis in the Non-Endemic Hilly Region of Uttarakhand, India." *Transactions of the Royal Society of Tropical Medicine and Hygiene* 101 (7): 730–32.

Volpedo, G, T Pacheco-Fernandez, E A Holcomb, N Cipriano, B Cox, and A R Satoskar. 2021. "Mechanisms of Immunopathogenesis in Cutaneous Leishmaniasis and Post Kala-Azar Dermal Leishmaniasis (PKDL)." *Frontiers in Cellular and Infection Microbiology* 11: 685296.

Wanders, R J A, and H R Waterham. 2005. "Peroxisomal Disorders I: Biochemistry and Genetics of Peroxisome Biogenesis Disorders." *Clinical Genetics* 67 (2): 107–33.

Zaba, L C, J G Krueger, and M A Lowes. 2009. "Resident and 'Inflammatory' Dendritic Cells in Human Skin." *Journal of Investigative Dermatology* 129 (2): 302–8.

Zijlstra, E E. 2014. "PKDL and Other Dermal Lesions in HIV Co-Infected Patients with Leishmaniasis: Review of Clinical Presentation in Relation to Immune Responses." *PLoS Neglected Tropical Diseases* 8 (11https://doi.org/10.1371/journal.pntd.0003258.

Zijlstra, E E. 2016. "The Immunology of Post-Kala-Azar Dermal Leishmaniasis (PKDL)." *Parasites & Vectors* 9 (1): 464.

Zijlstra E E, and J Alvar. 2012. *The Post Kala-Azar Dermal Leishmaniasis (PKDL) Atlas A Manual for Health Workers.* Spain: World Health Organization (WHO).

Zijlstra, E E, and A M El-Hassan. 2001. "Leishmaniasis in Sudan. 4. Post Kala-Azar Dermal Leishmaniasis." *Transactions of the Royal Society of Tropical Medicine and Hygiene* 95 (Supplement_1): S59–S76.

Zijlstra, E E, A M Musa, E A G Khalil, I M El Hassan, and A M El-Hassan. 2003. "Post-Kala-Azar Dermal Leishmaniasis." *The Lancet Infectious Diseases* 3 (2): 87–98.

10 Immunopathogenesis of Visceral Leishmaniasis and Post-kala-azar Dermal Leishmaniasis

Dayakar Alti, Vikash Kumar, and Ashish Kumar*
Department of Immunology, ICMR-Rajendra Memorial Research Institute of Medical Sciences, Patna, Bihar, India
*Corresponding Author: Dr. Dayakar Alti, Scientist-B, Department of Immunology, ICMR-Rajendra Memorial Research Institute of Medical Sciences, Patna, Bihar, India

10.1 INTRODUCTION

The clearance of *Leishmania* parasites from the host body mainly depends on the balanced host immune response. However, this response is very complex during *Leishmania* infection and varies based on the parasite species and their virulence factors, host species and their genetic makeup, and demographic factors, etc. (Ehrlich et al., 2014; McMahon-Pratt and Alexander, 2004; Hurrell et al., 2016; Reiner and Locksley, 1995; Nashed et al., 2000; Launois et al., 1998; Kharazmi et al., 1999; Ganguly et al., 2010a, 2010b). Both arms of immunity, innate and adaptive responses, are critically important in offering protection against leishmaniasis (Dos-Santos et al., 2016). When exposed to the sandfly bite, the saliva-composed molecules can cause vasodilation, coagulation inhibition, and immune stimulation at the site of inoculation (Andrade et al., 2007). On the other hand, skin-resident macrophages, mast cells, keratinocytes, and Langerhans cells (LCs) can express a variety of pattern recognition receptors (PRRs), such as toll-like receptors (TLRs), which can recognize pathogen-associated molecular patterns (PAMPs) to trigger the phagocytosis of the *Leishmania* parasite. Simultaneously, chemokines such as monocyte chemoattractant protein-1 (MCP-1), chemokine (C-X-C motif) ligand 1 (CXCL1), and CXCL2 produced by the skin-resident cells can initiate further activation of immune components (Teixeira et al., 2006; Pacheco-Fernandez et al., 2021), including the recruitment of monocytes and neutrophils at the site of inoculation (Teixeira et al., 2006; Giraud et al., 2018). The activated cells undergo various structural and functional modifications, which increase *Leishmania* phagocytosis and oxidative burst. After the first encounter with neutrophils, the *Leishmania* parasites are repeatedly phagocytosed by antigen-presenting cells (APCs) such as monocytes, macrophages, and dendritic cells (DCs) (Goundry et al., 2018). The coordinated interaction between innate and adaptive immune cells is of utmost importance for the survival of the host against pathogen invasion and to prevent reinfection, as shown in Figure 10.1 (Dayakar et al., 2019).

The B- and T-lymphocytes, including their subsets and their interaction with APCs, are crucial in bridging the gap between the two arms of the immune system. However, cell-mediated immunity (CMI) is the ultimate response for clearing intracellular pathogens from host tissues.

 DOI: 10.1201/9781032643120-10

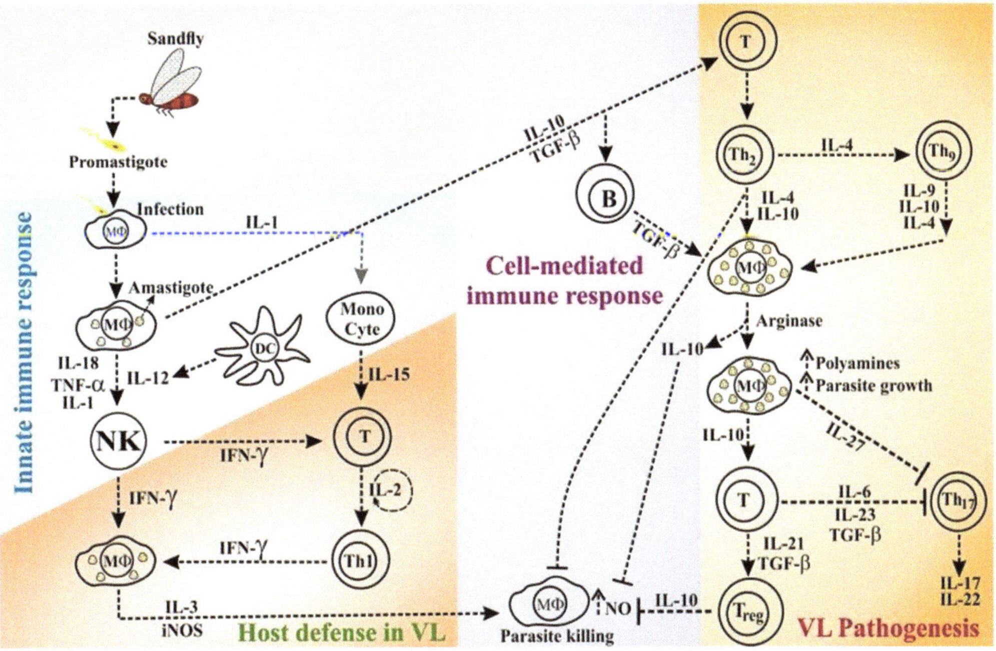

FIGURE 10.1 Innate and adaptive immune cell interaction via their cytokines during VL pathogenesis and host protection (Dayakar et al., 2019)

10.2 INNATE IMMUNITY EVENTS DURING *LEISHMANIA* INFECTION

Phagocytosis: Upon phagocytosis, the internalized promastigotes form an endosome or phagosome that fuses with the lysosome to form a phagolysosome, where they transform into amastigotes. To resist hostile conditions such as acidic pH, high temperature, and oxidative stress in the phagolysosome, promastigotes avoid the fusion of the phagosome with the lysosome, resulting in the inhibition of endosomal maturation, indicated by the delayed expression of Ras-related protein Rab7 and lysosome-associated membrane protein (LAMP)-1 (Olivier et al., 2005; Scianimanico et al., 1999). The survival mechanism of *L. donovani* inside the early endosomal compartment of macrophages is shown in Figure 10.2 (Verma et al., 2017).

Lipophosphoglycan (LPG) expressed on the parasite surface prevents the endosome maturation (Holm et al., 2001) and acidification of the phagosome by interfering with the vacuolar-type ATPase (V-ATPase) pump, resulting in the development of resistant amastigotes (Vinet et al., 2009). To counteract the acidic environment and survive inside the phagolysosome, the *Leishmania* amastigotes scavenge host sphingolipids (Ali et al., 2012). As iron is an important metabolite for intracellular amastigotes' growth (Huynh and Andrews, 2008), *Leishmania* upregulates the expression of iron transporters LIT 1 and LIT 2 to increase their iron acquisition by competing with the iron-sequestering mechanisms of the host. As a counter to this action of the parasite, the host cells induce the expression of natural-resistance-associated macrophage protein 1 (Nrampl), which translocates iron (Fe^{+2}) from the phagolysosome to the cytosol, making iron unavailable to the parasite (Kaye and Scott, 2011). L-arginine is another essential metabolite for the intracellular parasites for their polyamine biosynthesis and is simultaneously essential to the host for the production of nitric oxide (NO), a potential innate antileishmanial factor. The *Leishmania*-encoded arginase enzyme competes with the nitric oxide synthase 2 (NOS2) of the host for L-arginine metabolism, resulting in the attenuation of NO-mediated death of the intracellular amastigotes by the infected macrophages (Gaur et al., 2007; Reguera et al., 2009). These evasion mechanisms of the parasite contribute to their survival inside the phagosomes and help to establish the infection in host cells.

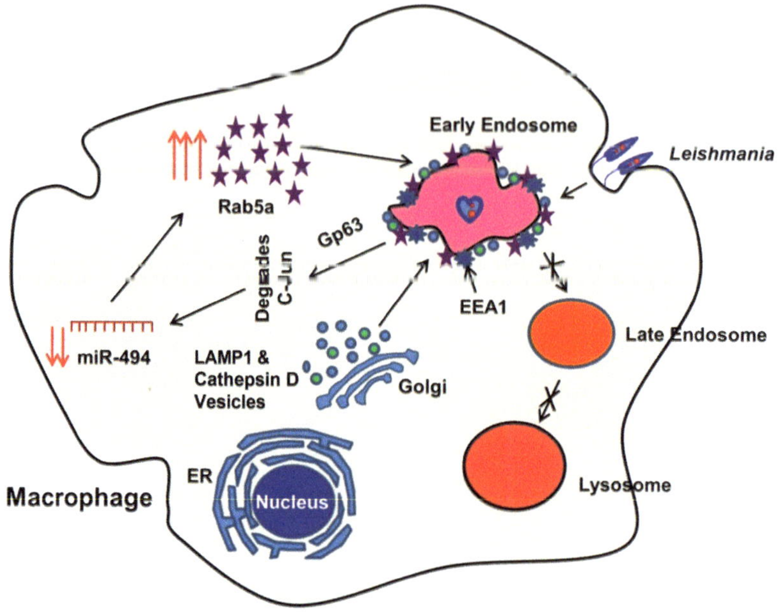

FIGURE 10.2 Survival mechanism of *L. donovani* inside the early endosomal compartment of macrophages after phagocytosis. Parasite-secreted gp63 induces the degradation of c-Jun, resulting in the downregulation of miR-494 and subsequent upregulation of Rab5a. The Rab5a and early endosome antigen 1 (EEA1) recruited in parasitophorous vacuole inhibits the lysosomal fusion. *Leishmania* exploits the inactive cathepsin D and LAMP-1 to thrive inside the endosomal compartment (Verma et al., 2017)

Complement system: While the parasite is ingested into the blood, the opsonization (complement-mediated phagocytosis) of metacyclic promastigotes occurs rapidly and results in the lysis of more than 90% of ingested parasites. However, *Leishmania* protein kinases phosphorylate the serum complement proteins C3, C5, and C9, leading to their inactivation (Hermoso et al., 1991; Nunes et al., 1997). The complement C3b is deposited on the surface of the parasite with the help of factor H and factor I from human serum (Pereira Filho et al., 2021) and inactivated upon interaction with the membrane protease gp63 (63 kDa glycoprotein) and LPG, which resists the complement-mediated lysis and further facilitates the internalization of the parasites into macrophages or neutrophils via complement receptor (CR)2 (Domínguez et al., 2003; Nylén and Gautam, 2010). The inactivated C3b facilitates the CR1- and CR3-mediated 'silent entry' of the parasite; hence, the establishment of infection occurs due to the incompetent oxidative burst triggered by the host cells (Blackwell, 1985; Laurenti et al., 1996). Amastigotes, compared to promastigotes (Sacks and Perkins, 1984; Ueno et al., 2009), and strains causing visceral leishmaniasis (VL), compared to strains causing cutaneous leishmaniasis (CL), are more resistant to the complement-mediated lysis, due to the differential antigenic expression on their surface (Hoover et al., 1984). The *L. donovani* inhibitor of serine peptidase 2 (*Ld*ISP2) upregulates C5aR-mediated signaling in host cells, resulting in the inhibition of the lectin-mediated complement pathway and membrane attacking complex (MAC) formation, which supports parasite endurance, as shown in Figure 10.3 (Verma et al., 2018; Ingale et al., 2020).

10.3 INNATE IMMUNE CELL RESPONSE TO *LEISHMANIA* INFECTION

Neutrophils: As first-line innate immune cells, neutrophils infiltrate the site of infection in the dermis in response to the factors derived from the host, parasite, and sandfly (Ribeiro-Gomes and Sacks, 2012; Carlsen et al., 2015). The anti-parasitic activity of neutrophils is due to their phagocytosis via TLRs, reactive oxygen/nitrogen species production in phagosomes, neutrophils

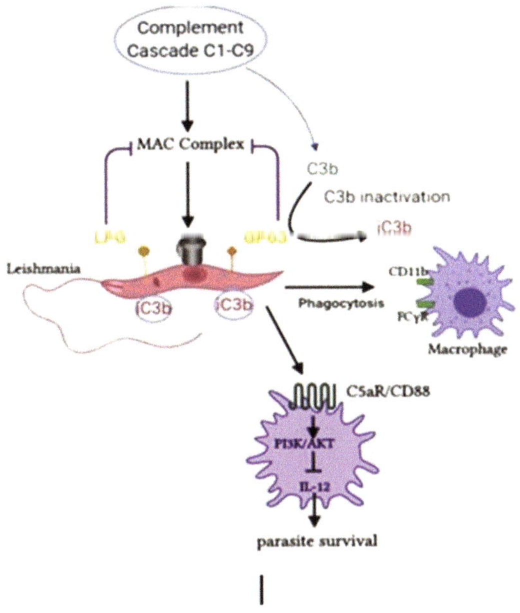

FIGURE 10.3 Evasion of the complement system by the *Leishmania* parasites (Ingale et al., 2020)

extracellular trap (NET) formation, and release of toxic mediators upon degranulation at the site of infection (Mollinedo et al., 2010; Peters et al., 2009; Pillay et al., 2010; Charmoy et al., 2007; Guimarães-Costa et al., 2009; Hurrell et al., 2016). Neutrophil granule-derived substances regulate the T-cell response, where interleukin (IL)-12, interferon (IFN)-γ, tumor necrosis factor (TNF)-α, and neutrophils elastase are involved in T-cell activation, while IL-10, transforming growth factor (TGF)-β, and thromboxane A2 are involved in T-cell suppression (Kalyan et al., 2014; Borregaard et al., 2007; Yang and Unanue, 2013; Ohms et al., 2021). The role of neutrophils varies in response to different *Leishmania* species that cause VL. For example, interaction with lipoxin A4 receptor by *L. donovani* causes the infiltration, survival, and deactivation of neutrophils (Ribeiro-Gomes and Sacks, 2012; van Zandbergen et al., 2002), whereas *L. infantum* promotes neutrophil activation and NET formation (Wei et al., 2019). Neutrophils may perform diverse functions in response to vector saliva containing sialogenins and yellow salivary protein (Guimarães-Costa et al., 2021; Prates et al., 2011). The neutrophil's phenotype (Nα or Nβ) determines their protective or detrimental function in VL. Nα phenotypic neutrophils are highly susceptible to VL infection in mice (Bhattacharya et al., 2020), and human leukocyte antigen (HLA)-DR$^+$ low-density neutrophils in human VL show higher expression of arginase-I, IL-10, and programmed death ligand 1 (PD-L1), which causes T-cell exhaustion (Kumar and Nylén, 2012; Sharma et al., 2016). In in vitro, neutrophils behave as 'Trojan Horses,' serving as intermediate cells for *Leishmania* before infecting the definitive host cells (van Zandbergen et al., 2004; Peters et al., 2008). The interaction during *Leishmania* infection between neutrophils and other immune cells, particularly macrophages and dentritic cells (DCs), varies depending on the parasite species, host genetics, and the apoptotic or necrotic behavior of neutrophils. Macrophages and dendritic cells can acquire parasites from neutrophils. During immune cells cross-talk with infected neutrophils, macrophages produce TGF-β and prostaglandin E2 (Teixeira et al., 2018), and DCs show impaired activation and cross-presentation to CD8$^+$ T-cells. Upon interaction with CD11b of neutrophils, DCs may induce the secretion of TNF-α in VL (Kupani et al., 2021). Neutrophil depletion in the host impairs the protective CD4$^+$ T-cell response against centrin gene knock-out *L. donovani* challenge (Bhattacharya et al., 2020).

Monocytes: After neutrophils, monocytes are the earliest cells recruited at the site of *Leishmania* infection (Beil et al., 1992). The monocyte subset recruitment depends on the *Leishmania* species and the localized cytokine and chemokine response (Terrazas et al., 2017; Geissmann et al., 2003). Monocytes exist in three different subsets with discrete functions: classical ($CD14^{++}CD16^{-}$), intermediate ($CD14^{++}CD16^{+}$), and non-classical ($CD14^{+}CD16^{++}$) (Ziegler-Heitbrock et al., 2010). The ingestion of *Leishmania* along with sandfly saliva causes C-C motif chemokine receptor 2 (CCR2)-mediated infiltration of inflammatory monocytes to the site of infection (Serbina and Pamer, 2006), which plays a contrasting role in different infections and is detrimental in VL infection (Vono et al., 2017). Monocytes interact with *Leishmania* parasites via TLRs and produce either pro- or anti-inflammatory cytokine responses depending on the infecting species (Viana et al., 2018; Campos et al., 2014; Fromm et al., 2012; Steinbrink et al., 2000). *L. donovani* infection in monocytes downregulates TLR2 and TLR4 signaling, resulting in the inhibition of oxidative burst and antigen presentation and high expression of IL-10 (Roy et al., 2015; Chandra and Naik, 2008). The inflammatory monocytes recruited in the spleen during *L. donovani* infection acquire a suppressor cell phenotype, with the help of hypoxia-inducible factor (HIF)-1α, and cause chronic infection (Hammami et al., 2017). In contrast, monocytes are involved in granuloma formation, a key event in the clearance of parasites from the liver and spleen, during *L. donovani* infection in a murine model (Terrazas et al., 2016). Active VL patients in India show higher $CD14^{+}$ monocyte populations, IL-10 production (Roy et al., 2018; Peruhype-Magalhães et al., 2005), and anti-inflammatory monocyte responses in peripheral blood mononuclear cells (PBMCs) (Roy et al., 2015; Singh et al., 2018). Extramedullary hematopoiesis of myeloid cells in the spleen of hamsters during *L. donovani* infection causes the accumulation of monocytes, which promotes parasite replication (Osorio et al., 2020). The early response of monocytes in *L. donovani* infection determines the clinical outcome. In post-kala-azar dermal leishmaniasis (PKDL), there is no apparent difference in the distribution of monocyte subsets in peripheral blood. As reported in VL, the $CD14^{+}$ monocytes in the peripheral blood of PKDL patients show reduced TLR2 and TLR4 expression, resulting in reduced phosphorylation of mitogen-activated protein kinase (MAPK), CD40 signaling, oxidative burst, and redox imbalance in monocytes. This creates an anti-inflammatory pro-parasitic microenvironment that helps in parasite survival (Mukhopadhyay et al., 2015).

Macrophages: Macrophages are the definitive host cells for *Leishmania* after phagocytosis of infected apoptotic neutrophils (van Zandbergen et al., 2004). Neutrophil-derived chemokines such as MCP-1 (CCL2), macrophage inflammatory protein (MIP)-1α, MIP-1β, and MIP-2 (CXCL2) facilitate the recruitment of macrophages to the site of infection (Gregory et al., 2008). The macrophage response to infection depends on the cross-talk between *Leishmania* PAMPs and host PRRs. *Leishmania* PAMPs, such as surface gp63, glycosylphosphatidylinositol (GPI), and LPG, can generally be recognized by TLRs (Tuon et al., 2008), fibronectin, mannose receptor (MRC)-1, complement, and Fc receptors of the host (Rostami and Khamesipour, 2021). *Leishmania*'s interaction with complement or Fc receptors delays the phagosome maturation (Polando et al., 2013) and enhances IL-10 secretion (Rostami and Khamesipour, 2021; Kane and Mosser, 2001; Sutterwala et al., 1998). The interplay of *Leishmania* gp63 with host TLRs activates nuclear factor (NF)-κB signaling (Hawn et al., 2002), which negatively regulates reactive oxygen species (ROS) production (Isnard et al., 2012), nod-like receptor family pyrin domain containing 3 (NLRP3) inflammasome activation, and IL-1β secretion (Kumar et al., 2018; Saha et al., 2018). To establish subclinical infection, *Leishmania* escapes host resistance mechanisms and survives throughout life (Gupta et al., 2013). To do this, *Leishmania* evades the complement system by inhibiting MAC formation and lowering the expression of B7 and CD40 on macrophages, which are essential to promote the anti-parasitic activity of T-cells and the upregulation of iron transporters. *Leishmania* exploits the TLR-2/4 signaling in macrophages to suppress the Janus kinase/signal transducers and activators of the transcription (JAK-STAT) pathway, in turn affecting cytokine signaling (Hawn et al., 2002). In response, the host counters the infected macrophages by inducing FasL-mediated apoptosis (Huang et al., 1998).

Macrophages are double-edged swords in function, and can act as either host or foe for *Leishmania* depending on their phenotype (Bogdan, 2020). This dual function also depends on the *Leishmania* species, their recognition pattern, and microenvironmental signals in the host (Italiani and Boraschi, 2014). Macrophage polarization towards either classical (M1-type) or alternative (M2-type) phenotype can produce inflammatory or anti-inflammatory effects, respectively, which can determine the disease outcome (Ribeiro-Gomes et al., 2004; Loría-Cervera and Andrade-Narvaez, 2020; Pace, 2014; Bronte and Zanovello, 2005). Visceral infection causes *L. infantum* lipid mediators promote alternative activation of macrophages by inhibiting IFN-γ-induced M1 polarization (Paloque et al., 2019). M2 macrophages act as niche reservoirs for *Leishmania* parasites, possibly due to unusual expression of specific receptors for parasite uptake or impaired antileishmanial activity. M2 polarization is a hallmark of Indian PKDL patients. After successful antileishmanial drug therapy, M2→M1 plasticity is observed in the host (Peruhype-Magalhães et al., 2005). *Leishmania* infection redistributes metabolites in macrophages. Lipid droplets accumulate in and around parasitophorous vacuoles, and providing the carbon source for amastigotes (Saunders and McConville, 2020). The metacyclic stage of *L. infantum* infection causes M2 polarization of macrophages by inducing docosahexaenoic acid levels, a polyunsaturated fatty acid derived from resolvin D1 (Titos et al., 2011). M1-type macrophage activation depends on IFN-γ stimuli and TLR signaling (Bronte and Zanovello, 2005; Orecchioni et al., 2019), whereas M2-type activation depends on IL-13 and IL-4 stimuli and fatty acid oxidation, but not on TLR signaling (Ganeshan and Chawla, 2014; Van den Bossche et al., 2017). Upregulation of the pentose phosphate pathway in M1 macrophages (Baardman et al., 2018) synthesizes adenosine triphosphate (ATP) and nicotinamide adenine dinucleotide phosphate hydrogen (NADPH), facilitating ROS production for parasite killing. Higher mitochondrial respiration in M2 macrophages makes them more permissive for amastigotes survival (Saunders and McConville, 2020; Moreira et al., 2015).

Macrophages and parasites compete for L-arginine for the production of NO and polyamines, respectively, using NOS2 and arginase-I enzymes, respectively. NO is a strong antileishmanial innate immune component, whereas polyamines support parasite growth. In M1 macrophages, the increased L-arginine transportation produces abundant NO and is IFN-γ dependent. With low IFN-γ production, L-arginine availability increases for parasite polyamine biosynthesis (Saunders and McConville, 2020; Wanasen et al., 2007). *L. donovani* infection in macrophages hampers STAT1 and STAT4 phosphorylation and the signal transduction of IL-12 and IFN-γ (Bhardwaj et al., 2010; Nandan and Reiner, 1995). The surface LPG of *L. donovani* inhibits MAPK activation, in turn affecting IL-12 and ROS production (Bhardwaj et al., 2010). Predominant M2 macrophage activity during human VL produces IL-10 and IL-4, preventing NO production (Rossi and Fasel, 2018; Loría-Cervera and Andrade-Narvaez, 2020; Nylén and Sacks, 2007). IL-10 opposes IFN-γ-mediated activation of macrophages (Rostami and Khamesipour, 2021). The M1-M2 dichotomy is not always applicable in VL pathogenesis, as IL-4 and IL-13 derived from M2 macrophages contribute to collagen production and granuloma formation, favoring the host (Stäger et al., 2003). As professional APCs, macrophages are crucial for *Leishmania* antigenic epitopes-mediated induction of T-cell immunity. However, this depends on the cholesterol distribution and lipid raft integrity in the macrophage membrane. During *L. donovani* infection, cholesterol depletion in the macrophage membrane disturbs membrane fluidity, affecting antigen presentation and T-cell activation (Chakraborty et al., 2005; Ghosh et al., 2014; Roy et al., 2016).

Decreased production of reactive oxygen and nitrogen species in monocytes of PKDL patients suggests alternative activation of macrophages (Mukhopadhyay et al., 2015). The IL-4 and IL-10 signaling in these macrophages induces the expression of certain genes, including mannose receptor (CD206), arginase 1, peroxisome proliferator-activated receptor (*PPAR*)-γ, and those found in the inflammatory zone (*Fizz*)-1 (Locati et al., 2020), skewing them towards an M2 phenotype. The CD68$^+$ macrophages highly expressing these gene products are recruited in PKDL patients' dermal lesions. Ultraviolet (UV)-light exposure induces lesion development in PKDL and breaks subcutaneous cholesterol deposits into vitamin D. Vitamin D receptor signaling in PKDL patients alternatively activates peripheral monocytes and dermal macrophages to an M2-phenotype, maintaining disease chronicity (Mukhopadhyay et al., 2015).

Dendritic cells: Like macrophages, dendritic cells are professional phagocytes and APCs, which can uptake the antigens via Fc receptors, C-type lectin receptors (CLRs), and TLRs (Geijtenbeek et al., 2004). *Leishmania* DC interaction occurs via TLR-2, TLR-4, and TLR-9 (Tibúrcio et al., 2019). TLR-9-mediated DC activation during *L. infantum* infection induces the secretion of neutrophil chemotactic factors and IL-12 (Sacramento et al., 2015). Upon encountering *Leishmania* infection, DCs mature and express, expressing major histocompatibility complex (MHC)-II, CD80, CD86, and CD40. Then they migrate to secondary lymphoid tissues to present antigens to naïve T-cells (Liu and Uzonna, 2012). DCs are the main source of IL-12 production, which can polarize the naïve T cells to the IFN-γ^+ Th1 subset (Gorak et al., 1998). Early in DC infection by *L. donovani*, cells produce IL-12, IL-23, and IL-27, with IL-12 primarily derived from the CD8α^+ DC subset (Maroof and Kaye, 2008). However, chronic *L. donovani* infection causes functional impairment of CD11c+ splenic DCs, reducing surface expression of MHC-II and IL-12 production (Basu et al., 2000). IL-12 expression in DCs can also be regulated by HIF-1α (Hammami et al., 2018), vascular cell adhesion molecule-1 (VCAM) and very late antigen (VLA)-4 interaction (Stanley et al., 2008), miR-21 (Varikuti et al., 2021a), and miR-155 (Varikuti et al., 2021b). HIF-1α expression in DCs impairs the CD4+Th1 (Hammami et al., 2018) and CD8+ T-cell (Hammami et al., 2015) expansion during *L. donovani* infection. In contrast, in macrophages, HIF-1α expression limits susceptibility to *L. donovani* infection, suggesting a host cell-specific HIF-1α role in glycolysis regulation (Mesquita et al., 2020). *L. infantum* secretory proteins can reduce IL-12p70 production from human DCs (Markikou-Ouni et al., 2015). *L. donovani* exosome-derived factors inhibit myeloid DC's ability to prime naïve CD4+ T-cells to the IFN-γ-producing Th1 subset (Silverman et al., 2010). Similarly, the bone marrow DCs-derived exosomes during *L. donovani* infection inhibit IL-12 production via miR-21 (Gannavaram et al., 2019). DCs can prime the Th17 response against *L. donovani* infection, and *Leishmania* elongation factor 2 and Nucleoside hydrolase NH36 antigen-primed DCs could serve as vaccine candidates by inducing host-protective CD4 and CD8 T-cell response (Agallou et al., 2018; Banerjee et al., 2018; Nico et al., 2018).

In the PKDL patients' peripheral blood, myeloid and plasmacytoid DC distribution is comparable to that of the healthy controls. However, at the PKDL lesion site, dermal DCs such as Langerhans cells and migratory dermal DCs (dDCs) show significant decreases in CD1a, MHC-II, and IL-12 expression (Ismail et al., 2006a; Mukherjee et al., 2015). LCs reduction at the lesion site differentiates PKDL from lepromatous leprosy, which shares other overlapping symptoms with PKDL. This reduction in LCs and dDCs is more prominent in polymorphic than macular lesions (Sengupta et al., 2019). The HLA-DR expression in DC decreases significantly in both forms of PKDL cases (Sengupta et al., 2019; Mukherjee et al., 2015). UVB exposure damages the epidermal LCs profoundly in polymorphic lesion PKDL patients (Clydesdale et al., 2001), decreasing B7-1 expression and causing defective antigen presentation and immunosuppression (Mukhopadhyay et al., 2014; Clydesdale et al., 2001).

Mast cells: Mast cells are important innate immune cells that play a key role in the CL and VL pathogenesis. As they reside mostly in the skin, their mechanism of action is well understood in CL. Although the literature on mast cell interaction with *L. donovani* is limited, it has been demonstrated that their role is specific to *Leishmania* (Naqvi et al., 2017) and host species (Saha et al., 2004). For instance, the mast cell's uptake of *L. tropica* and *L. donovani* differs. During this process, mast cells produce extracellular traps (MCETs), analogous to NETs of neutrophils. However, *L. donovani* is more resistant to MCET-based uptake by mast cells than *L. tropica*. Mast cell-induced inflammation (von Stebut et al., 2003) and Th1 polarization (Fu et al., 2019) may offer protection against *L. donovani* infection.

Natural killer and NKT cells: As a part of innate immunity, natural killer (NK) and NKT cells play a vital role in VL pathogenesis. NK cells are identified as $CD3^-CD56^+$ phenotype in humans, producing cytotoxicity and cytokine responses (Björkström et al., 2022). Three NK cell subsets ($CD56^-CD161^+$, $CD56^+CD161^-$, and $CD56^+CD161^+$) were found in VL patients, with $CD56^+CD161^+$ NK cells occurring in lower numbers in VL patients compared to healthy controls (Rai et al., 2018).

NK cell interaction with *Leishmania* metacyclic promastigotes via TLR-2 strongly induces the NF-kB translocation to the nucleus and IFN-γ and TNF-α secretion (Becker et al., 2003). TLR-9-based interaction regulates NK cell cytotoxicity and IFN-γ production (Liese et al., 2007). NK cells are crucial for establishing Th1 response against *L. donovani* amastigotes (Kirkpatrick and Farrell, 1982), but not against *L. major* and *L. tropica* infection (Satoskar et al., 1999).

Although NKT cells are involved in early immune responses, they are considered part of innate immunity due to their lack of immunological memory. Their functions bridge the two arms of immunity. NKT cells are generally CD1d$^+$-restricted T-cells with properties of both NK (CD161 marker) and T-cells. Studies on NKT cells in VL are inadequate, but their phenotype appears to determine whether they are protective or detrimental in VL. *Leishmania* interaction with CD1d$^+$ NKT cells via surface glycoconjugates offers protection against VL in the early stages (Amprey et al., 2004). CD8$^+$ NKT cells upregulate killer cell immunoglobulin-like receptors (KIRs) and secrete IFN-γ, benefiting the host without migrating to the infection site. In contrast, CD4$^+$ NKT cells migrate to the infection site, upregulating CD25 and forkhead box protein (FoxP)3 expression and secreting IL-10, which is pathogenic to the host (Kumari et al., 2015).

In conclusion, active human VL exhibits a dichotomy of innate immune cells regarding cytokine expression, including increased IL-4$^+$ neutrophils, NK cells, and IL-10$^+$ monocytes, alongside decreased IL-12$^+$ and IFN-γ $^+$ eosinophils and TNF-α $^+$ and IL-12$^+$ monocytes. Additionally, a mixed profile of innate immune cells is observed, featuring increased IFN-γ $^+$ neutrophils/eosinophils and NK cells, IL-12$^+$ eosinophils/monocytes, IL-4$^+$ neutrophils and NK cells, and IL-10$^+$ eosinophils/monocytes (Peruhype-Magalhães et al., 2005).

10.4 ADAPTIVE IMMUNE CELL RESPONSE TO *LEISHMANIA* INFECTION

B-lymphocytes: While B-cells are classically antibody-producing cells involved in humoral immune response (Figure 10.4), they also act as professional APCs, facilitate co-stimulatory signals, and are involved in cytokine production and immune regulation.

Different subsets of B-cells, such as naïve, immature, regulatory (Bregs), plasma, and memory B-cells, are classified based on their lineage and differentiation markers. B-cell-mediated humoral immunity is under the control of IL-10, a regulatory cytokine (Rousset et al., 1992). B-cell proliferation, differentiation, apoptosis, and homeostasis are regulated by molecules such as Blimp-1 (B-lymphocyte induced maturation protein-1), BCMA (B-cell maturation antigen), and TACI (transmembrane activator and calcium modulator and cyclophilin ligand interactor) (Rickert et al., 2011). The decisive role of B-cells in intracellular infections, including VL and PKDL pathogenesis, is not yet fully understood. In VL cases, B-cells exhibit various phenotypes with different functions. Throughout VL progression, effector B-cells become exhausted and resting memory B-cells are depleted. The exhausted B-cells, under the control of Blimp-1, show defective antigen presentation due to their compromised HLA-DR expression. Since the antibodies are major immune regulators produced by B-cells, they potentially influence VL pathogenesis by regulating plasma cell growth and differentiation with the help of IL-10, Blimp-1, BCMA, and TACI (Singh et al., 2019). Most data on B-cells' contribution to VL pathogenesis comes from animal models, with scarce information from human VL.

In active human VL, the percentage of B-cells gradually decreases, and they acquire CD21$^-$ CD27$^-$ phenotype markers, which are known for tissue-like memory (TLM) B-cells with functional impairment. TLM B-cells are compromised in antigen processing and presentation and induce the production of anti-inflammatory cytokines and VL-specific antibodies that increase disease severity through increased parasite load in the spleen. These B-cell outcomes in VL may have lethal consequences on T-cell responses, resulting in an overall imbalance in antileishmanial immunity. Antileishmanial treatment could restore the effector functions of B-cells. Therefore, a better understanding of their role in human VL pathogenesis may open a new avenue for immunotherapy (Singh et al., 2019).

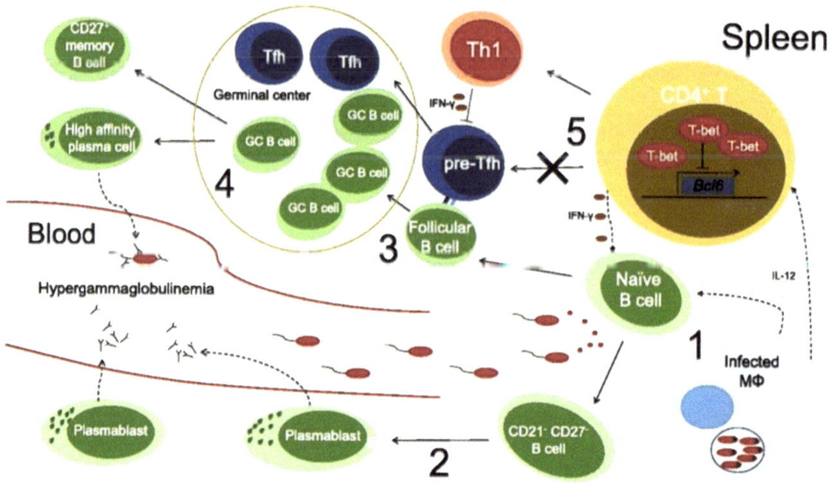

FIGURE 10.4 Impaired B-cell-mediated humoral immunity during VL. (1) Non-specific activation of B-cells in response to the soluble factors of *Leishmania* and inflammatory components produced during infection. (2) Expansion of unconventional B-cells with $CD21^-$ $CD27^-$ phenotype to produce plasmablast cells, which secretes abundant levels of antibodies, results in hypergammaglobulinemia. (3) Specific activation of B-cells through the B-cell receptor (BCR) leads to a follicular pathway interaction with pre-T follicular helper cells (Tfh) cells. (4) Productive interaction between B-cells and Tfh cells forms a germinal center (GC), causing affinity maturation and clonal selection of high-affinity antigen-specific B-cells. After exiting the GC, CD27+ phenotypic memory B-cells and plasma cells produce antileishmanial antibodies. (5) During chronic VL infection in the spleen, the inflammatory Th1 response mediated by the over-expression of T-bet inhibits the Bcl-6 transcription factor of Tfh cells, resulting in decreased Tfh cell numbers (Rodrigues et al., 2016)

T-lymphocytes: T-cells and their subpopulations are differentiated by the specific expression of cell surface markers such as CD3, CD4, and CD8, and transcription factors such as GATA-3, T-bet, retinoic-acid-receptor-related orphan nuclear receptor (ROR)γt, and FoxP3, among others. Mature T-cells can recognize the antigens presented by MHC molecules, which are expressed on the surface of APCs or infected host cells. The antigen's nature determines the interaction between the T-cell type and MHC molecule. For example, exogenous antigens are recognized by $CD4^+$ T-helper cells via MHC-II, while endogenous antigens are recognized by $CD8^+$ T-cytotoxic cells via MHC-I (João et al., 2006; Mingari and Moretta, 1982). This results in the activation of T-cells, which in turn activates B-cells and other immune cells to produce the effector and memory immune response (Rabb, 2002).

CD4+ T-cells: These comprise an abundant fraction of total T-cells that express the membrane glycoprotein CD4 molecule and function as helper T-cells. After recognizing the antigen presented by MHC-II, these naïve T-cells differentiate into either Th1 or Th2 subsets based on the microenvironment (Luckheeram et al., 2012). This dichotomy of $CD4^+$ T-cells in leishmaniasis, studied a few decades ago (Mosmann et al., 1986), is now completely understood and has expanded to include Th1, Th2, Th17, and regulatory T-cell (Treg) subsets based on their mechanism of action and cytokine profile (Dayakar et al., 2019). During host protection against VL, the $CD4^+$ T-cells produce IL-12, IFN-γ, and TNF-α (Jawed et al., 2018), though this is not always the case. During VL progression in hamsters, $CD4^+$ T-cells in the spleen show higher inhibitory receptor programmed cell death protein (PD)-1 expression and a mixed phenotype of Th1 and Th2 cytokine production. Inhibiting PD-1 results in decreased arginase-1 expression and splenic parasite load (Medina-Colorado et al., 2017). In mouse models of VL, the increase in IFN-γ^+ $CD4^+$ T-cells causes irreversible damage to the bone marrow functions (Pinto et al., 2017). This suggests that an increase in $CD4^+$ T-cell number alone does not ensure host protection against *L. donovani* infection; rather, their polarization and proinflammatory status determine the outcome. During *L. donovani* infection, increased $CD4^+$ Th2 and Treg subsets provide disease progression by inducing the secretion of immunosuppressive cytokines such as IL-10 and TGF-β. In contrast, increased $CD4^+$ Th1 and Th17 subsets offer protection via proinflammatory cytokine response.

In Indian PKDL patients, $CD4^+$ T-cell recruitment to the lesion site is almost absent (Mukherjee et al., 2019), whereas in self-healing African PKDL, the number of CD4+ T-cells far exceeds $CD8^+$ T-cells (Ismail et al., 2006b). The absence of $CD4^+$ T-cells leads to the alternative activation of macrophages via STAT-6 signaling, which favors the disease progression (Huber et al., 2010; Badley et al., 1997). These M2 macrophages induce the expression of CCL17 and CCL22 chemokines at the lesion site, interacting with the CCR4 receptor and resulting in Th2 cell skin-homing and production of IL-4, IL-13, and IL10 cytokines (Mukhopadhyay et al., 2015), promoting PKDL progression.

CD8+ T-cells: The cytotoxic cells express the CD8 surface marker, a dimeric glycoprotein. When naïve $CD8^+$ T-cells recognize antigens via MHC-1, they activate to clear the infection by releasing cytotoxic granules such as perforin and granzyme, secreting IFN-γ and TNF-α, or upregulating the Fas-FasL-mediated death of infected cells (Wherry et al., 2003). $CD8^+$ T-cells confer protection against acute and chronic *L. donovani* infection through IFN-γ, perforin, and granzyme secretion. They are also promising in vaccine-induced protection against VL in mouse models, where they upregulate levels of CXCL-10, a chemokine, against organ-resident parasites (Majumder et al., 2012). Upon the healing of VL, the elevated $CD8^+$ T-cells resist the *L. donovani* infection via granzyme B secretion (Kaushal et al., 2014). Long-lasting VL infection in humans drives $CD8^+$ T-cell exhaustion, hindering the ability to combat the parasite's dominance (Gautam et al., 2014).

In Indian PKDL patients, the increase in IL-10-producing $CD8^+$ T-cells is a key observation in immunopathogenesis (Ganguly et al., 2008). The dermal homing of $CD8^+$ T-cells is more common in polymorphic than macular forms (Sengupta et al., 2019; Mukherjee et al., 2019). These $CD8^+CCR^+$ T-cells cannot produce perforin, granzyme, and zap-70, and overexpress PD-1, IL-10, and IL-5, contributing to PKDL progression (Mukherjee et al., 2019). Antigen-specific memory cell responses are absent in PKDL patients' peripheral blood, indicating immune anergy characterized by a lack of T-cell propagation and IFN-γ secretion (Ganguly et al., 2008; Belkaid, 2007). A prominent decrease in the expression of CD86 on APCs (Linsley and Ledbetter, 1993) and CD28 on $CD8^+$ T-cells (Belkaid, 2007) leads to a decrease in the co-stimulation signal, resulting in peripheral immune anergy in PKDL, as shown in Figure 10.5 (Chatterjee et al., 2020). PD-1^{High} $CD127^{low}$ peripheral $CD8^+$ T-cells mimic the Tregs phenotype in PKDL cases (Mukherjee et al., 2019).

Th17 subset: These are a distinct lineage of $CD4^+$ T-cells, with the transcription factor RORγt responsible for their differentiation. The differentiation of the Th17 subset remains debatable, as previously it was thought that IL-1β, IL-6, and/or IL-23 promote and TGF-β1 inhibits this process (Crome et al., 2010, Acosta-Rodriguez et al., 2007). However, it was later shown that TGF-β1 plays an important role in IL-17 cytokine production. The Th17 lineage is primarily characterized by IL-17 secretion, showing pleiotropic effects (Crome et al., 2010). Th17 cells are stimulated by several cytokines and chemokines due to the presence of receptors on their surface (Sato et al., 2007), and produce various cytokines, such as IL-17A, IL-17F, IL-21, IL-22, TNF-α, IL-6, and IL-26 (Harrington et al., 2005). These cytokines exhibit both beneficial and detrimental effects on the host. An acute Th17 cytokine response can help in microbial clearance, but persistent IL-17 release can cause chronic inflammation and tissue damage, leading to autoimmunity (Ouyang et al., 2008). In VL, the Th17 subset shows a host protective immune response via IL-17 production and neutrophil recruitment (Gonçalves-de-Albuquerque et al., 2017). High levels of IL-17 in serum are associated with resistance to *L. donovani* infection and fast recovery from VL in human subjects (Tiwananthagorn et al., 2012). IL-17 produced against *Leishmania* antigen-primed dendritic cell vaccine offers protection against VL (Jawed et al., 2016). In contrast, IL-17A-mediated collapse of IFN-γ secretion and neutrophil recruitment increases susceptibility to *L. donovani* infection (Terrazas et al., 2016). Overall, the presence of IL-17 alone cannot determine disease outcome; the cross-talk between other Th17 cytokines is also essential in this context.

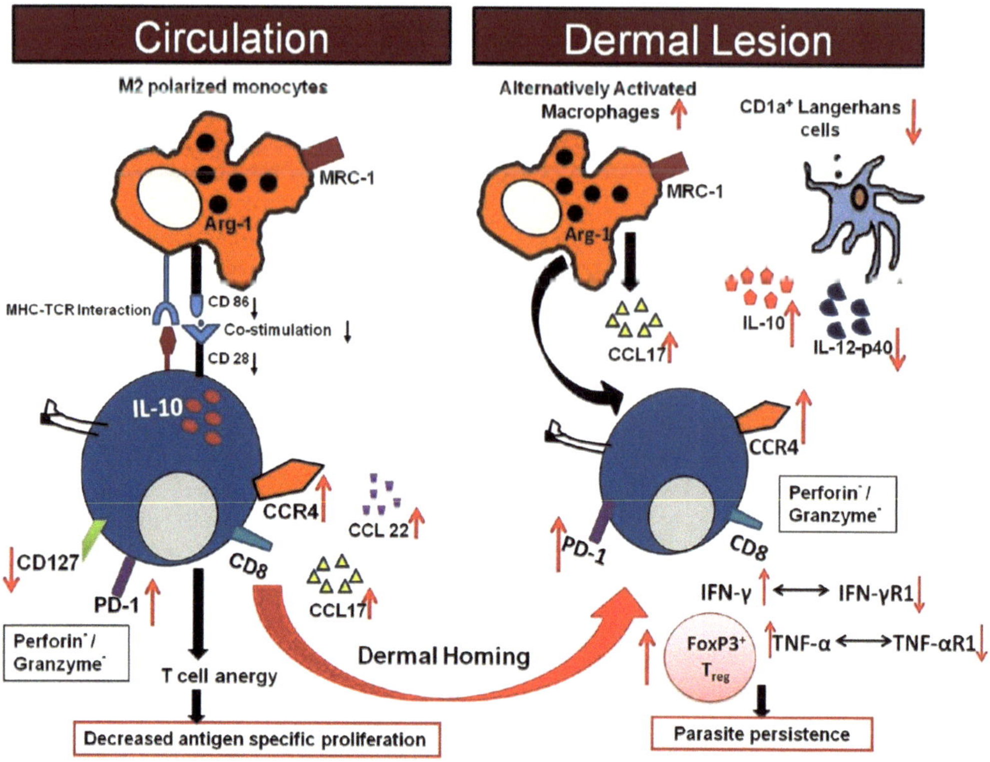

FIGURE 10.5 Pattern of immune cell response and their interaction between peripheral blood and skin lesions of PKDL patients (Chatterjee et al., 2020)

Regulatory T-cells: Tregs are also a subset of $C4^+$ T-cells with important immunoregulatory functions, constituting only 5–10% of the total $CD4^+$ T-cell population. Tregs are actively involved in controlling allergic reactions, autoimmunity, and maintaining self-tolerance. They are identified by specific cell surface markers, including CD25, cytotoxic T-lymphocyte associated antigen (CTLA)-4, CD127, and specific transcription factors like FoxP3 (Allos et al., 2019). Treg suppressor activity is attributed to IL-10 and TGF-β production (Allos et al., 2019; Togashi et al., 2019). In acute infection, Tregs protect the host from immune damage after parasite clearance, whereas in chronic infection, they benefit the parasite due to their prolonged immunosuppressive effects (Jonuleit et al., 2001). During human and experimental VL, Tregs are associated with severe pathogenesis due to their production of IL-10 and TGF-β (Rai et al., 2012). $IL\text{-}10^+$ Tregs in infected hosts further stimulate IL-10-secreting $CD4^+$ T-cells, causing immune suppression (Leveque et al., 2009). IL-2 levels in serum determine the balance between Th17 and Tregs population, positively correlating with Th17 and negatively correlating with Tregs (Karmakar et al., 2012). $CD4^+$ $Foxp3^+$ Tregs but not $CD8^+$ $Foxp3^+$ Tregs increase susceptibility to *L. donovani* infection through IL-10 secretion (Tiwananthagorn et al., 2012). This indicates that Treg phenotype and regulation by other cytokines are crucial in determining VL susceptibility or tolerance. The percentage of $CD4^+CD25^+$ Tregs in peripheral blood is comparable between healthy individuals and PKDL patients. However, FoxP3+ Tregs are abundant in PKDL lesions, with higher IFN-γ and IL-10 mRNA expression (Belkaid, 2007).

Natural killer T-cells: NKT cells are an unusual subset of immune cells, characterized by the absence of typical T-cell receptors and the presence of NK 1.1, NK 1.2, CD16, and CD56 markers on their surface. These T-cells act like $CD8^+$ T-cells when they encounter antigens, secreting huge amounts of cytokines. Their activation does not require MHC-based antigen presentation (Wu and

Van Kaer, 2009). Activated NKT cells produce a variety of immunoregulatory cytokines to combat parasite invasion. During leishmaniasis, NKT cells secrete IFN-γ, TNF-α, IL-23, and IL-17 cytokines, aiding in immune restoration and parasite elimination. NKT cell IL-17 production is independent of IL-6 action (Karmakar et al., 2012). Similar to Tregs, NKT cell phenotype determines *L. donovani* infection outcome, with $CD4^+CD56^+$ NKT cells promoting disease progression via TGF-β secretion and $CD8^+CD56^-$ NKT cells contributing to host protection (Kumari et al., 2018).

γ δ T-cells: These are unconventional T-cells with γ and δ chains in their T-cell receptor instead of the α and β chains found in conventional T-cells (Saha et al., 1999). This less abundant subset of T-cells does not require traditional antigen presentation for their activation (Lawand et al., 2017). They are activated by phospho-antigens or stress antigens and produce large amounts of TNF-α, IFN-γ, and granzyme (Vantourout and Hayday, 2013), resulting in inflammation, pathogen clearance, and tissue homeostasis (Bonneville et al., 2010). To create immunosuppressive conditions in the host, the *L. donovani* infection increases the γ δ T-cell population, which is associated with low HLA-DR expression (Saha et al., 1999). Elevated γ δ T-cells cause abnormal humoral immunity and hypergammaglobulinemia by inducing secretion of B-cell growth and differentiation factors associated with VL pathogenesis (Raziuddin et al., 1992).

10.5 DISCUSSION

The humoral immune response mediated by B-cells and antibodies in VL pathogenesis remains quite unclear. Active VL patients show elevated levels of IgE, IgM, and IgG antileishmanial antibodies (Ghosh et al., 1995; Anam et al., 1999; da Matta et al., 2000; Ryan et al., 2002), which correlate positively with disease pathology but not resolution (Galvao-Castro et al., 1984). While antileishmanial antibodies present in serum may predict disease occurrence, they cannot differentiate between current, past, or subclinical infections. Studies in endemic areas of Bihar show that only 3.5% out of the 33% of cases that are seropositive are transformed into active disease, which is similar to the percentage (2.6%) of seronegative cases converted into disease within two years. This suggests that the serological profile may not reflect disease conversion in the healthy group (Gidwani et al., 2009; Ostyn et al., 2011). B-cell deficient mice are resistant to *L. donovani* infection (Smelt et al., 2000). High antibody titers in VL induce IL-10 production, which promotes B-cell and plasma cell survival and proliferation, as well as IgG class-switching to IgG1 and IgG3 subtypes (Caldas et al., 2005). The persistence of antileishmanial antibodies long after cure suggests their indispensable role in host protection against VL (Gidwani et al., 2011). In a polymorphic group of PKDL patients, the humoral immune response is much stronger than in the macular variant, characterized by greater recruitment of $CD20^+$ B-cells at the lesions (Sengupta et al., 2019) and increased production of IgG1 and IgG3 subclasses (Singh et al., 2005). These levels normalize after treatment. Differences in the immunopathology of these two PKDL forms might be attributed to the variation in parasite load at lesions. Higher parasite load in the polymorphic form of PKDL is supported by higher IgG avidity (Singh et al., 2005; Mukhopadhyay et al., 2012).

CMI in leishmaniasis is routinely discussed in terms of the Th1-Th2 dichotomy, first described in CL in a murine model, where Th1 and Th2 cytokine responses are associated with host protection and disease progression, respectively (Kemp et al., 1997; Lohoff et al., 1989). This phenomenon is ambiguous in VL due to overlapping Th1 and Th2 responses. During active VL in humans, CMI is characterized by absence of T-cell proliferation and low or no production of IFN-γ and IL-2 in response to *Leishmania* antigen stimulation. Additionally, the abundant presence of Th2 and Treg cells and their respective cytokines IL-4, TGF-β, and IL-10 are associated with the disease progression through inhibition of M1 macrophage's antileishmanial activity and Th1 response (Murray et al., 2002; Wilson et al., 2002; Gomez et al., 2009).

Notably, IL-10 abundance is critical in determining the clinical progression of VL rather than the presence or absence of IFN-γ (Murphy et al., 2001). This is because IL-10 negatively regulates IFN-γ production and completely shuts down IFN-γ-mediated activation of macrophages (Ghalib et al., 1993; Bhattacharyya et al., 2001; Melby et al., 2001). In contrast, IL-12-producing DCs and macrophage-mediated Th1 stimulation aid in recovery from VL (Ghalib et al., 1995; Bacellar et al., 1996) via IFN-γ production, which in turn activates the macrophages to kill the intracellular parasites oxidatively through NO production (Engwerda et al., 1998; Bacellar et al., 2000). Similar findings were observed in a murine model of VL, where IL-10 levels determined disease establishment (Mesquita et al., 2018). *L. donovani* infection in the spleen induces constitutive expression of CCL2, which activates the IL4$^+$ Th2 subset, in turn mediating the alternative activation of macrophages. These M2 macrophages are a potential source of arginase 1 expression and polyamine biosynthesis, supporting parasite growth (Bhattacharya and Ali, 2013). The exhaustion of T-cells during chronic VL is characterized by elevated expression of immunosuppressive markers such as PD-1 and CTLA-4 and increased production of TGF-β (Gomes and Dosreis, 2001). The Th1-Th2 dichotomy alone is not sufficient to determine disease outcome; other T-cell subsets and their interplay also potentially influence this. For example, the disease progression in VL patients is linked to IL-27 levels in serum and spleen (Ansari et al., 2011), where CD4$^+$ Th17 cells are unable to produce IL-17 and IL-22 in the presence of IL-27 (Murugaiyan et al., 2009). IL-27 promotes the expansion of IL-10$^+$ T-cells, facilitating parasite dissemination.

The CMI in PKDL is transitional between Th1 and Th2 responses. As PKDL develops in VL patients after treatment, the systemic illness is mitigated in terms of the absence of fever and hepatosplenomegaly. However, parasites remain in the skin for a longer time, developing lesions (Gasim et al., 2000). The immune dissociation between the skin and viscera is the key rationale for the development of PKDL lesions after VL treatment (Zijlstra et al., 2003). The Th2-based IL-10 response continues in the skin of PKDL patients after a VL episode, whereas the systemic Th1 based IFN-γ response persists after VL therapy (Ismail et al., 2006; Ansari et al., 2006; Ismail et al., 1999). PKDL lesions develop in sun-exposed regions of the skin, such as the face and neck, where UV light damages resident DCs, resulting in a surge of Th2 response and immune devastation (Ismail et al., 2006a; Ganguly et al., 2008; Katara et al., 2011; Mukherjee et al., 2015). UV exposure induces 1α,25-dihydroxy vitamin D3 levels in the skin, which mediates the unusual differentiation of macrophages (Mukhopadhyay et al., 2015). IL-10$^+$ CD8$^+$ regulatory T-cell infiltration at the lesion site is increased in polymorphic PKDL patients (Ganguly et al., 2010a). Additionally, keratinocyte-produced TGF-β, TNF-α, IL-10, and IL-12 play key roles in PKDL pathogenesis. Though TNF-α levels are high, TNF-α receptor 1 expression is low in PKDL and VL patients (Ansari et al., 2008). The TNF-α to IL-10 ratio of 2.66 in PKDL patient skin biopsies is twice as high as the 1.18 ratio in VL patient bone marrow aspirates, suggesting its prognostic value (Ansari et al., 2008). Higher TNF-α levels positively correlate with matrix metalloproteinase (MMP)-9, and a high ratio of MMP-9 to TIMP1 (tissue inhibitor of matrix metalloproteinase 1) is associated with active PKDL (Ansari et al., 2008; Islam et al., 2013). A higher level of TIMP3 in PKDL compared to VL helps confine the parasites to skin lesions (Ansari et al., 2008). In contrast to VL, IL-17 upregulation in PKDL patients' lesions and blood is associated with pathogenesis and normalizes after chemotherapy. The differentiation of the Th17 subset into IL-17-producing cells is mediated by TGF-β, IL-23, IL-6, and IL-1 (Katara et al., 2012).

Though CMI varies among individuals over time, acute PKDL exhibits stronger CMI than chronic PKDL (Haldar et al., 1983). Likewise, the macular variant shows stronger CMI than the papulo-nodular variant of PKDL, as the former has fewer parasites and low antibody titers (except IgG1), whereas the latter, with low CMI, has a high parasite load, increased levels of IgG1 and IgG3, and abundant production of TGF-β, IL-10, and Tregs markers (Katara et al., 2011; Haldar et al., 1983; Saha et al., 2007; Mukhopadhyay et al., 2012).

There is no distinctive immunological evidence influencing the time required for the conversion of VL into PKDL. In African patients, this interval is shorter, while in Asian patients, it is about 1–3 years. Understanding the events occurring in the VL-to-PKDL transition in relation to immune reconstitution would help in controlling PKDL occurrence. VL and PKDL are more opportunistic in immunosuppressive HIV patients and organ transplant recipients with steroid therapy (van Griensven et al., 2014; El Hassan et al., 2013). About 95% of HIV/PKDL co-infection patients have more than 200 $CD4^+$ cells per mm^3 of blood (Zijlstra, 2014). HIV targets the $CD4^+$ DCs and macrophages directly, whereas VL targets them indirectly by promoting viral replication and Th2 response, characterized by low IL-12, IL-18, and IFN-γ, creating an environment favoring PKDL development (Wolday et al., 2000). In HIV co-infection, Th17 exhaustion occurs before that of the Th1 subset (Bixler and Mattapallil, 2013). Immune reconstitution during antiretroviral therapy and antileishmanial therapy may lead to PKDL development in HIV-VL co-infection cases.

10.6 CONCLUSION

The parasite persistence-driven inflammatory response in the host damages the architecture of lymphoid tissues. In this situation, certain proinflammatory molecules mediate self-immune tolerance via regulatory T-cells, acting as a negative feedback response. For instance, parasite-driven IFN-γ, along with IL-1β, acts on DCs and macrophages to produce IL-27, which blocks the production of the effector Th17 response and elevates the IL-10 mediated regulatory T-cell response. IL-10 impairs the antigen presentation ability of DCs and macrophages, rendering them insensitive to activation signals. Although the regulatory T-cell response restores tissue injury, its persistence further favors parasite dissemination. Consequently, the parasite-driven inflammation continues until the occurrence of CD4+ T-cell exhaustion and immune failure.

ACKNOWLEDGMENT

The authors are grateful to the ICMR-Rajendra Memorial Research Institute of Medical Sciences, Patna, for providing the necessary facilities for this research work.

REFERENCES

Acosta-Rodriguez, E. V., G. Napolitani, A. Lanzavecchia, and F. Sallusto. "Interleukins 1β and 6 but not transforming growth factor-β are essential for the differentiation of interleukin 17–producing human T helper cells." *Nature Immunology* 8, no. 9 (2007): 942–949.

Agallou, M., E. Pantazi, E. Tsiftsaki, D. K. Toubanaki, C. Gaitanaki, D. Smirlis, and E. Karagouni. "Induction of protective cellular immune responses against experimental visceral leishmaniasis mediated by dendritic cells pulsed with the N-terminal domain of Leishmania infantum elongation factor-2 and CpG oligodeoxynucleotides." *Molecular Immunology* 103 (2018): 7–20.

Ali, H. Z., C. R. Harding, and P. W. Denny. "Endocytosis and sphingolipid scavenging in Leishmania mexicana amastigotes." *Biochemistry Research International* 2012 (2012): 691363. doi: 10.1155/2012/691363. Epub 2011 Sep 21.

Allos, H., B. S. Al Dulaijan, J. Choi, and J. Azzi. "Regulatory T cells for more targeted immunosuppressive therapies." *Clinics in Laboratory Medicine* 39, no. 1 (2019): 1–13.

Amprey, J. L., J. S. Im, S. J. Turco, H. W. Murray, P. A. Illarionov, G. S. Besra, S. A. Porcelli, and G. F. Späth. "A subset of liver NK T cells is activated during *Leishmania donovani* infection by CD1d-bound lipophosphoglycan." *The Journal of Experimental Medicine* 200, no. 7 (2004): 895–904.

Anam, K., F. Afrin, D. Banerjee, N. Pramanik, S. K. Guha, R. P. Goswami, P. N. Gupta, S. K. Saha, and N. Ali. "Immunoglobulin subclass distribution and diagnostic value of *Leishmania donovani* antigen-specific immunoglobulin G3 in Indian kala-azar patients." *Clinical Diagnostic Laboratory Immunology* 6, no. 2 (1999): 231–235.

Andrade, B. de B., C. I. De Oliveira, C. Ida Brodskyn, A. Barral, and M. Barral-Netto. "Role of sand fly saliva in human and experimental leishmaniasis: current insights." *Scandinavian Journal of Immunology* 66, no. 2-3 (2007): 122–127.

Ansari, N. A., G. K. Katara, V. Ramesh, and P. Salotra. "Evidence for involvement of TNFR1 and TIMPs in pathogenesis of post-kala-azar dermal leishmaniasis." *Clinical & Experimental Immunology* 154, no. 3 (2008): 391–398.

Ansari, N. A., R. Kumar, S. Gautam, S. Nylén, O. P. Singh, S. Sundar, and D. Sacks. "IL-27 and IL-21 are associated with T cell IL-10 responses in human visceral leishmaniasis." *The Journal of Immunology* 186, no. 7 (2011): 3977–3985.

Ansari, N. A., S. Saluja, and P. Salotra. "Elevated levels of interferon-γ, interleukin-10, and interleukin-6 during active disease in Indian kala azar." *Clinical Immunology* 119, no. 3 (2006): 339–345.
Baardman, J., S. G. S. Verberk, K. H. M. Prange, M. van Weeghel, S. van der Velden, D. G. Ryan, R. C. I. Wüst, et al. "A defective pentose phosphate pathway reduces inflammatory macrophage responses during hypercholesterolemia." *Cell Reports* 25, no. 8 (2018): 2044–2052.
Bacellar, O., C. Brodskyn, J. Guerreiro, M. Barral-Netto, C. H. Costa, R. L. Coffman, W. D. Johnson, and E. M. Carvalho. "Interleukin-12 restores interferon-γ production and cytotoxic responses in visceral leishmaniasis." *Journal of Infectious Diseases* 173, no. 6 (1996): 1515–1518.
Bacellar, O., A. D'Oliveira Jr, S. Jerônimo, and E. M. Carvalho. "IL-10 and IL-12 are the main regulatory cytokines in visceral leishmaniasis." *Cytokine* 12, no. 8 (2000): 1228–1231.
Badley, A. D., D. Dockrell, M. Simpson, R. Schut, D. H. Lynch, P. Leibson, and C. V. Paya. "Macrophage-dependent apoptosis of CD4+ T lymphocytes from HIV-infected individuals is mediated by FasL and tumor necrosis factor." *The Journal of Experimental Medicine* 185, no. 1 (1997): 55–64.
Banerjee, A., P. Bhattacharya, P. K. Dagur, S. Karmakar, N. Ismail, A. B. Joshi, A. D. Akue, et al. "Live attenuated *Leishmania donovani* centrin gene–deleted parasites induce IL-23–dependent IL-17–protective immune response against visceral leishmaniasis in a murine model." *The Journal of Immunology* 200, no. 1 (2018): 163–176.
Basu, A., G. Chakrabarti, A. Saha, and S. Bandyopadhyay. "Modulation of CD11C+ splenic dendritic cell functions in murine visceral leishmaniasis: correlation with parasite replication in the spleen." *Immunology* 99, no. 2 (2000): 305–313.
Becker, I., N. Salaiza, M. Aguirre, J. Delgado, N. Carrillo-Carrasco, L. G. Kobeh, A. Ruiz, et al. "Leishmania lipophosphoglycan (LPG) activates NK cells through toll-like receptor-2." *Molecular and Biochemical Parasitology* 130, no. 2 (2003): 65–74.
Beil, W. J., G. Meinardus-Hager, D. -C. Neugebauer, and C. Sorg. "Differences in the onset of the inflammatory response to cutaneous leishmaniasis in resistant and susceptible mice." *Journal of Leukocyte Biology* 52, no. 2 (1992): 135–142.
Belkaid, Y. "Regulatory T cells and infection: a dangerous necessity." *Nature Reviews Immunology* 7, no. 11 (2007): 875–888.
Bhardwaj, S., N. Srivastava, R. Sudan, and B. Saha. "Leishmania interferes with host cell signaling to devise a survival strategy." *BioMed Research International* 2010 (2010): 109189. doi: 10.1155/2010/109189.
Bhattacharya, P., and N. Ali. "Involvement and interactions of different immune cells and their cytokines in human visceral leishmaniasis." *Revista da sociedade brasileira de medicina tropical* 46 (2013): 128–134.
Bhattacharya, P., R. Dey, A. Saxena, S. Karmakar, N. Ismail, S. Gannavaram, P. K. Dagur, et al. "Essential role of neutrophils in the protective immune response induced by a live attenuated leishmania vaccine." *The Journal of Immunology* 205, no. 12 (2020): 3333–3347.
Bhattacharyya, S., S. Ghosh, P. L. Jhonson, S. K. Bhattacharya, and S. Majumdar. "Immunomodulatory role of interleukin-10 in visceral leishmaniasis: defective activation of protein kinase C-mediated signal transduction events." *Infection and Immunity* 69, no. 3 (2001): 1499–1507.
Bixler, S. L., and J. J. Mattapallil. "Loss and dysregulation of Th17 cells during HIV infection." *Clinical and Developmental Immunology* 2013 (2013): 852418. doi: 10.1155/2010/109189.
Björkström, N. K., B. Strunz, and H. -G. Ljunggren. "Natural killer cells in antiviral immunity." *Nature Reviews Immunology* 22, no. 2 (2022): 112–123.
Blackwell, J. M. "Role of macrophage complement and lectin-like receptors in binding Leishmania parasites to host macrophages." *Immunology Letters* 11, no. 3–4 (1985): 227–232.
Bogdan, C. "Macrophages as host, effector and immunoregulatory cells in leishmaniasis: Impact of tissue micro-environment and metabolism." *Cytokine: X* 2, no. 4 (2020): 100041.
Bonneville, M., R. L. O'brien, and W. K. Born. "γ δ T cell effector functions: a blend of innate programming and acquired plasticity." *Nature Reviews Immunology* 10, no. 7 (2010): 467–478.
Borregaard, N., O. E. Sørensen, and K. Theilgaard-Mönch. "Neutrophil granules: a library of innate immunity proteins." *Trends in Immunology* 28, no. 8 (2007): 340–345.
Bronte, V., and P. Zanovello. "Regulation of immune responses by L-arginine metabolism." *Nature Reviews Immunology* 5, no. 8 (2005): 641–654.
Caldas, A., C. Favali, D. Aquino, V. Vinhas, J. Van Weyenbergh, C. Brodskyn, J. Costa, M. Barral-Netto, and A. Barral. "Balance of IL-10 and interferon-γ plasma levels in human visceral leishmaniasis: implications in the pathogenesis." *BMC Infectious Diseases* 5 (2005): 1–9.

Campos, T. M., S. T. Passos, F. O. Novais, D. P. Beiting, R. S. Costa, A. Queiroz, D. Mosser, P. Scott, E. M. Carvalho, and L. P. Carvalho. "Matrix metalloproteinase 9 production by monocytes is enhanced by TNF and participates in the pathology of human cutaneous Leishmaniasis." *PLoS Neglected Tropical Diseases* 8, no. 11 (2014): e3282.

Carlsen, E. D., Y. Liang, T. R. Shelite, D. H. Walker, P. C. Melby, and L. Soong. "Permissive and protective roles for neutrophils in leishmaniasis." *Clinical & Experimental Immunology* 182, no. 2 (2015): 109–118.

Chakraborty, D., S. Banerjee, A. Sen, K. K. Banerjee, P. Das, and S. Roy. "*Leishmania donovani* affects antigen presentation of macrophage by disrupting lipid rafts." *The Journal of Immunology* 175, no. 5 (2005): 3214–3224.

Chandra, D., and S. Naik. "*Leishmania donovani* infection down-regulates TLR2-stimulated IL-12p40 and activates IL-10 in cells of macrophage/monocytic lineage by modulating MAPK pathways through a contact-dependent mechanism." *Clinical & Experimental Immunology* 154, no. 2 (2008): 224–234.

Charmoy, M., R. Megnekou, C. Allenbach, C. Zweifel, C. Perez, K. Monnat, M. Breton, C. Ronet, P. Launois, and F. Tacchini-Cottier. "Leishmania major induces distinct neutrophil phenotypes in mice that are resistant or susceptible to infection." *Journal of Leucocyte Biology* 82, no. 2 (2007): 288–299.

Chatterjee, M., R. Sengupta, D. Mukhopadhyay, S. Mukherjee, A. Dighal, S. Moulik, and S. Sengupta. "Immune responses in post kala-azar dermal leishmaniasis." *Indian Journal of Dermatology* 65, no. 6 (2020): 452.

Clydesdale, G. J., G. W. Dandie, and H. Konrad Muller. "Ultraviolet light induced injury: immunological and inflammatory effects." *Immunology and Cell Biology* 79, no. 6 (2001): 547–568.

Crome, S. Q., A. Y. Wang, and M. K. Levings. "Translational mini-review series on Th17 cells: function and regulation of human T helper 17 cells in health and disease." *Clinical & Experimental Immunology* 159, no. 2 (2010): 109–119.

da Matta, V. L. R., S. Hoshino-Shimizu, R. Dietze, and C. E. P. Corbett. "Detection of specific antibody isotypes and subtypes before and after treatment of American visceral leishmaniasis." *Journal of Clinical Laboratory Analysis* 14, no. 1 (2000): 5–12.

Dayakar, A., S. Chandrasekaran, S. V. Kuchipudi, and S. K. Kalangi. "Cytokines: key determinants of resistance or disease progression in visceral leishmaniasis: opportunities for novel diagnostics and immunotherapy." *Frontiers in Immunology* 10 (2019): 670.

Domínguez, M., I. Moreno, C. Aizpurua, and A. Toraño. "Early mechanisms of Leishmania infection in human blood." *Microbes and Infection* 5, no. 6 (2003): 507–513.

Dos-Santos, A. L. A., L. F. Carvalho-Kelly, C. F. Dick, and J. R. Meyer-Fernandes. "Innate immunomodulation to trypanosomatid parasite infections." *Experimental Parasitology* 167 (2016): 67–75.

Ehrlich, A., T. M. Castilho, K. Goldsmith-Pestana, W. -J. Chae, A. L. M. Bothwell, T. Sparwasser, and D. McMahon-Pratt. "The immunotherapeutic role of regulatory T cells in Leishmania (Viannia) panamensis infection." *The Journal of Immunology* 193, no. 6 (2014): 2961–2970.

El Hassan, A. M., E. A. Gasim Khalil, W. M. Elamin, L. A. M. El Hassan, M. E. Ahmed, and A. M. Musa. "Misdiagnosis and mistreatment of post-kala-azar dermal leishmaniasis." *Case Reports in Medicine* 2013 (2013): 351579. doi: 10.1155/2013/351579. Epub 2013 Feb 21.

Engwerda, C. R., M. L. Murphy, S. E. J. Cotterell, S. C. Smelt, and P. M. Kaye. "Neutralization of IL-12 demonstrates the existence of discrete organ-specific phases in the control of *Leishmania donovani*." *European Journal of Immunology* 28, no. 2 (1998): 669–680.

Fromm, P. D., J. Kling, M. Mack, J. D. Sedgwick, and H. Koerner. "Loss of TNF signaling facilitates the development of a novel Ly-6Clow macrophage population permissive for Leishmania major infection." *The Journal of Immunology* 188, no. 12 (2012): 6258–6266.

Fu, Z., S. Akula, M. Thorpe, and L. Hellman. "Highly selective cleavage of TH2-promoting cytokines by the human and the mouse mast cell tryptases, indicating a potent negative feedback loop on TH2 immunity." *International Journal of Molecular Sciences* 20, no. 20 (2019): 5147.

Galvao-Castro, B., J. A. Sa Ferreira, K. F. Marzochi, M. C. Marzochi, S. G. Coutinho, and P. H. Lambert. "Polyclonal B cell activation, circulating immune complexes and autoimmunity in human American visceral leishmaniasis." *Clinical and Experimental Immunology* 56, no. 1 (1984): 58.

Ganeshan, K., and A. Chawla. "Metabolic regulation of immune responses." *Annual Review of Immunology* 32 (2014): 609–634.

Ganguly, S., N. K. Das, J. N. Barbhuiya, and M. Chatterjee. "Post-kala-azar dermal leishmaniasis–an overview." *International Journal of Dermatology* 49, no. 8 (2010b): 921–931.

Ganguly, S., N. K. Das, M. Panja, S. Pal, D. Modak, M. Rahaman, S. Mallik, et al. "Increased levels of interleukin-10 and IgG3 are hallmarks of Indian post-kala-azar dermal leishmaniasis." *Journal of Infectious Diseases* 197, no. 12 (2008): 1762–1771.

Ganguly, S., D. Mukhopadhyay, N. K. Das, M. Chaduvula, S. Sadhu, U. Chatterjee, M. Rahman, et al. "Enhanced lesional Foxp3 expression and peripheral anergic lymphocytes indicate a role for regulatory T cells in Indian post-kala-azar dermal leishmaniasis." *Journal of Investigative Dermatology* 130, no. 4 (2010a): 1013–1022.

Gannavaram, S., P. Bhattacharya, A. Siddiqui, N. Ismail, S. Madhavan, and H. L. Nakhasi. "miR-21 expression determines the early vaccine immunity induced by LdCen–/– immunization." *Frontiers in Immunology* 10 (2019): 2273.

Gasim, S., A. M. Elhassan, A. Kharazmi, E. A. G. Khalil, A. Ismail, and T. G. Theander. "The development of post-kala-azar dermal leishmaniasis (PKDL) is associated with acquisition of Leishmania reactivity by peripheral blood mononuclear cells (PBMC)." *Clinical & Experimental Immunology* 119, no. 3 (2000): 523–529.

Gaur, U., S. C. Roberts, R. P. Dalvi, I. Corraliza, B. Ullman, and M. E. Wilson. "An effect of parasite-encoded arginase on the outcome of murine cutaneous leishmaniasis." *The Journal of Immunology* 179, no. 12 (2007): 8446–8453.

Gautam, S., R. Kumar, N. Singh, A. K. Singh, M. Rai, D. Sacks, S. Sundar, and S. Nylén. "CD8 T cell exhaustion in human visceral leishmaniasis." *The Journal of Infectious Diseases* 209, no. 2 (2014): 290–299.

Geijtenbeek, T. B. H., S. J. van Vliet, A. Engering, B. A. 't Hart, and Y. van Kooyk. "Self-and nonself-recognition by C-type lectins on dendritic cells." *Annual Review of Immunology* 22 (2004): 33–54.

Geissmann, F., S. Jung, and D. R. Littman. "Blood monocytes consist of two principal subsets with distinct migratory properties." *Immunity* 19, no. 1 (2003): 71–82.

Ghalib, H. W., M. R. Piuvezam, Y. A. Skeiky, M. Siddig, F. A. Hashim, A. M. El-Hassan, D. M. Russo, and S. G. Reed. "Interleukin 10 production correlates with pathology in human *Leishmania donovani* infections." *The Journal of Clinical Investigation* 92, no. 1 (1993): 324–329.

Ghalib, H. W., J. A. Whittle, M. Kubin, F. A. Hashim, A. M. El-Hassan, K. H. Grabstein, G. Trinchieri, and S. G. Reed. "IL-12 enhances Th1-type responses in human *Leishmania donovani* infections." *Journal of Immunology (Baltimore, Md.: 1950)* 154, no. 9 (1995): 4623–4629.

Ghosh, A. K., S. Dasgupta, and A. C. Ghose. "Immunoglobulin G subclass-specific antileishmanial antibody responses in Indian kala-azar and post-kala-azar dermal leishmaniasis." *Clinical Diagnostic Laboratory Immunology* 2, no. 3 (1995): 291–296.

Ghosh, J., R. Guha, S. Das, and S. Roy. "Liposomal cholesterol delivery activates the macrophage innate immune arm to facilitate intracellular *Leishmania donovani* killing." *Infection and Immunity* 82, no. 2 (2014): 607–617.

Gidwani, K., A. Picado, B. Ostyn, S. Prakash Singh, R. Kumar, B. Khanal, V. Lejon, F. Chappuis, M. Boelaert, and S. Sundar. "Persistence of *Leishmania donovani* antibodies in past visceral leishmaniasis cases in India." *Clinical and Vaccine Immunology* 18, no. 2 (2011): 346–348.

Gidwani, K., M. Rai, J. Chakravarty, M. Boelaert, and S. Sundar. "Evaluation of leishmanin skin test in Indian visceral leishmaniasis." *The American Journal of Tropical Medicine and Hygiene* 80, no. 4 (2009): 566–567.

Giraud, E., T. Lestinova, T. Derrick, O. Martin, R. J. Dillon, P. Volf, I. Műller, P. A. Bates, and M. E. Rogers. "Leishmania proteophosphoglycans regurgitated from infected sand flies accelerate dermal wound repair and exacerbate leishmaniasis via insulin-like growth factor 1-dependent signalling." *PLoS Pathogens* 14, no. 1 (2018): e1006794.

Gomes, N. A., and G. A. DosReis. "The dual role of CTLA-4 in Leishmania infection." *Trends in Parasitology* 17, no. 10 (2001): 487–491.

Gomez, M. A., I. Contreras, M. Hallé, M. L. Tremblay, R. W. McMaster, and M. Olivier. "Leishmania GP63 alters host signaling through cleavage-activated protein tyrosine phosphatases." *Science Signaling* 2, no. 90 (2009): ra58–ra58.

Gonçalves-de-Albuquerque, S. da C., R. Pessoa-e-Silva, L. A. M. Trajano-Silva, T. C. de Goes, R. C. S. de Morais, C. N. da C. Oliveira, V. M. B. de Lorena, and M. de Paiva-Cavalcanti. "The equivocal role of Th17 cells and neutrophils on immunopathogenesis of leishmaniasis." *Frontiers in Immunology* 8 (2017): 1437.

Gorak, P. M. A., C. R. Engwerda, and P. M. Kaye. "Dendritic cells, but not macrophages, produce IL-12 immediately following *Leishmania donovani* infection." *European Journal of Immunology* 28, no. 2 (1998): 687–695.

Goundry, A., A. Romano, A. P. C. A. Lima, J. C. Mottram, and E. Myburgh. "Inhibitor of serine peptidase 2 enhances Leishmania major survival in the skin through control of monocytes and monocyte-derived cells." *The FASEB Journal* 32, no. 3 (2018): 1315.

Gregory, D. J., M. Godbout, I. Contreras, G. Forget, and M. Olivier. "A novel form of NF- κB is induced by Leishmania infection: Involvement in macrophage gene expression." *European Journal of Immunology* 38, no. 4 (2008): 1071–1081.

Guimarães-Costa, A. B., M. T. C. Nascimento, G. S. Froment, R. P. P. Soares, F. N. Morgado, F. Conceição-Silva, and E. M. Saraiva. "Leishmania amazonensis promastigotes induce and are killed by neutrophil extracellular traps." *Proceedings of the National Academy of Sciences* 106, no. 16 (2009): 6748–6753.

Guimarães-Costa, A. B., J. P. Shannon, I. Waclawiak, J. Oliveira, C. Meneses, W. de Castro, X. Wen, et al. "A sand fly salivary protein acts as a neutrophil chemoattractant." *Nature Communications* 12, no. 1 (2021): 3213.

Gupta, G., S. Oghumu, and A. R. Satoskar. "Mechanisms of immune evasion in leishmaniasis." *Advances in Applied Microbiology* 82 (2013): 155–184.

Haldar, J. P., S. Ghose, K. C. Saha, and A. C. Ghose. "Cell-mediated immune response in Indian kala-azar and post-kala-azar dermal leishmaniasis." *Infection and Immunity* 42, no. 2 (1983): 702–707.

Hammami, A., B. M. Abidin, T. Charpentier, A. Fabié, A. -P. Duguay, K. M. Heinonen, and S. Stäger. "HIF-1α is a key regulator in potentiating suppressor activity and limiting the microbicidal capacity of MDSC-like cells during visceral leishmaniasis." *PLoS Pathogens* 13, no. 9 (2017): e1006616.

Hammami, A., B. M. Abidin, K. M. Heinonen, and S. Stäger. "HIF-1α hampers dendritic cell function and Th1 generation during chronic visceral leishmaniasis." *Scientific Reports* 8, no. 1 (2018): 3500.

Hammami, A., T. Charpentier, M. Smans, and S. Stäger. "IRF-5-mediated inflammation limits CD8+ T cell expansion by inducing HIF-1α and impairing dendritic cell functions during Leishmania infection." *PLoS Pathogens* 11, no. 6 (2015): e1004938.

Hawn, T. R., A. Ozinsky, D. M. Underhill, F. S. Buckner, S. Akira, and A. Aderem. "Leishmania major activates IL-1α expression in macrophages through a MyD88-dependent pathway." *Microbes and Infection* 4, no. 8 (2002): 763–771.

Hermoso, T., Z. Fishelson, S. I. Becker, K. Hirschberg, and C. L. Jaffe. "Leishmanial protein kinases phosphorylate components of the complement system." *The EMBO Journal* 10, no. 13 (1991): 4061–4067.

Holm, Å., K. Tejle, K. -E. Magnusson, A. Descoteaux, and B. Rasmusson. "*Leishmania donovani* lipophosphoglycan causes periphagosomal actin accumulation: correlation with impaired translocation of PKCα and defective phagosome maturation." *Cellular Microbiology* 3, no. 7 (2001): 439–447.

Hoover, D. A. V. I. D. L., M. E. L. V. I. N. Berger, C. A. R. O. L. A. Nacy, W. A. Y. N. E. T. Hockmeyer, and M. S. Meltzer. "Killing of Leishmania tropica amastigotes by factors in normal human serum." *Journal of Immunology (Baltimore, Md.: 1950)* 132, no. 2 (1984): 893–897.

Huang, F. -P., D. Xu, E. -O. Esfandiari, W. Sands, X. -q. Wei, and F. Y. Liew. "Cutting edge: mice defective in Fas are highly susceptible to Leishmania major infection despite elevated IL-12 synthesis, strong Th1 responses, and enhanced nitric oxide production." *The Journal of Immunology* 160, no. 9 (1998): 4143–4147.

Huber, S., R. Hoffmann, F. Muskens, and D. Voehringer. "Alternatively activated macrophages inhibit T-cell proliferation by Stat6-dependent expression of PD-L2." *Blood, The Journal of the American Society of Hematology* 116, no. 17 (2010): 3311–3320.

Hurrell, B. P., I. B. Regli, and F. Tacchini-Cottier. "Different Leishmania species drive distinct neutrophil functions." *Trends in Parasitology* 32, no. 5 (2016): 392–401.

Huynh, C., and N. W. Andrews. "Iron acquisition within host cells and the pathogenicity of Leishmania." *Cellular Microbiology* 10, no. 2 (2008): 293–300.

Ingale, P., R. Kabra, and S. Singh. "Structural sequence evolution and computational modeling approaches of the complement system in leishmaniasis." *Advances in Protein Chemistry and Structural Biology* 120 (2020): 409–424.

Islam, S., E. Kenah, M. A. Alam Bhuiyan, K. M. Rahman, B. Goodhew, C. M. Ghalib, M. M. Zahid, et al. "Clinical and immunological aspects of post–kala-azar dermal leishmaniasis in Bangladesh." *The American Journal of Tropical Medicine and Hygiene* 89, no. 2 (2013): 345.

Ismail, A., A. M. El Hassan, K. Kemp, S. Gasim, A. El Gadir, M. Y. Kadaru, T. Møller, A. Kharazmi, and T. G. Theander. "Immunopathology of post kala-azar dermal leishmaniasis (PKDL): T-cell phenotypes and cytokine profile." *The Journal of Pathology* 189, no. 4 (1999): 615–622.

Ismail, A., A. F. A. Gadir, T. G. Theander, A. Kharazmi, and A. M. El Hassan. "Pathology of post-kala-azar dermal leishmaniasis: a light microscopical, immunohistochemical, and ultrastructural study of skin lesions and draining lymph nodes." *Journal of Cutaneous Pathology* 33, no. 12 (2006b): 778–787.

Ismail, A., E. A. G. Khalil, A. M. Musa, I. M. EL Hassan, M. E. Ibrahim, T. G. Theander, and A. M. EL Hassan. "The pathogenesis of post kala-azar dermal leishmaniasis from the field to the molecule: does ultraviolet light (UVB) radiation play a role?." *Medical Hypotheses* 66, no. 5 (2006): 993–999.

Isnard, A., M. T. Shio, and M. Olivier. "Impact of Leishmania metalloprotease GP63 on macrophage signaling." *Frontiers in Cellular and Infection Microbiology* 2 (2012): 72.

Italiani, P., and D. Boraschi. "From monocytes to M1/M2 macrophages: phenotypical vs. functional differentiation." *Frontiers in Immunology* 5 (2014): 514.

Jawed, J. J., S. Banerjee, S. Bandyopadhyay, S. Parveen, B. P. Chowdhury, P. Saini, and S. Majumdar. "Immunomodulatory effect of Arabinosylated lipoarabinomannan restrict the progression of visceral leishmaniasis through NOD2 inflammatory pathway: Functional regulation of T cell subsets." *Biomedicine & Pharmacotherapy* 106 (2018): 724–732.

Jawed, J. J., S. Majumder, S. Bandyopadhyay, S. Biswas, S. Parveen, and S. Majumdar. "SLA-PGN-primed dendritic cell-based vaccination induces Th17-mediated protective immunity against experimental visceral leishmaniasis: a crucial role of PKCÎ²." *FEMS Pathogens and Disease* 74, no. 5 (2016): ftw041.

João, C., B. M. Ogle, and S. Geyer. "Immunoglobulin promotes the diversity and the function of T cells." *European Journal of Immunology* 36, no. 7 (2006): 1718–1728.

Jonuleit, H., E. Schmitt, M. Stassen, A. Tuettenberg, J. Knop, and A. H. Enk. "Identification and functional characterization of human CD4+ CD25+ T cells with regulatory properties isolated from peripheral blood." *The Journal of Experimental Medicine* 193, no. 11 (2001): 1285–1294.

Kalyan, S., and D. Kabelitz. "When neutrophils meet T cells: beginnings of a tumultuous relationship with underappreciated potential." *European Journal of Immunology* 44, no. 3 (2014): 627–633.

Kane, M. M., and D. M. Mosser. "The role of IL-10 in promoting disease progression in leishmaniasis." *The Journal of Immunology* 166, no. 2 (2001): 1141–1147.

Karmakar, S., S. K. Bhaumik, J. Paul, and T. De. "TLR4 and NKT cell synergy in immunotherapy against visceral leishmaniasis." *PLoS Pathogens* 8, no. 4 (2012): e1002646.

Katara, G. K., N. A. Ansari, A. Singh, V. Ramesh, and P. Salotra. "Evidence for involvement of Th17 type responses in post kala azar dermal leishmaniasis (PKDL)." *PLoS Neglected Tropical Diseases* 6, no. 6 (2012): e1703.

Katara, G. K., N. A. Ansari, S. Verma, V. Ramesh, and P. Salotra. "Foxp3 and IL-10 expression correlates with parasite burden in lesional tissues of post kala azar dermal leishmaniasis (PKDL) patients." *PLoS Neglected Tropical Diseases* 5, no. 5 (2011): e1171.

Kaushal, H., R. Bras-Gonçalves, N. S. Negi, J. -L. Lemesre, G. Papierok, and P. Salotra. "Role of CD8+ T cells in protection against *Leishmania donovani* infection in healed Visceral Leishmaniasis individuals." *BMC Infectious Diseases* 14, no. 1 (2014): 1–7.

Kaye, P., and P. Scott. "Leishmaniasis: complexity at the host–pathogen interface." *Nature Reviews Microbiology* 9, no. 8 (2011): 604–615.

Kemp, K., L. Hviid, A. Kharazmi, and M. Kemp. "Interferon- γ production by human T cells and natural killer cells in vitro in response to antigens from the two intracellular pathogens Mycobacterium tuberculosis and leishmania major." *Scandinavian Journal of Immunology* 46, no. 5 (1997): 495–499.

Kharazmi, A., K. Kemp, A. Ismail, S. Gasim, A. Gaafar, J. A. L. Kurtzhals, A. M. El Hassan, T. G. Theander, and M. Kemp. "T-cell response in human leishmaniasis." *Immunology Letters* 65, no. 1–2 (1999): 105–108.

Kirkpatrick, C. A. R. L. E., and J. P. Farrell. "Leishmaniasis in beige mice." *Infection and Immunity* 38, no. 3 (1982): 1208–1216.

Kumar, A., S. Das, A. Mandal, S. Verma, K. Abhishek, A. Kumar, V. Kumar, A. K. Ghosh, and P. Das. "Retracted: Leishmania infection activates host mTOR for its survival by M2 macrophage polarization." *Parasite Immunology* 40, no. 11 (2018): e12586.

Kumar, R., and S. Nylén. "Immunobiology of visceral leishmaniasis." *Frontiers in Immunology* 3 (2012): 251.

Kumari, S., F. Jamal, P. Shivam, A. Thakur, M. Kumar, S. Bimal, V. N. R. Das, et al. "*Leishmania donovani* skews the CD56+ Natural Killer T cell response during human visceral leishmaniasis." *Cytokine* 73, no. 1 (2015): 53–60.

Kumari, S., P. Shivam, S. Kumar, F. Jamal, M. Kumar Singh, S. Bimal, S. Narayan, et al. "*Leishmania donovani* mediated higher expression of CCL4 induces differential accumulation of CD4+ CD56+ NKT and CD8+ CD56+ NKT cells at infection site." *Cytokine* 110 (2018): 306–315.

Kupani, M., R. K. Pandey, and S. Mehrotra. "Neutrophils and Visceral Leishmaniasis: Impact on innate immune response and cross-talks with macrophages and dendritic cells." *Journal of Cellular Physiology* 236, no. 4 (2021): 2255–2267.

Launois, P., F. Conceição-Silva, H. Himmerlich, C. Parra-López, F. Tacchini-Cottier, and J. A. Louis. "Setting in motion the immune mechanisms underlying genetically determined resistance and susceptibility to infection with Leishmania major." *Parasite Immunology* 20, no. 5 (1998): 223–230.

Laurenti, M. D., C. E. Pereira Corbett, M. N. Sotto, I. L. Sinhorini, and H. Goto. "The role of complement in the acute inflammatory process in the skin and in host–parasite interaction in hamsters inoculated with Leishmania (Leishmania) chagasi." *International Journal of Experimental Pathology* 77, no. 1 (1996): 15–24.

Lawand, M., J. Déchanet-Merville, and M. -C. Dieu-Nosjean. "Key features of gamma-delta T-cell subsets in human diseases and their immunotherapeutic implications." *Frontiers in Immunology* 8 (2017): 761.

Leveque, L., F. Deknuydt, G. Bioley, L. J. Old, J. Matsuzaki, K. Odunsi, M. Ayyoub, and D. Valmori. "Interleukin 2-mediated conversion of ovarian cancer-associated CD4+ regulatory T cells into pro-inflammatory interleukin 17-producing helper T cells." *Journal of Immunotherapy* 32, no. 2 (2009): 101–108.

Liese, J., U. Schleicher, and C. Bogdan. "TLR9 signaling is essential for the innate NK cell response in murine cutaneous leishmaniasis." *European Journal of Immunology* 37, no. 12 (2007): 3424–3434.

Linsley, P. S., and J. A. Ledbetter. "The role of the CD28 receptor during T cell responses to antigen." *Annual Review of Immunology* 11, no. 1 (1993): 191–212.

Liu, D., and J. E. Uzonna. "The early interaction of Leishmania with macrophages and dendritic cells and its influence on the host immune response." *Frontiers in Cellular and Infection Microbiology* 2 (2012): 83.

Locati, M., G. Curtale, and A. Mantovani. "Diversity, mechanisms, and significance of macrophage plasticity." *Annual Review of Pathology: Mechanisms of Disease* 15 (2020): 123–147.

Lohoff, M., F. Sommer, W. Solbach, and M. Röllinghoff. "Coexistence of antigen-specific TH1 and TH2 cells in genetically susceptible BALB/c mice infected with Leishmania major." *Immunobiology* 179, no. 4–5 (1989): 412–421.

Loría-Cervera, E. N., and F. Andrade-Narvaez. "The role of monocytes/macrophages in Leishmania infection: A glance at the human response." *Acta Tropica* 207 (2020): 105456.

Luckheeram, R. V., R. Zhou, A. D. Verma, and B. Xia. "CD4+ T cells: differentiation and functions." *Clinical and Developmental Immunology* 2012 (2012): 925135. doi: 10.1155/2012/925135. Epub 2012 Mar 14.

Majumder, S., S. Bhattacharjee, B. P. Chowdhury, and S. Majumdar. "CXCL10 is critical for the generation of protective CD8 T cell response induced by antigen pulsed CpG-ODN activated dendritic cells." *PLoS One* 7, no. 11 (2012): e48727.

Markikou-Ouni, W., S. Drini, N. Bahi-Jaber, M. Chenik, and A. Meddeb-Garnaoui. "Immunomodulatory effects of four Leishmania infantum potentially excreted/secreted proteins on human dendritic cells differentiation and maturation." *PLoS One* 10, no. 11 (2015): e0143063.

Maroof, A., and P. M. Kaye. "Temporal regulation of interleukin-12p70 (IL-12p70) and IL-12-related cytokines in splenic dendritic cell subsets during *Leishmania donovani* infection." *Infection and Immunity* 76, no. 1 (2008): 239–249.

McMahon-Pratt, D., and J. Alexander. "Does the Leishmania major paradigm of pathogenesis and protection hold for New World cutaneous leishmaniases or the visceral disease?" *Immunological Reviews* 201, no. 1 (2004): 206–224.

Medina-Colorado, A. A., E. Y. Osorio, O. A. Saldarriaga, B. L. Travi, F. Kong, H. Spratt, L. Soong, and P. C. Melby. "Splenic CD4+ T cells in progressive visceral leishmaniasis show a mixed effector-regulatory phenotype and impair macrophage effector function through inhibitory receptor expression." *PLoS One* 12, no. 1 (2017): e0169496.

Melby, P. C., B. Chandrasekar, W. Zhao, and J. E. Coe. "The hamster as a model of human visceral leishmaniasis: progressive disease and impaired generation of nitric oxide in the face of a prominent Th1-like cytokine response." *The Journal of Immunology* 166, no. 3 (2001): 1912–1920.

Mesquita, I., C. Ferreira, A. M. Barbosa, C. M. Ferreira, D. Moreira, A. Carvalho, C. Cunha, et al. "The impact of IL-10 dynamic modulation on host immune response against visceral leishmaniasis." *Cytokine* 112 (2018): 16–20.

Mesquita, I., C. Ferreira, D. Moreira, G. E. Gabriel Kluck, A. M. Barbosa, E. Torrado, R. J. Dinis-Oliveira, et al. "The absence of HIF-1α increases susceptibility to *Leishmania donovani* infection via activation of BNIP3/mTOR/SREBP-1c axis." *Cell Reports* 30, no. 12 (2020): 4052–4064.

Mingari, M. C., and L. Moretta. "Surface markers of human T lymphocytes." *Ricerca in clinica e in laboratorio* 12 (1982): 439–448.

Mollinedo, F., H. Janssen, J. de la Iglesia-Vicente, J. A. Villa-Pulgarin, and J. Calafat. "Selective fusion of azurophilic granules with Leishmania-containing phagosomes in human neutrophils." *Journal of Biological Chemistry* 285, no. 45 (2010): 34528–34536.

Moreira, D., V. Rodrigues, M. Abengozar, L. Rivas, E. Rial, M. Laforge, X. Li, et al. "Leishmania infantum modulates host macrophage mitochondrial metabolism by hijacking the SIRT1-AMPK axis." *PLoS Pathogens* 11, no. 3 (2015): e1004684.

Mosmann, T. R., Cherwinski, H., Bond, M. W., Giedlin, M. A., and Coffman, R. L. "Two types of murine helper T cell clone. I. Definition according to profiles of lymphokine activities and secreted proteins." *J Immunol.* 136, no. 7 (1 April 1986): 2348–2357.

Mukherjee, S., D. Mukhopadhyay, C. Braun, J. N. Barbhuiya, N. K. Das, U. Chatterjee, E. von Stebut, and M. Chatterjee. "Decreased presence of Langerhans cells is a critical determinant for Indian Post kala-azar dermal leishmaniasis." *Experimental Dermatology* 24, no. 3 (2015): 232–234.

Mukherjee, S., R. Sengupta, D. Mukhopadhyay, C. Braun, S. Mitra, S. Roy, N. Kanti Das, U. Chatterjee, E. von Stebut, and M. Chatterjee. "Impaired activation of lesional CD8+ T-cells is associated with enhanced expression of Programmed Death-1 in Indian Post Kala-azar Dermal Leishmaniasis." *Scientific Reports* 9, no. 1 (2019): 762.

Mukhopadhyay, D., J. E. Dalton, P. M. Kaye, and M. Chatterjee. "Post kala-azar dermal leishmaniasis: an unresolved mystery." *Trends in Parasitology* 30, no. 2 (2014): 65–74.

Mukhopadhyay, D., N. K. Das, S. De Sarkar, A. Manna, D. N. Ganguly, J. N. Barbhuiya, A. K. Maitra, A. Hazra, and M. Chatterjee. "Evaluation of serological markers to monitor the disease status of Indian post kala-azar dermal leishmaniasis." *Transactions of the Royal Society of Tropical Medicine and Hygiene* 106, no. 11 (2012): 668–676.

Mukhopadhyay, D., S. Mukherjee, S. Roy, J. E. Dalton, S. Kundu, A. Sarkar, N. K. Das, P. M. Kaye, and M. Chatterjee. "M2 polarization of monocytes-macrophages is a hallmark of Indian post kala-azar dermal leishmaniasis." *PLoS Neglected Tropical Diseases* 9, no. 10 (2015): e0004145.

Murphy, M. L., U. Wille, E. N. Villegas, C. A. Hunter, and J. P. Farrell. "IL-10 mediates susceptibility to *Leishmania donovani* infection." *European Journal of Immunology* 31, no. 10 (2001): 2848–2856.

Murray, H. W., C. M. Lu, S. Mauze, S. Freeman, A. L. Moreira, G. Kaplan, and R. L. Coffman. "Interleukin-10 (IL-10) in experimental visceral leishmaniasis and IL-10 receptor blockade as immunotherapy." *Infection and Immunity* 70, no. 11 (2002): 6284–6293.

Murugaiyan, G., A. Mittal, R. Lopez-Diego, L. M. Maier, D. E. Anderson, and H. L. Weiner. "IL-27 is a key regulator of IL-10 and IL-17 production by human CD4+ T cells." *The Journal of Immunology* 183, no. 4 (2009): 2435–2443.

Nandan, D., and N. E. Reiner. "Attenuation of gamma interferon-induced tyrosine phosphorylation in mononuclear phagocytes infected with *Leishmania donovani*: selective inhibition of signaling through Janus kinases and Stat1." *Infection and Immunity* 63, no. 11 (1995): 4495–4500.

Naqvi, N., K. Ahuja, A. Selvapandiyan, R. Dey, H. Nakhasi, and N. Puri. "Role of Mast Cells in clearance of Leishmania through extracellular trap formation." *Scientific Reports* 7, no. 1 (2017): 13240.

Nashed, B. F., Y. Maekawa, M. Takashima, T. Zhang, K. Ishii, T. Dainichi, H. Ishikawa, T. Sakai, H. Hisaeda, and K. Himeno. "Different cytokines are required for induction and maintenance of the Th2 type response in DBA/2 mice resistant to infection with Leishmania major." *Microbes and Infection* 2, no. 12 (2000): 1435–1443.

Nico, D., F. M. Almeida, J. M. Motta, F. S. dos Santos Cardoso, C. G. Freire-de-Lima, L. Freire-de-Lima, P. M. De Luca, A. M. Blanco Martinez, A. Morrot, and C. B. Palatnik-de-Sousa. "NH36 and F3 antigen-primed dendritic cells show preserved migrating capabilities and CCR7 expression and F3 is effective in immunotherapy of visceral leishmaniasis." *Frontiers in Immunology* 9 (2018): 967.

Nunes, A. C., F. R. Almeida-Campos, M. F. Horta, and F. J. Ramalho-Pinto. "Leishmania amazonensis promastigotes evade complement killing by interfering with the late steps of the cascade." *Parasitology* 115, no. 6 (1997): 601–609.

Nylén, S., and S. Gautam. "Immunological perspectives of leishmaniasis." *Journal of Global Infectious Diseases* 2, no. 2 (2010): 135.

Nylén, S., and D. Sacks. "Interleukin-10 and the pathogenesis of human visceral leishmaniasis." *Trends in immunology* 28, no. 9 (2007): 378–384.

Ohms, M., C. Ferreira, H. Busch, I. Wohlers, A. C. G. de Souza, R. Silvestre, and T. Laskay. "Enhanced glycolysis is required for antileishmanial functions of neutrophils upon infection with *Leishmania donovani*." *Frontiers in Immunology* 12 (2021): 632512.

Olivier, M., D. J. Gregory, and G. Forget. "Subversion mechanisms by which Leishmania parasites can escape the host immune response: a signaling point of view." *Clinical Microbiology Reviews* 18, no. 2 (2005): 293–305.

Orecchioni, M., Y. Ghosheh, A. B. Pramod, and K. Ley. "Macrophage polarization: different gene signatures in M1 (LPS+) vs. classically and M2 (LPS–) vs. alternatively activated macrophages." *Frontiers in Immunology* 10 (2019): 1084.

Osorio, E. Y., A. A. Medina-Colorado, B. L. Travi, and P. C. Melby. "In-situ proliferation contributes to the accumulation of myeloid cells in the spleen during progressive experimental visceral leishmaniasis." *PLoS One* 15, no. 11 (2020): e0242337.

Ostyn, B., K. Gidwani, B. Khanal, A. Picado, F. Chappuis, S. Prakash Singh, S. Rijal, S. Sundar, and M. Boelaert. "Incidence of symptomatic and asymptomatic *Leishmania donovani* infections in high-endemic foci in India and Nepal: a prospective study." *PLoS Neglected Tropical Diseases* 5, no. 10 (2011): e1284.

Ouyang, W., J. K. Kolls, and Y. Zheng. "The biological functions of T helper 17 cell effector cytokines in inflammation." *Immunity* 28, no. 4 (2008): 454–467.

Pace, D. "Leishmaniasis. vol. 20." *Journal of Information Security* (2014): 1–9.

Pacheco-Fernandez, T., G. Volpedo, C. Verma, and A. R. Satoskar. "Understanding the immune responses involved in mediating protection or immunopathology during leishmaniasis." *Biochemical Society Transactions* 49, no. 1 (2021): 297–311.

Paloque, L., T. Perez-Berezo, A. Abot, J. Dalloux-Chioccioli, S. Bourgeade-Delmas, P. Le Faouder, J. Pujo, et al. "Polyunsaturated fatty acid metabolites: biosynthesis in Leishmania and role in parasite/host interaction." *Journal of Lipid Research* 60, no. 3 (2019): 636–647.

Pereira Filho, A. A., A. A. de Sousa Nascimento, N. A. Araújo Saab, R. T. Fugiwara, G. C. D'Ávila Pessoa, L. B. Koerich, M. H. Pereira, R. N. Araújo, M. R. Viana Sant'Anna, and N. F. Gontijo. "Evasion of the complement system by Leishmania through the uptake of factor H, a complement regulatory protein." *Acta Tropica* 224 (2021): 106152.

Peruhype-Magalhães, V., O. A. Martins-Filho, A. Prata, L. De A. Silva, A. Rabello, A. Teixeira-Carvalho, R. M. Figueiredo, S. F. Guimarães-Carvalho, T. C. A. Ferrari, and R. Correa-Oliveira. "Immune response in human visceral leishmaniasis: analysis of the correlation between innate immunity cytokine profile and disease outcome." *Scandinavian Journal of Immunology* 62, no. 5 (2005): 487–495.

Peters, N. C., J. G. Egen, N. Secundino, A. Debrabant, N. Kimblin, S. Kamhawi, P. Lawyer, M. P. Fay, R. N. Germain, and D. Sacks. "In vivo imaging reveals an essential role for neutrophils in leishmaniasis transmitted by sand flies." *Science* 321, no. 5891 (2008): 970–974.

Peters, N. C., N. Kimblin, N. Secundino, S. Kamhawi, P. Lawyer, and D. L. Sacks. "Vector transmission of leishmania abrogates vaccine-induced protective immunity." *PLoS Pathogens* 5, no. 6 (2009): e1000484.

Pillay, J., I. Den Braber, N. Vrisekoop, L. M. Kwast, R. J. De Boer, J. A. M. Borghans, K. Tesselaar, and L. Koenderman. "In vivo labeling with $2H_2O$ reveals a human neutrophil lifespan of 5.4 days." *Blood, The Journal of the American Society of Hematology* 116, no. 4 (2010): 625–627.

Pinto, A. I., N. Brown, O. Preham, J. S. P. Doehl, H. Ashwin, and P. M. Kaye. "TNF signalling drives expansion of bone marrow CD4+ T cells responsible for HSC exhaustion in experimental visceral leishmaniasis." *PLoS Pathogens* 13, no. 7 (2017): e1006465.

Polando, R., U. G. Dixit, C. R. Carter, B. Jones, J. P. Whitcomb, W. Ballhorn, M. Harintho, C. L. Jerde, M. E. Wilson, and M. A. McDowell. "The roles of complement receptor 3 and Fcγ receptors during Leishmania phagosome maturation." *Journal of Leukocyte Biology* 93, no. 6 (2013): 921–932.

Prates, D. B., T. Araújo-Santos, N. Farias Luz, B. B. Andrade, J. França-Costa, L. Afonso, J. Clarêncio, et al. "Lutzomyia longipalpis saliva drives apoptosis and enhances parasite burden in neutrophils." *Journal of Leukocyte Biology* 90, no. 3 (2011): 575–582.

Rabb, H. "The T cell as a bridge between innate and adaptive immune systems: implications for the kidney." *Kidney International* 61, no. 6 (2002): 1935–1946.

Rai, A. K., C. Prasad Thakur, P. Kumar, S. Saini, A. K. Kureel, S. Kumari, T. Seth, and D. K. Mitra. "Decrease in the frequency of circulating CD56+ CD161+ NK cells in human visceral leishmaniasis." *Immunological Investigations* 47, no. 2 (2018): 125–134.

Rai, A. K., C. P. Thakur, A. Singh, T. Seth, S. K. Srivastava, P. Singh, and D. K. Mitra. "Regulatory T cells suppress T cell activation at the pathologic site of human visceral leishmaniasis." *PLoS One* 7, no. 2 (2012): e31551.

Raziuddin, S., A. W. Telmasani, M. El-Hag El-Awad, O. Al-Amari, and M. Al-Janadi. "γ δ T cells and the immune response in visceral leishmaniasis." *European Journal of Immunology* 22, no. 5 (1992): 1143–1148.

Reguera, R. M., R. Balaña-Fouce, M. Showalter, S. Hickerson, and S. M. Beverley. "Leishmania major lacking arginase (ARG) are auxotrophic for polyamines but retain infectivity to susceptible BALB/c mice." *Molecular and Biochemical Parasitology* 165, no. 1 (2009): 48–56.
Reiner, S. L., and R. M. Locksley. "The regulation of immunity to Leishmania major." *Annual Review of Immunology* 13, no. 1 (1995): 151–177.
Ribeiro-Gomes, F. L., A. C. Otero, N. A. Gomes, M. C. A. Moniz-de-Souza, L. Cysne-Finkelstein, A. C. Arnholdt, V. L. Calich, S. G. Coutinho, M. F. Lopes, and G. A. DosReis. "Macrophage interactions with neutrophils regulate Leishmania major infection." *The Journal of Immunology* 172, no. 7 (2004): 4454–4462.
Ribeiro-Gomes, F. L., and D. Sacks. "The influence of early neutrophil-Leishmania interactions on the host immune response to infection." *Frontiers in Cellular and Infection Microbiology* 2 (2012): 59.
Rickert, R. C., J. Jellusova, and A. V. Miletic. "Signaling by the tumor necrosis factor receptor superfamily in B-cell biology and disease." *Immunological Reviews* 244, no. 1 (2011): 115–133.
Rodrigues, V., A. Cordeiro-da-Silva, M. Laforge, R. Silvestre, and J. Estaquier. "Regulation of immunity during visceral Leishmania infection." *Parasites & Vectors* 9 (2016): 1–13.
Rossi, M., and N. Fasel. "How to master the host immune system? Leishmania parasites have the solutions!." *International Immunology* 30, no. 3 (2018): 103–111.
Rostami, M. N., and A. Khamesipour. "Potential biomarkers of immune protection in human leishmaniasis." *Medical Microbiology and Immunology* 210, no. 2–3 (2021): 81–100.
Rousset, F., E. Garcia, T. Defrance, C. Peronne, N. Vezzio, D.-H. Hsu, R. Kastelein, K. W. Moore, and J. Banchereau. "Interleukin 10 is a potent growth and differentiation factor for activated human B lymphocytes." *Proceedings of the National Academy of Sciences* 89, no. 5 (1992): 1890–1893.
Roy, K., S. Mandloi, S. Chakrabarti, and S. Roy. "Cholesterol corrects altered conformation of MHC-II protein in *Leishmania donovani* infected macrophages: implication in therapy." *PLoS Neglected Tropical Diseases* 10, no. 5 (2016): e0004710.
Roy, S., D. Mukhopadhyay, S. Mukherjee, S. Ghosh, S. Kumar, K. Sarkar, D. Pal, et al. "A defective oxidative burst and impaired antigen presentation are hallmarks of human visceral leishmaniasis." *Journal of Clinical Immunology* 35 (2015): 56–67.
Roy, S., D. Mukhopadhyay, S. Mukherjee, S. Moulik, S. Chatterji, N. Brahme, N. Pramanik, R. P. Goswami, B. Saha, and M. Chatterjee. "An IL-10 dominant polarization of monocytes is a feature of Indian Visceral Leishmaniasis." *Parasite Immunology* 40, no. 7 (2018): e12535.
Ryan, J. R., A. M. Smithyman, G.-H. Rajasekariah, L. Hochberg, J. M. Stiteler, and S. K. Martin. "Enzyme-linked immunosorbent assay based on soluble promastigote antigen detects immunoglobulin M (IgM) and IgG antibodies in sera from cases of visceral and cutaneous leishmaniasis." *Journal of Clinical Microbiology* 40, no. 3 (2002): 1037–1043.
Sacks, D. L., and P. V. Perkins. "Identification of an infective stage of Leishmania promastigotes." *Science* 223, no. 4643 (1984): 1417–1419.
Sacramento, L., S. C. Trevelin, M. S. Nascimento, D. S. Lima-Jùnior, D. L. Costa, R. P. Almeida, F. Q. Cunha, J. S. Silva, and V. Carregaro. "Toll-like receptor 9 signaling in dendritic cells regulates neutrophil recruitment to inflammatory foci following Leishmania infantum infection." *Infection and Immunity* 83, no. 12 (2015): 4604–4616.
Saha, A., M. Basu, and A. Ukil. "Recent advances in understanding *Leishmania donovani* infection: the importance of diverse host regulatory pathways." *IUBMB Life* 70, no. 7 (2018): 593–601.
Saha, B., A. M. D. J. Tonkal, S. Croft, and S. Roy. "Mast cells at the host–pathogen interface: host-protection versus immune evasion in leishmaniasis." *Clinical & Experimental Immunology* 137, no. 1 (2004): 19–23.
Saha, C., and S. Bandyopadhyay. "*Leishmania donovani* Parasites Interact with γ/δ+ Human Peripheral Blood T Cells and Induce Susceptibility to NK Cell-Mediated Lysis." *Scandinavian Journal of Immunology* 50, no. 6 (1999): 588–595.
Saha, S., S. Mondal, R. Ravindran, S. Bhowmick, D. Modak, S. Mallick, M. Rahman, et al. "IL-10-and TGF-β-mediated susceptibility in kala-azar and post-kala-azar dermal leishmaniasis: the significance of amphotericin B in the control of *Leishmania donovani* infection in India." *The Journal of Immunology* 179, no. 8 (2007): 5592–5603.
Sato, W., T, Aranami and T. Yamamura. "Cutting edge: human Th17 cells are identified as bearing CCR2+ CCR5− phenotype." *The Journal of Immunology* 178, no. 12 (2007): 7525–7529.
Satoskar, A. R., L. M. Stamm, X. Zhang, A. A. Satoskar, M. Okano, C. Terhorst, J. R. David, and B. Wang. "Mice lacking NK cells develop an efficient Th1 response and control cutaneous Leishmania major infection." *The Journal of Immunology* 162, no. 11 (1999): 6747–6754.

Saunders, E. C., and M. J. McConville. "Immunometabolism of Leishmania granulomas." *Immunology and Cell Biology* 98, no. 10 (2020): 832–844.

Scianimanico, S., M. Desrosiers, J.-F. Dermine, S. Méresse, A. Descoteaux, and M. Desjardins. "Impaired recruitment of the small GTPase rab7 correlates with the inhibition of phagosome maturation by *Leishmania donovani* promastigotes." *Cellular Microbiology* 1, no. 1 (1999): 19–32.

Sengupta, R., S. Mukherjee, S. Moulik, S. Mitra, S. J. Chaudhuri, N. Kanti Das, U. Chatterjee, and M. Chatterjee. "In-situ immune profile of polymorphic vs. macular Indian post kala-azar dermal leishmaniasis." *International Journal for Parasitology: Drugs and Drug Resistance* 11 (2019): 166–176.

Serbina, N. V., and E. G. Pamer. "Monocyte emigration from bone marrow during bacterial infection requires signals mediated by chemokine receptor CCR2." *Nature Immunology* 7, no. 3 (2006): 311–317.

Sharma, S., R. E. Davis, S. Srivastva, S. Nylén, S. Sundar, and M. E. Wilson. "A subset of neutrophils expressing markers of antigen-presenting cells in human visceral leishmaniasis." *The Journal of Infectious Diseases* 214, no. 10 (2016): 1531–1538.

Silverman, J. M., J. Clos, E. Horakova, A. Y. Wang, M. Wiesgigl, I. Kelly, M. A. Lynn, et al. "Leishmania exosomes modulate innate and adaptive immune responses through effects on monocytes and dendritic cells." *The Journal of Immunology* 185, no. 9 (2010): 5011–5022.

Singh, B., O. Prakash Singh, N. Singh, S. Sankar Singh, and S. Sundar. "Abnormal B-Cell Subset and blimp-1–mediated humoral responses associated with visceral Leishmaniasis pathogenesis." *The American Journal of Tropical Medicine and Hygiene* 100, no. 4 (2019): 816.

Singh, N., R. Kumar, S. B. Chauhan, C. Engwerda, and S. Sundar. "Peripheral blood monocytes with an antiinflammatory phenotype display limited phagocytosis and oxidative burst in patients with visceral leishmaniasis." *The Journal of Infectious Diseases* 218, no. 7 (2018): 1130–1141.

Singh, R., B. V. Subba Raju, R. K. Jain, and P. Salotra. "Potential of direct agglutination test based on promastigote and amastigote antigens for serodiagnosis of post-kala-azar dermal leishmaniasis." *Clinical and Vaccine Immunology* 12, no. 10 (2005): 1191–1194.

Smelt, S. C., S. E. J. Cotterell, C. R. Engwerda, and P. M. Kaye. "B cell-deficient mice are highly resistant to *Leishmania donovani* infection, but develop neutrophil-mediated tissue pathology." *The Journal of Immunology* 164, no. 7 (2000): 3681–3688.

Stäger, S., J. Alexander, K. Christine Carter, F. Brombacher, and P. M. Kaye. "Both interleukin-4 (IL-4) and IL-4 receptor α signaling contribute to the development of hepatic granulomas with optimal antileishmanial activity." *Infection and Immunity* 71, no. 8 (2003): 4804–4807.

Stanley, A. C., J. E. Dalton, S. H. Rossotti, K. P. MacDonald, Y. Zhou, F. Rivera, W. A. Schroder, et al. "VCAM-1 and VLA-4 modulate dendritic cell IL-12p40 production in experimental visceral leishmaniasis." *PLoS Pathogens* 4, no. 9 (2008): e1000158.

Steinbrink, K., F. SchÖnlau, U. Rescher, U. Henseleit, T. Vogel, C. Sorg, and C. Sunderötter. "Ineffective elimination of Leishmania major by inflammatory (MRP 14-positive) subtype of monocytic cells." *Immunobiology* 202, no. 5 (2000): 442–459.

Sutterwala, F. S., G. J. Noel, P. Salgame, and D. M. Mosser. "Reversal of proinflammatory responses by ligating the macrophage Fcγ receptor type I." *The Journal of Experimental Medicine* 188, no. 1 (1998): 217–222.

Teixeira, C. R., C. da S. Santos, D. B. Prates, R. T. Dos Santos, T. Araújo-Santos, S. M. de Souza-Neto, V. M. Borges, M. Barral-Netto, and C. I. Brodskyn. "Lutzomyia longipalpis saliva drives interleukin-17-induced neutrophil recruitment favoring Leishmania infantum infection." *Frontiers in Microbiology* 9 (2018): 881.

Teixeira, M. J., C. R. Teixeira, B. B. Andrade, M. Barral-Netto, and A. Barral. "Chemokines in host–parasite interactions in leishmaniasis." *Trends in Parasitology* 22, no. 1 (2006): 32–40.

Terrazas, C., S. Varikuti, J. Kimble, E. Moretti, P. N. Boyaka, and A. R. Satoskar. "IL-17A promotes susceptibility during experimental visceral leishmaniasis caused by *Leishmania donovani*." *The FASEB Journal* 30, no. 3 (2016): 1135.

Terrazas, C., S. Varikuti, S. Oghumu, H. M. Steinkamp, N. Ardic, J. Kimble, H. Nakhasi, and A. R. Satoskar. "Ly6Chi inflammatory monocytes promote susceptibility to *Leishmania donovani* infection." *Scientific Reports* 7, no. 1 (2017): 14693.

Tibúrcio, R., S. Nunes, I. Nunes, M. R. Ampuero, I. Bonyek Silva, R. Lima, N. M. Tavares, and C. Brodskyn. "Molecular aspects of dendritic cell activation in leishmaniasis: an immunobiological view." *Frontiers in Immunology* 10 (2019): 227.

Titos, E., B. Rius, A. González-Périz, C. López-Vicario, E. Morán-Salvador, M. Martínez-Clemente, V. Arroyo, and J. Claria. "Resolvin D1 and its precursor docosahexaenoic acid promote resolution of adipose tissue inflammation by eliciting macrophage polarization toward an M2 like phenotype." *The Journal of Immunology* 187, no. 10 (2011): 5408–5418.

Tiwananthagorn, S., K. Iwabuchi, M. Ato, T. Sakurai, H. Kato, and K. Katakura. "Involvement of CD4+ Foxp3+ regulatory T cells in persistence of *Leishmania donovani* in the liver of alymphoplastic aly/aly mice." *PLoS Negl Trop Dis* 6, no. 8 (2012): e1798. doi: 10.1371/journal.pntd.0001798. Epub 2012 Aug 21

Togashi, Y., K. Shitara, and H. Nishikawa. "Regulatory T cells in cancer immunosuppression—implications for anticancer therapy." *Nature Reviews Clinical Oncology* 16, no. 6 (2019): 356–371.

Tuon, F. F., V. S. Amato, H. A. Bacha, T. AlMusawi, M. I. Duarte, and V. Amato Neto. "Toll-like receptors and leishmaniasis." *Infection and Immunity* 76, no. 3 (2008): 866–872.

Ueno, N., C. L. Bratt, N. E. Rodriguez, and M. E. Wilson. "Differences in human macrophage receptor usage, lysosomal fusion kinetics and survival between logarithmic and metacyclic Leishmania infantum chagasi promastigotes." *Cellular Microbiology* 11, no. 12 (2009): 1827–1841.

Van den Bossche, J., L. A. O'Neill, and D. Menon. "Macrophage immunometabolism: where are we (going)?" *Trends in Immunology* 38, no. 6 (2017): 395–406.

Van Griensven, J., E. Carrillo, R. López-Vélez, L. Lynen, and J. Moreno. "Leishmaniasis in immunosuppressed individuals." *Clinical Microbiology and Infection* 20, no. 4 (2014): 286–299.

Van Zandbergen, G., N. Hermann, H. Laufs, W. Solbach, and T. Laskay. "Leishmania promastigotes release a granulocyte chemotactic factor and induce interleukin-8 release but inhibit gamma interferon-inducible protein 10 production by neutrophil granulocytes." *Infection and Immunity* 70, no. 8 (2002): 4177–4184.

Van Zandbergen, G., M. Klinger, A. Mueller, S. Dannenberg, A. Gebert, W. Solbach, and T. Laskay. "Cutting edge: neutrophil granulocyte serves as a vector for Leishmania entry into macrophages." *The Journal of Immunology* 173, no. 11 (2004): 6521–6525.

Vantourout, P., and A. Hayday. "Six-of-the-best: unique contributions of γ δ T cells to immunology." *Nature Reviews Immunology* 13, no. 2 (2013): 88–100.

Varikuti, S., C. Verma, E. Holcomb, B. K. Jha, A. Viana, R. Maryala, F. Lamenza, et al. "MicroRNA-21 deficiency promotes the early Th1 immune response and resistance toward visceral leishmaniasis." *The Journal of Immunology* 207, no. 5 (2021a): 1322–1332.

Varikuti, S., C. Verma, G. Natarajan, S. Oghumu, and A. R. Satoskar. "MicroRNA155 plays a critical role in the pathogenesis of cutaneous leishmania major infection by promoting a Th2 response and attenuating dendritic cell activity." *The American Journal of Pathology* 191, no. 5 (2021b): 809–816.

Verma, J. K., R. Rastogi, and A. Mukhopadhyay. "*Leishmania donovani* resides in modified early endosomes by upregulating Rab5a expression via the downregulation of miR-494." *PLoS Pathogens* 13, no. 6 (2017): e1006459.

Verma, S., A. Mandal, M. Y. Ansari, A. Kumar, K. Abhishek, A. K. Ghosh, A. Kumar, V. Kumar, S. Das, and P. Das. "*Leishmania donovani* inhibitor of serine peptidases 2 mediated inhibition of lectin pathway and upregulation of C5aR signaling promote parasite survival inside host." *Frontiers in Immunology* 9 (2018): 63.

Viana, A. G., L. M. Dias Magalhães, R. C. Giunchetti, W. O. Dutra, and K. J. Gollob. "Infection of human monocytes with Leishmania infantum strains induces a downmodulated response when compared with infection with Leishmania braziliensis." *Frontiers in Immunology* 8 (2018): 1896.

Vinet, A. F., M. Fukuda, S. J. Turco, and A. Descoteaux. "The *Leishmania donovani* lipophosphoglycan excludes the vesicular proton-ATPase from phagosomes by impairing the recruitment of synaptotagmin V." *PLoS Pathogens* 5, no. 10 (2009): e1000628.

von Stebut, E., M. Metz, G. Milon, J. Knop, and M. Maurer. "Early macrophage influx to sites of cutaneous granuloma formation is dependent on MIP-1α/β released from neutrophils recruited by mast cell–derived TNFα." *Blood, The Journal of the American Society of Hematology* 101, no. 1 (2003): 210–215.

Vono, M., A. Lin, A. Norrby-Teglund, R. A. Koup, F. Liang, and K. Loré. "Neutrophils acquire the capacity for antigen presentation to memory CD4+ T cells in vitro and ex vivo." *Blood, The Journal of the American Society of Hematology* 129, no. 14 (2017): 1991–2001.

Wanasen, N., C. L. MacLeod, L. G. Ellies, and L. Soong. "L-arginine and cationic amino acid transporter 2B regulate growth and survival of Leishmania amazonensis amastigotes in macrophages." *Infection and Immunity* 75, no. 6 (2007): 2802–2810.

Wei, F., W. Gong, J. Wang, Y. Yang, J. Liu, Y. Wang, and J. Cao. "Role of the lipoxin A4 receptor in the development of neutrophil extracellular traps in Leishmania infantum infection." *Parasites & Vectors* 12, no. 1 (2019): 1–10.

Wherry, E. J., V. Teichgräber, T. C. Becker, D. Masopust, S. M. Kaech, R. Antia, U. H. Von Andrian, and R. Ahmed. "Lineage relationship and protective immunity of memory CD8 T cell subsets." *Nature Immunology* 4, no. 3 (2003): 225–234.

Wilson, M. E., T. J. Recker, N. E. Rodriguez, B. M. Young, K. K. Burnell, J. A. Streit, and J. N. Kline. "The TGF- β response to Leishmania chagasi in the absence of IL-12." *European Journal of Immunology* 32, no. 12 (2002): 3556–3565.

Wolday, D., N. Berhe, S. Britton, and H. Akuffo. "HIV-1 alters T helper cytokines, interleukin-12 and interleukin-18 responses to the protozoan parasite *Leishmania donovani*." *AIDS* 14, no. 8 (2000): 921–929.

Wu, L., and L. V. Kaer. "Natural killer T cells and autoimmune disease." *Current Molecular Medicine* 9, no. 1 (2009): 4–14.

Yang, C. -W., and E. R. Unanue. "Neutrophils control the magnitude and spread of the immune response in a thromboxane A2-mediated process." *Journal of Experimental Medicine* 210, no. 2 (2013): 375–387.

Ziegler-Heitbrock, L., P. Ancuta, S. Crowe, M. Dalod, V. Grau, D. N. Hart, P. J. M. Leenen, et al. "Nomenclature of monocytes and dendritic cells in blood." *Blood, The Journal of the American Society of Hematology* 116, no. 16 (2010): e74–e80.

Zijlstra, E. E. "PKDL and other dermal lesions in HIV co-infected patients with leishmaniasis: review of clinical presentation in relation to immune responses." *PLoS Neglected Tropical Diseases* 8, no. 11 (2014): e3258.

Zijlstra, E. E., A. M. Musa, E. A. G. Khalil, I. M. El Hassan, and A. M. El-Hassan. "Post-kala-azar dermal leishmaniasis." *The Lancet Infectious Diseases* 3, no. 2 (2003): 87–98.

11 A Recent Update on Visceral Leishmaniasis and Post-kala-azar Dermal Leishmaniasis

Epidemiology, Diagnosis, and Treatments

*Saurav Kumar Mishra and John J. Georrge**
Department of Bioinformatics, University of North Bengal, Raja Rammohunpur, West Bengal, India
*Corresponding Author: John J Georrge, Department of Bioinformatics, University of North Bengal, Raja Rammohunpur, West Bengal, India

11.1 INTRODUCTION

Leishmaniasis is a vector-borne disease caused by protozoa parasites from over 20 *Leishmania* species that affect populations worldwide. The transmission of *Leishmania* parasites to humans occurs through the bite of infected female sandflies. Currently,, approximately 90 sandfly species are known to transmit *Leishmania* parasites to their respective hosts, and the clinical manifestations depend on the causative *Leishmania* species. There are three primary forms of leishmaniasis: visceral leishmaniasis (VL), also referred to as kala-azar; cutaneous leishmaniasis (CL); and mucocutaneous leishmaniasis, as shown in Figure 11.1 (Ready 2013; Mahto et al. 2023; Alvar et al. 2006). According to the WHO, 90% of new VL cases were reported among various forms of leishmaniasis in different geographical locations, especially in South Sudan, Sudan, Kenya, Ethiopia, Eritrea, India, Brazil, China, Somalia, and Yemen (www.who.int/news-room/fact-sheets/detail/leishmaniasis). VL is the most severe form of the disease and is transmitted by two species: *Labrus donovani* and *L. infantum* (*L. chagasi infantum*). While symptoms vary depending on the geographical area, VL commonly affects internal organs such as the spleen, liver, and bone marrow, and is transmitted disease occurs by the bite of female sandflies (Diptera, Psychodidae) (Lari et al. 2022; Chappuis et al. 2007; Ready 2013; Mann et al. 2021; Safavi et al. 2021). Early detection and treatment are crucial for cure; otherwise, VL can lead to death. Due to its clinical complications, the mortality rate of VL is higher than that of other forms of leishmaniasis. These species and their transmission are mainly distributed throughout tropical and subtropical areas and can progress easily from normal to severe. While VL is transmitted by *L. donovani* and *L. infantum* (*L. chagasi infantum*), there are differences in transmission patterns: *L. donovani* infects all age groups, whereas *L. infantum* mainly affects children under five years old (Jancar et al. 2023; Cavalcante et al. 2022; Alvar et al. 2006). Due to the diverse and severe nature of species transmission, VL has become a growing health concern worldwide. To combat this disease, rapid case detection, treatment, and local vector control are the main aspects that can be utilised for improvement and to reduce transmission rates. In the Indian subcontinent, *L. donovani* is the main species that transmits VL in its severe form in the target host (https://ncvbdc.mohfw.gov.in/). As the species and their symptoms vary in different regions,

DOI: 10.1201/9781032643120-11

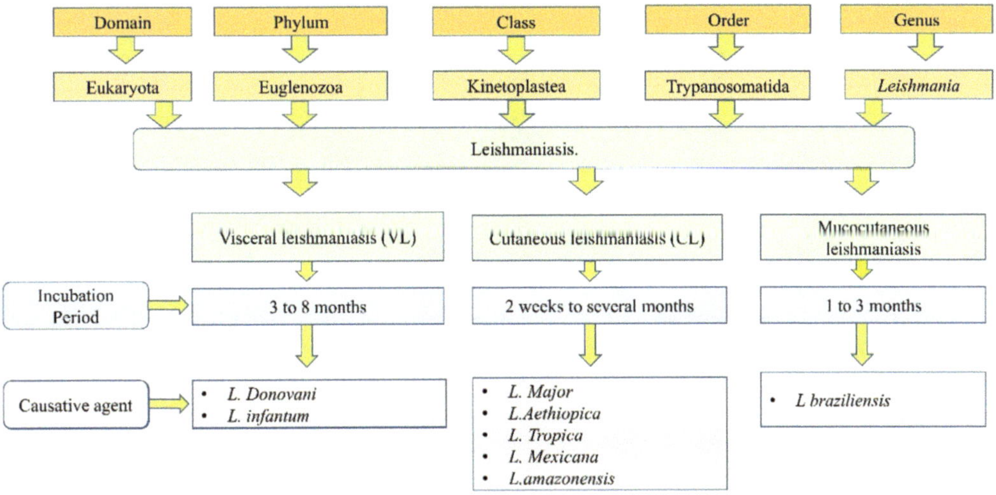

FIGURE 11.1 The graphical representation of *leishmaniasis*

several factors influence the clinical outcomes, including the parasite species involved. Co-infection cases can enhance clinical complications in infected patients, such as hypergammaglobulinemia and weight loss, and can be fatal if left untreated (Kumar and Nylen 2012; Kumar et al. 2023).

Moreover, post-kala-azar dermal leishmaniasis (PKDL) is usually a sequel of VL that appears after an apparent cure of VL, but sometimes occurs even without a prior history of VL in some regions. It manifests as a macular, papular, or nodular rash on the face, upper arms, trunk, and other areas of the body (Mahto et al. 2023; Zijlstra 2016; Sundar and Singh 2018). PKDL complications occur in 5–10% of kala-azar patients, primarily in East Africa (mainly in Sudan) and on the Indian subcontinent (www.who.int/news-room/fact-sheets/detail/leishmaniasis). This late cutaneous manifestation of VL plays a crucial role in transmitting the disease and severely threatens VL elimination in affected populations. PKDL commonly manifests after apparent recovery from VL, raising concerns about eliminating infection from the host. Other species, such as *L. infantum* or *L. tropica*, can also lead to PKDL complications in some locations (Mediterranean countries and Latin America). While this transmission occurs after VL treatment, it has virtually no mortality compared to other species' transmission. However, it can lead to new VL infections as it can serve as a parasite reservoir. To address these challenges, active surveillance, early detection, vector control, the development of therapeutics, and a better understanding of immune responses in VL and its further complications leading to PKDL are required. This chapter will provide insights into the epidemiology, life cycle, prevention, treatments, and other significant information that can help in understanding VL and PKDL (Figure 11.1) (Kumar et al. 2023; Zijlstra 2016; Volpedo et al. 2021; Koley et al. 2013).

11.2 EPIDEMIOLOGY AND LIFE CYCLE

VL is a potentially fatal vector-borne disease that varies by region and is influenced by various factors. It ranks second and seventh among tropical diseases in mortality and loss of disability-adjusted life years, respectively, due to its significant complications worldwide. However, VL is considered a neglected disease. Case distribution depends on the distribution of the *Leishmania* parasite, the presence of competent sandfly vectors, geographical distribution, environmental conditions, and host behaviour. According to the WHO (www.who.int/news-room/fact-sheets/detail/leishmaniasis), more than 90% of worldwide VL cases are reported in Brazil, Ethiopia, India, Kenya, Somalia, South Sudan, and Sudan. Most infections occur in Brazil, East Africa, and India,

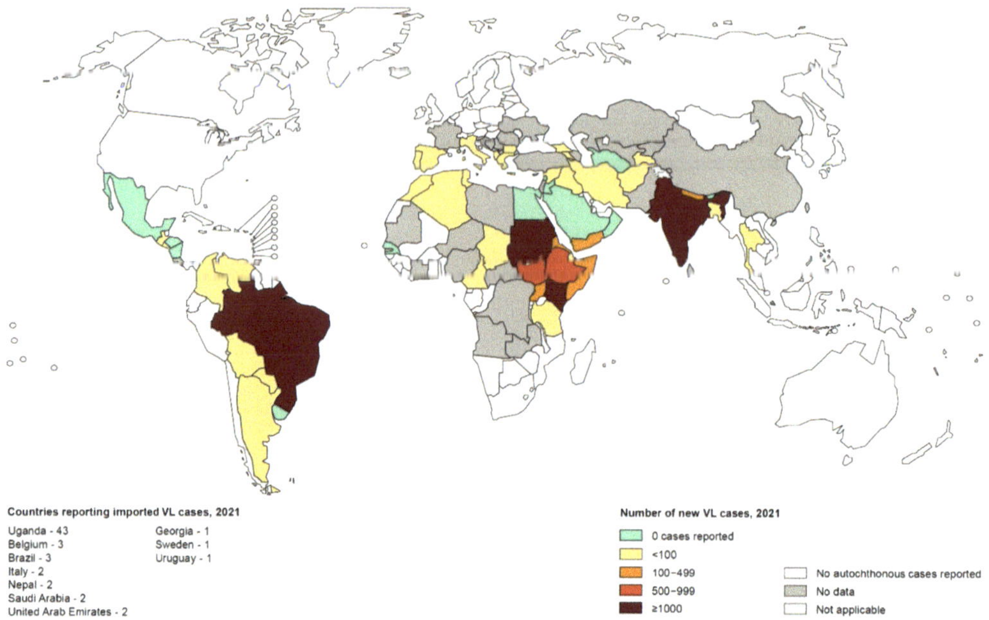

FIGURE 11.2 Graphical representation of the geographical region of infection cases

with approximately 50,000 to 90,000 new cases of VL occurring annually worldwide; only 25–45% are reported to WHO. PKDL also occurs in these regions, especially in East Africa (mainly Sudan) and the Indian subcontinent. Approximately 5–10% of patients treated for VL develop this condition, and due to its diverse classification, it is considered an epidemiological marker for the kala-azar elimination programme (Chastonay and Chastonay 2022; Kumar et al. 2022; Pacheco-Fernandez et al. 2021). Understanding how the number of cases varies from region to region is essential. In the host-pathogen interaction, once *Leishmania* enters the human host, it starts to produce protease, which decreases macrophage activity. Macrophages play a crucial role in this infection by attacking the parasite and its essential proteins with reactive oxygen and nitrogen molecules to kill the transmission process.

PKDL mainly occurs due to the persistence of *Leishmania* parasites in the skin, although the transition from VL to PKDL is not fully understood. It mainly manifests as skin lesions and can lead to further infection. In PKDL caused by *L. donovani*, symptoms may appear months to years after VL treatment. The clinical manifestation of VL, from mild to moderate, can last as long as 24 months; the incubation phase usually lasts between 2 and 6 months and includes fever, weight loss, hypergammaglobulinemia, pancytopenia, and hepatosplenomegaly as symptoms in infected patients. Further clinical complications lead to PKDL symptoms such as macular, papular, or nodular rashes on the face, upper arms, trunk, and other body areas in the infected patients (Kumar and Nylen 2012; Zijlstra 2016; Pal et al. 2023; Kumar et al. 2022; Sundar and Singh 2018). Figure 11.2 shows the graphical representation of the endemic situation of VL, retrieved from the WHO data programme website (https://apps.who.int/neglected_diseases/ntddata/leishmaniasis/leishmaniasis.html).

Tables 11.1 and 11.2 illustrate cases in different locations (www.who.int/data/gho/data/indicators/indicator-details/GHO/number-of-cases-of-visceral-leishmaniasis-reported) and endemic status (www.who.int/data/gho/data/indicators/indicator-details/GHO/status-of-endemicity-of-visceral-leishmaniasis), respectively. Figure 11.3 represents VL's basic life cycle transmission. The major hosts are the phlebotomine sandfly and a vertebrate host (such as a dog or human) (www.who.int/news-room/fact-sheets/detail/leishmaniasis). The phlebotomine sandfly is responsible for the most

TABLE 11.1
Representation of a Location-Based Number of Cases of Visceral Leishmaniasis from 2015 to 2021

Location	2021	2020	2019	2018	2017	2016	2015
Afghanistan	20	12	20	23	14	9	8
Albania	20	6	7	45	22	15	15
Algeria	30	33	46	40	34	74	38
Angola	N/A	N/A	N/A	N/A	N/A	N/A	N/A
Argentina	16	11	6	2	9	11	8
Armenia	10	15	17	17	17	18	9
Azerbaijan	N/A	N/A	61	40	51	44	28
Bangladesh	35	47	97	124	210	258	544
Bhutan	0	4	2	3	1	N/A	N/A
Bolivia (Plurinational State of)	2	2	1	0	0	0	0
Bosnia and Herzegovina	N/A	N/A	N/A	N/A	0	0	0
Brazil	1492	1954	2526	3460	4098	3123	3223
Bulgaria	1	0	3	1	0	3	5
Cameroon	6	33	13	3	0	N/A	N/A
Central African Republic	N/A	N/A	N/A	N/A	N/A	N/A	N/A
Chad	59	74	0	2	0	0	N/A
China	N/A	215	166	180	190	321	514
Colombia	7	8	7	15	29	37	21
Costa Rica	N/A	0	0	0	0	0	0
Croatia	N/A	N/A	0	0	N/A	0	0
Cyprus	N/A	0	0	2	1	1	1
Democratic Republic of the Congo	N/A	N/A	N/A	N/A	N/A	N/A	N/A
Djibouti	23	9	1	12	34	9	10
Egypt	0	0	0	0	0	0	0
El Salvador	1	0	0	3	2	0	0
Eritrea	268	1034	514	502	N/A	N/A	N/A
Ethiopia	794	1077	1360	1828	1490	1593	1593
France	N/A	26	22	17	15	8	8
Gambia	N/A	N/A	N/A	N/A	N/A	N/A	N/A
Georgia	15	42	47	52	38	49	54
Greece	24	34	22	51	83	57	64
Guatemala	1	0	1	4	2	2	2
Honduras	0	0	3	8	8	7	6
India	1187	2033	2822	4360	5758	6249	8500
Iran (Islamic Republic of)	19	N/A	72	67	60	61	59
Iraq	66	61	170	259	172	183	427
Israel	N/A	0	N/A	1	N/A	1	3
Italy	75	29	7	72	19	49	55
Jordan	0	0	2	1	0	0	0
Kazakhstan	N/A	0	0	0	0	0	0
Kenya	1746	1178	1621	590	950	692	894
Kyrgyzstan	N/A	N/A	N/A	N/A	N/A	N/A	N/A
Lebanon	N/A	N/A	N/A	N/A	N/A	N/A	N/A
Libya	N/A	24	28	34	18	10	0
Malta	N/A	1	0	2	5	0	N/A
Mauritania	N/A	N/A	N/A	N/A	N/A	N/A	N/A

(*Continued*)

TABLE 11.1 (CONTINUED)
Representation of a Location-Based Number of Cases of Visceral Leishmaniasis from 2015 to 2021

Location	2021	2020	2019	2018	2017	2016	2015
Mexico	0	0	1	0	1	0	1
Monaco	N/A	N/A	N/A	0	1	N/A	N/A
Montenegro	N/A	N/A	N/A	N/A	4	6	5
Morocco	40	67	91	108	160	92	81
Nepal	240	206	185	208	244	237	217
Nicaragua	0	0	0	0	0	0	0
Niger	N/A	N/A	N/A	N/A	N/A	N/A	N/A
Nigeria	N/A	0	0	0	0	0	0
North Macedonia (Republic of)	N/A	N/A	N/A	12	9	5	4
Oman	0	0	0	0	2	N/A	1
Pakistan	N/A	N/A	N/A	N/A	N/A	N/A	N/A
Paraguay	55	26	22	19	34	64	92
Portugal	2	2	2	5	4	8	0
Romania	N/A	N/A	N/A	0	1	1	1
Saudi Arabia	0	0	0	1	4	4	3
Senegal	0	N/A	0	N/A	N/A	N/A	N/A
Serbia	N/A	N/A	N/A	N/A	0	0	0
Slovenia	N/A	0	0	0	0	0	0
Somalia	831	471	294	411	857	780	1165
South Sudan	792	827	1013	1867	3567	4285	2840
Spain	52	49	78	69	196	176	N/A
Sri Lanka	N/A	N/A	N/A	N/A	N/A	N/A	N/A
Sudan	3323	2563	2851	2711	3894	9810	2829
Syrian Arab Republic	29	22	47	38	55	25	20
Türkiye	N/A	N/A	N/A	N/A	N/A	N/A	N/A
Tajikistan	14	N/A	15	32	23	25	25
Thailand	2	0	0	1	1	0	1
Tunisia	14	28	27	23	23	17	30
Turkmenistan	0	0	0	0	0	1	0
Uganda	129	55	101	29	31	35	34
Ukraine	N/A	0	0	N/A	0	1	0
United Republic of Tanzania	1	N/A	N/A	N/A	N/A	N/A	N/A
Uruguay	N/A	0	2	0	1	0	0
Uzbekistan	N/A	47	41	34	49	38	34
Venezuela (Bolivarian Republic of)	30	6	23	43	40	33	37
Yemen	286	450	132	56	44	0	5
Zambia	N/A	N/A	N/A	N/A	0	0	0

significant vector-borne diseases. An infected phlebotomine sandfly regurgitates parasites when taking its blood meal, where the host's macrophage cells phagocytose the promastigotes. The promastigotes then transform into amastigotes and infect other phagocytic cells in the host, initiating complications (Gherbi et al. 2020; Kipp et al. 2020). When a sandfly ingests a blood meal from another host, it also ingests amastigotes, which develop into promastigotes in the gut and migrate further to continue the life cycle. Treated VL patients may still harbour the parasites, often in the amastigote form, which are adapted within host cells and can be present in skin lesions, making

TABLE 11.2
List of Location-Based Endemicity of Visceral Leishmaniasis

Location	Endemicity of VL Status
Afghanistan	Endemic
Albania	Endemic
Algeria	Endemic
Andorra	No autochthonous cases reported
Angola	Previously reported cases
Antigua and Barbuda	No autochthonous cases reported
Argentina	Endemic
Australia	No autochthonous cases reported
Austria	No autochthonous cases reported
Armenia	Endemic
Azerbaijan	Endemic
Bahamas	No autochthonous cases reported
Bahrain	No autochthonous cases reported
Bangladesh	Endemic
Barbados	No autochthonous cases reported
Belarus	No autochthonous cases reported
Belgium	No autochthonous cases reported
Belize	No autochthonous cases reported
Benin	No autochthonous cases reported
Bhutan	Endemic
Bolivia (Plurinational State of)	Endemic
Bosnia and Herzegovina	Endemic
Botswana	No autochthonous cases reported
Brazil	Endemic
Brunei Darussalam	No autochthonous cases reported
Bulgaria	Endemic
Burkina Faso	No autochthonous cases reported
Burundi	No autochthonous cases reported
Cabo Verde	No autochthonous cases reported
Cambodia	No autochthonous cases reported
Cameroon	Endemic
Canada	No autochthonous cases reported
Central African Republic	Previously reported cases
Chad	Endemic
Chile	No autochthonous cases reported
China	Endemic
Colombia	Endemic
Comoros	No autochthonous cases reported
Congo	No autochthonous cases reported
Cook Islands	No autochthonous cases reported
Costa Rica	Endemic
Croatia	Endemic
Cuba	No autochthonous cases reported
Cyprus	Endemic
Czechia	No autochthonous cases reported
Democratic People's Republic of Korea	No autochthonous cases reported
Democratic Republic of the Congo	Endemic
Denmark	No autochthonous cases reported

(Continued)

TABLE 11.2 (CONTINUED)
List of Location-Based Endemicity of Visceral Leishmaniasis

Location	Endemicity of VL Status
Djibouti	Endemic
Dominica	No autochthonous cases reported
Dominican Republic	No autochthonous cases reported
Ecuador	No autochthonous cases reported
Egypt	Endemic
El Salvador	Endemic
Equatorial Guinea	No autochthonous cases reported
Eritrea	Endemic
Estonia	No autochthonous cases reported
Eswatini	No autochthonous cases reported
Ethiopia	Endemic
Fiji	No autochthonous cases reported
Finland	No autochthonous cases reported
France	Endemic
Gabon	No autochthonous cases reported
Gambia	Previously reported cases
Georgia	Endemic
Germany	No autochthonous cases reported
Ghana	No autochthonous cases reported
Greece	Endemic
Grenada	No autochthonous cases reported
Guatemala	Endemic
Guinea	No autochthonous cases reported
Guinea-Bissau	No autochthonous cases reported
Guyana	No autochthonous cases reported
Haiti	No autochthonous cases reported
Honduras	Endemic
Hungary	No autochthonous cases reported
Iceland	No autochthonous cases reported
India	Endemic
Indonesia	No autochthonous cases reported
Iran (Islamic Republic of)	Endemic
Iraq	Endemic
Ireland	No autochthonous cases reported
Israel	Endemic
Italy	Endemic
Jamaica	No autochthonous cases reported
Japan	No autochthonous cases reported
Jordan	Endemic
Kazakhstan	Endemic
Kenya	Endemic
Kiribati	No autochthonous cases reported
Kuwait	No autochthonous cases reported
Kyrgyzstan	Endemic
Lao People's Democratic Republic	No autochthonous cases reported
Latvia	No autochthonous cases reported
Lebanon	Endemic
Lesotho	No autochthonous cases reported
Liberia	No autochthonous cases reported

(Continued)

TABLE 11.2 (CONTINUED)
List of Location-Based Endemicity of Visceral Leishmaniasis

Location	Endemicity of VL Status
Libya	Endemic
Lithuania	No autochthonous cases reported
Luxembourg	No autochthonous cases reported
Madagascar	No autochthonous cases reported
Malawi	No autochthonous cases reported
Malaysia	No autochthonous cases reported
Maldives	No autochthonous cases reported
Mali	No autochthonous cases reported
Malta	Endemic
Marshall Islands	No autochthonous cases reported
Mauritania	Endemic
Mauritius	No autochthonous cases reported
Mexico	Endemic
Micronesia (Federated States of)	No autochthonous cases reported
Monaco	Endemic
Mongolia	No autochthonous cases reported
Montenegro	Endemic
Morocco	Endemic
Mozambique	No autochthonous cases reported
Myanmar	No autochthonous cases reported
Namibia	No autochthonous cases reported
Nauru	No autochthonous cases reported
Netherlands	No autochthonous cases reported
New Zealand	No autochthonous cases reported
Nicaragua	Endemic
Niger	Endemic
Nigeria	Previously reported cases
Niue	No autochthonous cases reported
Norway	No autochthonous cases reported
North Macedonia (Republic of)	Endemic
Oman	Endemic
Pakistan	Endemic
Palau	No autochthonous cases reported
Panama	No autochthonous cases reported
Papua New Guinea	No autochthonous cases reported
Paraguay	Endemic
Peru	No autochthonous cases reported
Philippines	No autochthonous cases reported
Poland	No autochthonous cases reported
Portugal	Endemic
Qatar	No autochthonous cases reported
Republic of Korea	No autochthonous cases reported
Republic of Moldova	No autochthonous cases reported
Romania	Endemic
Russian Federation	No autochthonous cases reported
Rwanda	No autochthonous cases reported
Saint Kitts and Nevis	No autochthonous cases reported
Saint Lucia	No autochthonous cases reported

(Continued)

TABLE 11.2 (CONTINUED)
List of Location-Based Endemicity of Visceral Leishmaniasis

Location	Endemicity of VL Status
Saint Vincent and the Grenadines	No autochthonous cases reported
Samoa	No autochthonous cases reported
San Marino	No autochthonous cases reported
Sao Tome and Principe	No autochthonous cases reported
Saudi Arabia	Endemic
Senegal	Endemic
Serbia	Previously reported cases
Seychelles	No autochthonous cases reported
Sierra Leone	No autochthonous cases reported
Singapore	No autochthonous cases reported
Slovakia	No autochthonous cases reported
Slovenia	Endemic
Solomon Islands	No autochthonous cases reported
Somalia	Endemic
South Africa	No autochthonous cases reported
South Sudan	Endemic
Spain	Endemic
Sri Lanka	Endemic
Sudan	Endemic
Suriname	No autochthonous cases reported
Sweden	No autochthonous cases reported
Switzerland	No autochthonous cases reported
Syrian Arab Republic	Endemic
Tajikistan	Endemic
Thailand	Endemic
Timor-Leste	No autochthonous cases reported
Togo	No autochthonous cases reported
Tonga	No autochthonous cases reported
Trinidad and Tobago	No autochthonous cases reported
Tunisia	Endemic
Türkiye	Endemic
Turkmenistan	Endemic
Tuvalu	No autochthonous cases reported
Uganda	Endemic
Ukraine	Endemic
United Arab Emirates	No autochthonous cases reported
United Kingdom of Great Britain and Northern Ireland	No autochthonous cases reported
United Republic of Tanzania	Endemic
United States of America	No autochthonous cases reported
Uruguay	Endemic
Uzbekistan	Endemic
Vanuatu	No autochthonous cases reported
Venezuela (Bolivarian Republic of)	Endemic
Viet Nam	No autochthonous cases reported
Yemen	Endemic
Zambia	Endemic
Zimbabwe	No autochthonous cases reported

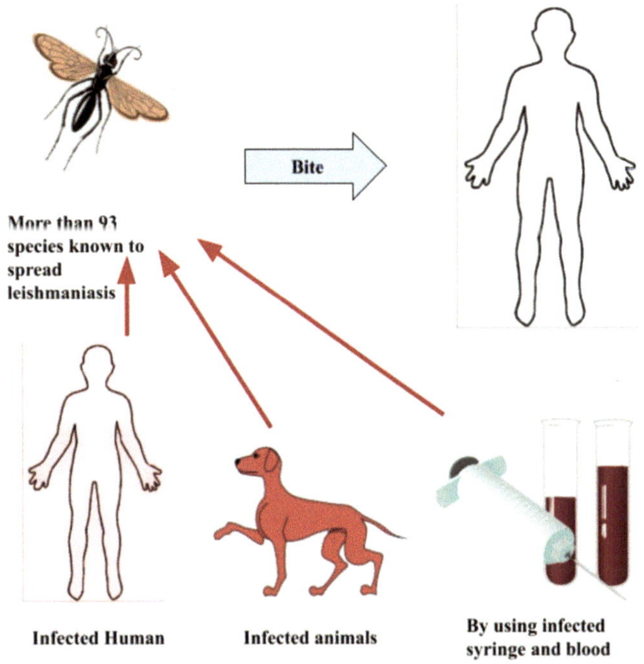

FIGURE 11.3 Representation of the basic life cycle of transmission of *leishmaniasis*

PKDL individuals infectious to sandfly vectors. Studies report that the domestic dog (*Canis familiaris*) is the primary reservoir for VL in urban and rural environments. Additionally, some studies report that the use of needles also contributes to parenteral and congenital transmission. Moreover, co-infection cases (with HIV, lymphatic filariasis, COVID-19, and malaria) have been reported, increasing clinical complications in infected patients (Figure 11.3; Tables 11.1 and 11.2) (Abbate et al. 2020; Ghimire et al. 2019; Burki 2022; Franssen et al. 2021; Kushwaha and Kaur 2021).

11.3 PREVENTION AND CONTROL

Due to the diverse nature and different mechanisms of action in different regions, accurate prevention and control of VL and PKDL remain challenging. Although this infection has a long history, no vaccine exists to protect against VL and PKDL diseases. However, as the mechanism of action and transmission are known, there are some preventive care practices for the human population to overcome these diseases, as shown in Figure 11.4 (Sundar and Singh 2018; Ibiapina et al. 2022; Li et al. 2022). Since PKDL is a complication occurring after the treatment of VL, similar prevention and control measures can lead to a low transmission rate. For VL, the primary and first line of prevention is to reduce the risk of sandfly bites, which are responsible for VL transmission. Insecticide-treated bed nets can help prevent sandfly interaction. Also, as pets can be reservoir hosts for the parasite, keeping them indoors during peak sandfly activity can help reduce transmission (Kumar et al. 2023; Santos Junior et al. 2023; Chapman et al. 2020). As Vl patients exhibit various symptoms, early diagnosis and treatment based on specific complications can be beneficial. In the absence of a vaccine, available antileishmanial medication can be utilised to prevent the disease. For controlling transmission, vector control is the most effective first line of defence. Indoor residual spraying and insecticide-treated bed nets can help control transmission. However, as the mechanisms and infection rates differ based on geographical regions, surveillance and monitoring of the mechanisms of action can help identify low and high infection areas, aiding in disease control. While medications are available, vaccine development is essential for worldwide control of transmission. Although the

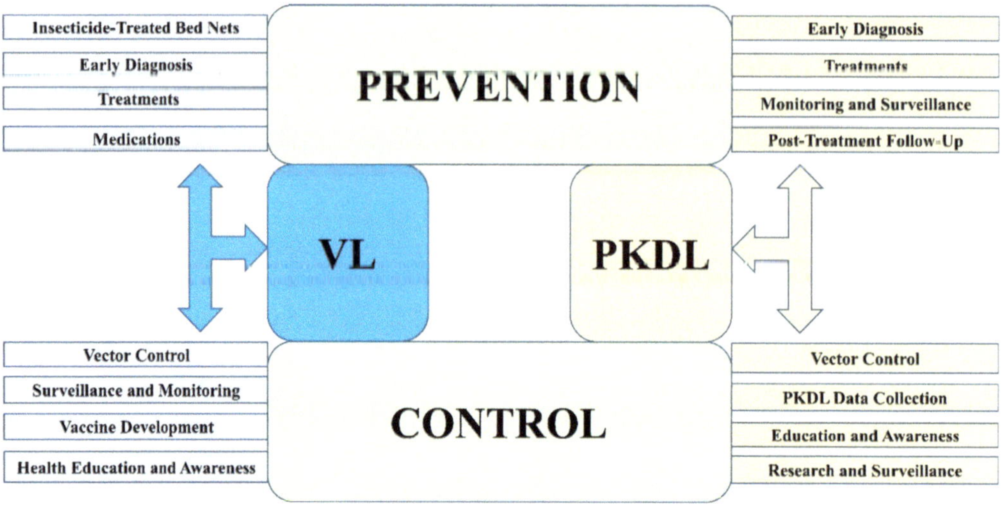

FIGURE 11.4 Representation of aspects of prevention and control for VL and PKDL

endemic situation differs from region to region, raising awareness can help locals understand how to control the initial stages of transmission (Berhe et al. 2022; Mahto et al. 2023).

For PKDL, prevention and control measures are the same as for VL, as PKDL occurs after the successful treatment of VL. The strategies mainly include early diagnosis and treatment based on the post-VL treatment symptoms. Monitoring and surveillance of previously infected patients can lead to a reduced transmission rate, as PKDL mainly occurs after VL. Patients who have been successfully treated for VL should undergo regular follow-up examinations to detect PKDL symptoms and signs. Vector control approaches used for VL can be similarly applied to control the transmission rate of PKDL. Due to varying endemic situations in different geographical regions, data collection, including early signs and symptoms and awareness of complications, can lead to better management (Li et al. 2022; Chapman et al. 2020; Pal et al. 2023). However, more focused research and analysis of transmission rates through accurate surveillance can be helpful in overcoming this disease (Figure 11.4).

11.4 DIAGNOSIS AND TREATMENTS FOR VL AND PKDL

People suspected of suffering from VL and PKDL should seek medical care and prevention immediately to avoid further complications. Several approaches are available for diagnosing VL and PKDL to enable early diagnosis and prevention. Although the severity and classification vary widely among geographical regions, some diagnostic methods and treatments are common for the first line of defence and further care (Mahto et al. 2023; Sundar et al. 2019). Figure 11.5 shows a representation of the diagnosis and treatments that can be applied to VL and PKDL for improved outcomes.

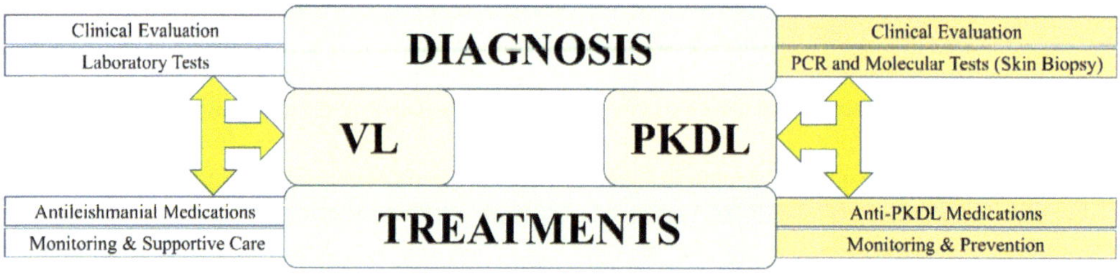

FIGURE 11.5 Representation of diagnosis and treatments for VL and PKDL

VL diagnosis involves two main approaches: direct (splenic aspirate, bone marrow aspirate, LN aspirate, and peripheral blood) and indirect diagnosis, including serological tests, rapid diagnostic tests (rK-39), direct agglutination tests, and other antibody tests, such as ELISA, indirect immunofluorescence, indirect agglutination tests, and immunochromatographic tests for more precise diagnosis. As VL often presents with non-specific symptoms, so clinical evaluation provides more accurate reports of changes and similar symptoms. Similarly, laboratory tests lead to better diagnosis, providing brief information about the patients. Various vaccine trials (Leishvaccine, Leishmune, ChAd63-KH, Leish-Tec, CaniLeish, LEISH-F1, F2, F3, SMTγ + NHμ) are ongoing for *Leishmania*, using different approaches, and are classified as the first, second, and third generations. However, no approved vaccines are available for VL. Some anti-VL drugs are available, such as pentavalent antimonials (sodium stibogluconate and meglumine antimoniate), amphotericin, lipid-associated liposomal, colloidal dispersion, lipid complex, pentamidine, miltefosine, and paromomycin, which are mostly in use. However, due to drug-resistant *Leishmania* strains, drug combinations may be used to enhance treatment efficacy, varying in different geographical regions (Sundar and Singh 2018; Aronson et al. 2017; Aronson and Joya 2019; Burza et al. 2018; Sundar et al. 2019; Mann et al. 2021; Teixeira et al. 2011; Moafi et al. 2019; Lage et al. 2020).

PKDL primarily affects the skin, and its diagnosis and treatment differ from VL, as it occurs as a sequel to VL treatment. PKDL complications, such as skin lesions, nodules, and rashes, are seen after the successful VL treatment, possibly months to years later. The main steps in diagnosing PKDL involve considering the patient's history of VL and analysing skin changes. To assess the patient's condition and determine whether they have PKDL symptoms, skin biopsies are used to identify the presence of *Leishmania* parasites in skin tissue. Additionally, these skin samples are further used to examine for *Leishmania* DNA through polymerase chain reaction tests. For PKDL treatments, some anti-VL drugs show promising results, such as miltefosine, sodium stibogluconate (SSG), and meglumine antimoniate (Desjeux et al. 2013; Dixit et al. 2020; Aronson et al. 2017; Goncalves and Costa 2018; Mann et al. 2021; Moafi et al. 2019; Lage et al. 2020). However, dosage and drug combinations differ based on the patients' geographical regions and stage of complication. Further, in-depth descriptions and other significant information about approved and investigational drugs can be retrieved from the DrugBank database (https://go.drugbank.com/) (Figure 11.5).

11.5 EARLIER STUDIES TO UNDERSTAND THE VACCINE DESIGN AND INHIBITOR IDENTIFICATION

In the present situation, computational-based learning is of keen interest to the research community in solving biological problems. Interestingly, during the COVID-19 pandemic, researchers conducted in silico studies and developed peptide-based vaccines, yielding promising results in identifying novel potential drug targets and inhibitors against the pathogen. These approaches proved to be cost-effective, time-efficient, and highly accurate. This section will help understand vaccine design and inhibitor identification through an in silico-based approach using a selected case study.

Immunoinformatics is the approach used for peptide-based vaccine design development. This approach mainly considers immune system aspects to identify potential B cell and T cell epitopes that encode B lymphocytes, T helper cells, and T cytotoxic cells, screening them through antigenicity, allergenicity, toxicity, and immunogenicity analysis to identify immunodominant epitopes. These epitopes are further fused using different linkers, and vaccine candidates are designed. Their validation is confirmed using molecular docking analysis between the signalling receptor and the developed vaccine to understand the immune response through their interaction. Several studies have been performed to develop vaccines targeting VL and PKDL, as illustrated in Table 11.3 (Oli et al. 2020; Thomas et al. 2022; Soleymani et al. 2021).

Recently, a study performed by Khatoon et al. used several secretory proteins (hypothetical, surface antigen, FG-GAP repeat, translocon-associated, and secretory lipase protein) to design B

TABLE 11.3
List of Selected Studies for Developing Peptide-Based Vaccine Design for VL and PKDL

Group Name	Target	Description	Reference
Khatoon et al. 2017	Hypothetical, surface antigen, FG-GAP repeat, translocon-associated and secretory lipase proteins	Identification of B and T cell epitopes for multiepitope subunit vaccine development	(Khatoon et al. 2017)
Khan et al. 2020	Amastigote secretory proteins	Identification of CD4+ and CD8+ T cell-specific epitopes for the subunit candidate vaccine development	(Khan et al. 2020)
Ojha et al. 2020	Membrane and secretory antigens	Identification of B cell, Helper T lymphocyte, and cytotoxic T lymphocyte epitopes for the multiepitope vaccine against VL	(Ojha et al. 2020)
Yadav et al. 2020	Hypothetical membrane proteins	Identification of helper T lymphocyte (HTL), cytotoxic T lymphocyte (CTL), and B cell epitopes for multiepitope subunit vaccine design	(Yadav et al. 2020)
Lari et al. 2021	Secretory proteins	Identification of CTL and HTL epitopes for the novel multiepitope vaccine development against VL	(Lari et al. 2022)
Saha et al. 2022	ATP-dependent zinc metallopeptidase, histidine secretory acid phosphatase, rhomboid-like protein, and amastin-like surface protein	Identification of CTL, HTL, and B cell epitopes for the multiepitope vaccine development	(Saha et al. 2022)
Onile et al. 2022	Disulfide isomerase protein	Identification of CTL and HTL epitopes to design a potential multiepitope peptide vaccine	(Onile et al. 2022)
Fayaz et al. 2022	Fusion protein	Identification of T cell epitopes in BALB/c mice and humans for the identification of vaccine candidates	(Fayaz et al. 2022)
Bhowmik et al. 2023	PrimPol protein	Identification of CTL, HTL, linear BCL epitopes, and discontinuous BCL epitopes for multiepitope subunit vaccine development against VL	(Bhowmik et al. 2023)
Bhattacharjee et al. 2023	Paraflagellar rod protein 2C, kinetoplast membrane protein 11 (KMP-11), cysteine proteinase A, leishmanolysin C1, antigen Lcr1 (fragment), hydrophilic acylated surface protein B, LACK protective antigen, surface membrane glycoprotein GP46/M-2, proteophosphoglycan, heat shock 70-related protein 1, tubulin beta chain, and eukaryotic translation initiation factor 5A	Identification of CTL and HTL epitopes for novel polyvalent multi-subunit peptide vaccine	(Bhattacharjee et al. 2023)

cell and T cell multiepitope subunit vaccines. In this study, they identified a total of 8 CTL and 15 HTL epitopes with promising profiles based on antigenicity and other properties. They also used toll-like receptor 4 (TLR-4) (PDB ID: 4G8A) agonist 50S ribosomal L7/L12 as an adjuvant (to enhance immune responses) for the final vaccine construction, employing AAY and GPGPG linkers. A vaccine construct composed of 528 amino acid residues was designed. Molecular docking of the designed vaccine with the immune receptor (TLR-4) was performed, and molecular dynamics (MD) simulation of the immune receptor-vaccine complex showed the stable nature of the docked complex. This study also performed codon optimisation of designed vaccine constructs for expression in the pET28a (+) vector. Overall, this study successfully designed a potential multiepitope subunit vaccine with a promising immune profile (Khatoon et al. 2017).

Khan et al. performed another study for an in silico-based subunit candidate vaccine targeting 26 secretory proteins (chaperonin heat shock protein [HSP] 60, mitochondrial; HSP70, putative; HSP70-kDa; eukaryotic translation initiation factor 5A; protein phosphatase-2C; stress-inducible protein STI1 homolog; proteasome endopeptidase complex; nucleoside diphosphate kinase, mitochondrial; glutathione peroxidase, uncharacterised proteins [containing META domain, alpha/beta hydrolase domain, complex1_LYR_1 motif, and RNA recognition motif RRM_8]; protein disulfide isomerase, putative; cysteine peptidase A [CPA]; cysteine peptidase C [CPC]; cysteine protease; thioredoxin-like protein; Mkiaa0324 protein-like protein [serine/arginine repetitive matrix protein 2]; leishmanolysin; fructose-1,6 bisphosphatase, cytosolic, putative; ubiquitin-conjugating enzyme e2, putative; lipophosphoglycan biosynthetic protein, putative). Through multiple screening steps, a total of 9 CTL and 14 HTL epitopes were utilised for vaccine candidate development, with these epitopes fused using the AAY, EAAK, and GPGPG linkers. A vaccine construct was designed, comprising 397 amino acid residues. To validate the immune response of the vaccine construct, molecular docking analysis was performed with TLR-4, and through MD simulation, the selected docked complex (vaccine-TLR-4) was found to be stable. The immune simulation study also found that the designed vaccine candidate has a lasting, robust immune profile that can help combat this infection (Khan et al. 2020).

A study performed by Ojha et al. used a vaccinomics strategy to develop a multiepitope subunit vaccine using apyrase, antigen 5-related, D7-related, pyrophosphatase, kinetoplastida membrane protein 11, LACK (*Leishmania* homolog of receptors for activated C kinase), cysteine proteinase, haemoglobin receptor (HBR), open reading frame, and glycoprotein (GP) 63. After utilising various rigorous vaccinomics steps, two different vaccine constructs, construct 1 (HTL-CTL-B-CELL) and construct 3 (B-cell-HTL-CTL), were designed. Construct 1 contained 483 amino acids, and construct 3 contained 486 amino acids. To understand the immune profile of the vaccine and its immune response, molecular docking analysis was performed using TLR-4 as a singling receptor, and the docked complex was found to be stable through the 100 ns MD simulation study. A comparative analysis of the immune simulation of the designed vaccine with the control (Leish-F3 vaccine sequence) showed that the designed vaccine has a promising profile and can be utilised to combat this pathogenic infection (Ojha, Pandey, and Prajapati 2020).

Similarly, Yadav et al. performed a study on the multiepitope subunit vaccine development using the three hypothetical membrane proteins of L. donovani. Through immunoinformatics-based screening, 6 HTL, 18 CTL, and 25 B cell epitopes were identified and utilised for vaccine construction with the help of different linkers. An adjuvant was also included in the designed vaccine to enhance its immune response. To confirm the target immune receptor response, molecular docking analysis was performed with the toll-like receptor-2 (TLR-2), and the docked complex's stability was evaluated using MD simulation. The expression of the designed multiepitope subunit vaccine in the expression vector was also analysed through in silico cloning. The overall screening from prediction to development and expression analysis showed that the designed multiepitope subunit vaccine has a promising profile and can potentially lead to vaccine development to reduce Vl pathogens (Yadav et al. 2020).

Lari et al. designed a novel subunit vaccine against VL using an immunoinformatics approach. They analysed amino acid sequences of three hypothetical proteins (HPs) of *L. donovani* (LDBPK_120390, LDBPK_300860, and LDBPK_312610) as secretory proteins, selecting five CTL epitopes and nine HTL epitopes. These peptides were linked with AAY and GPGPG linkers to develop a novel subunit vaccine against VL. The designed vaccine was then docked with TLR-4 to understand its feasibility in developing adaptive immunity, followed by MD simulation for stability analysis (Lari et al. 2022).

Saha et al. developed a multiepitope vaccine against *L. donovani* using an immunoinformatics approach. They identified six HTL epitopes, nine CTL epitopes, and four B cell epitopes through immunoinformatics-based screening for vaccine development. An adjuvant 50S ribosomal protein L7/L12 was used (to enhance the vaccine immune profile) and fused with GPGPG, AAY, and EAAAK linkers to construct the final vaccine. Molecular docking analysis between the designed vaccine and the TLR-4 receptor showed a robust profile based on the residue interaction. Through MD analysis over a 100 ns period, the docked complex of the vaccine-TLR-4 receptor was found to be stable. The study also included in silico immune simulation, codon optimisation, and in silico cloning to confirm the immunogenic profile of the designed vaccine. Interestingly, the designed multiepitope vaccine against *L. donovani* showed 100% world population coverage (Saha et al. 2022).

Onile et al. utilised disulfide isomerase protein for a potential multiepitope peptide vaccine through immunoinformatics studies. They identified 17 CTL epitopes and one potential HTL epitope for vaccine development. To enhance the profile of the vaccine candidate, TLR-4 agonist adjuvant (RS-09; sequence: APPHALS) and 6x histidine were also included and fused with EAAAK, GPGPG, and AAY linkers for the final vaccine design. The designed vaccine was found to have antigenic, non-allergenic, and immunogenic profiles. Molecular docking analysis between the designed vaccine and TLR-2 and TLR-4 showed a promising interaction, and its stability was confirmed using MD simulation through normal mode analysis (Onile et al. 2022).

Similarly, Fayaz et al. conducted a study utilising the fusion protein of *L. infantum* LiHyV and *Phlebotomus kandelakii* apyrase to identify vaccine candidates against VL. Through an immunoinformatics approach-based screening, they found epitopes consisting of truncated LiHyV (10–285) and full-length PkanAp (21–329) and named it Model 1. They identified 25 MHC-I and 141 MHC-II antigenic peptides, and based on homology-based analysis and various rigorous approaches, they concluded that the predicted Model 1 could be used as a vaccine candidate against VL (Fayaz et al. 2022).

Bhowmik et al. created a global multiepitope subunit vaccine against VL through an immunoinformatics-based approach using conserved epitopic regions of the target protein. They identified eight CTL epitopes, eight HTL epitopes, seven linear BCL epitopes, and one discontinuous BCL potential epitope. They designed a subunit vaccine with an adjuvant to establish the most promising profile for the designed vaccine. Further validation was performed to confirm the profile of the designed vaccine through molecular docking, molecular dynamics, immune stimulation, codon optimisation, and in silico cloning (Bhowmik et al. 2023).

Bhattacharjee et al. also utilised an immunoinformatics-based approach and successfully designed a novel polyvalent multi-subunit peptide vaccine leveraging cross-immunity against human visceral and cutaneous leishmaniasis using twelve proteins (paraflagellar rod protein 2C, kinetoplast membrane protein 11 [KMP-11], cysteine proteinase A, leishmanolysin C1, antigen Lcr1 [fragment], hydrophilic acylated surface protein B, LACK protective antigen, surface membrane GP46/M-2, proteophosphoglycan, heat shock 70-related protein 1, tubulin beta chain, and eukaryotic translation initiation factor 5A). They identified 26 CTL epitopes and three HTL epitopes based on their promising properties and fused them using AAY, EAAK, and GPGPG linkers. They also used human IL-12 (UniProt ID: P29459) with 219 amino acids and a 6XHis tag in this designed vaccine to enhance its effectiveness and facilitate protein identification and purification. After several rigorous steps such as molecular docking, simulation, immune simulation, and in silico cloning, the designed vaccine was found to have a promising profile that can play a major role in combating this infection (Bhattacharjee at al. 2023).

Several other studies on in silico-based vaccine development have successfully designed vaccine candidates and generated immune responses that can be useful for vaccine development. However, these in silico-based studies strongly require experimental validation through in vivo and in vitro analysis (Farzan et al. 2023; John et al. 2012; Mishra et al. 2023; Keshri et al. 2023; Khan et al. 2023).

As advanced computational approaches are utilised to identify potential peptides for vaccine development, the identification of potential targets and inhibitors is also ongoing to combat and overcome this pathogenesis. As the mechanism becomes known, several studies are performed daily for novel discoveries. This section will explore some selective studies to understand how in silico-based drug design is continuously evolving for target and inhibitor identification.

In drug design, selecting a potential drug target is the crucial first step. For ligand selection, several databases are available based on their data specifications, such as COCOCNUT (https://coconut.naturalproducts.net/), DrugBank (https://go.drugbank.com/), ChEMBL (www.ebi.ac.uk/chembl/), ZINC (https://zinc.docking.org/), and Dr Duke's phytochemical and ethnobotanical database (https://phytochem.nal.usda.gov/). These databases can be useful for identifying novel inhibitors for VL. They are regularly updated with a range of compound classes, and the collected datasets are carefully curated. The specifications and details of these compounds are typically verified and are available through the database interfaces (Vinjoda et al. 2024; Vaghasia et al. 2024; Vakhariya Sakina et al. 2023; Johny and Mishra 2023; Maru et al. 2023).

A study performed by Pasandideh et al. utilised mitochondrial iron transporter-1 (LMIT1), as it is essential for *Leishmania*'s replication, survival, and virulence, making it an ideal candidate for further exploration. They retrieved 1871 potential ligands from the DRUGBANK database for structure-based MD analysis with the target protein. As the 3D structure was unavailable, this study also performed 2D and 3D structure prediction, validation, and active pocket prediction. The DB06206 and DB04911 molecules showed promising results due to their better inhibitory orientation in interaction with the target protein, and the overall findings suggested that these compounds can be further utilised for developing new drugs against leishmaniasis in future studies (Pasandideh et al. 2020).

Sinha et al. conducted a study to identify valproate's potent analogues as drugs for *L. donovani* (Ld-MIPS) using MIP synthase as a target, aiming for minimal side effects and toxicity to the host. Through several computational approaches, such as 3D structure prediction and molecular docking, they identified compounds ZINC000001690029, ZINC000039101783, ZINC000004023366, ZINC000001758357, and ZINC000004728953 that could act as potent inhibitors for the selected target. In MD simulation, only the ZINC000001690029 and ZINC04023366 protein complexes showed better stability among the selected inhibitors. Additionally, the drug-likeness, absorption, distribution, excretion, metabolism, and toxicity profiles showed that the selected compounds have promising profiles, suggesting that the identified inhibitors can be helpful for drug development against VL (Sinha et al. 2022).

Tabrez et al. (2021) targeted the transferase enzyme pathway crucial for pathogenesis, i.e., sterol C-24 methyltransferase (LdSMT), to identify inhibitors. They retrieved 1355 purchasable drug compounds from the ZINC database in mol2 format, and the ligands library was prepared. As the target structure was unavailable, the structure of LdSMT was modelled through homology modelling. The structure was utilised for active binding pocket identification, which can help in understanding the strong residues involved in binding with the inhibitors. Molecular docking analysis revealed that simeprevir had the most significant profile based on the lowest binding energy of −10.6 Kcal/mol with LdSMT. The study also included antileishmanial assays and in vitro ROS assessment in parasites, indicating that simeprevir induced ROS generation in parasites in a dose-dependent manner. The authors suggested that further in vitro and in vivo validation is required to confirm its potential activity before using it as an antileishmanial drug (Tabrez et al. 2021).

Bhowmik et al. used an advanced computational approach utilising ligand-based virtual screening, docking, and molecular dynamics to evaluate potential drugs against VL targeting Ld-PriS (catalytic subunit) for the first time as a prospective target. The study designed a 3D structure of the target and performed ligand-based virtual screening using hybrid similarity recognition techniques. Ligands were retrieved from the ZINC database using sphingosine as a query. Through the molecular docking analysis and MD simulation, three leads, ZINC000009219046, ZINC000025998119, and ZINC000004677901, demonstrated strong binding properties with Ld-PriS. The ADMET predictions revealed a promising profile in terms of absorption, solubility, permeability, distribution, metabolism, toxicity, and good bioavailability. The study suggests that the final three compounds may serve as potential inhibitors and can lead to the development of new drugs against leishmaniasis (Bhowmik et al. 2021).

Exploring available data and drugs plays a crucial role in drug design, as it saves time and cost and has high accuracy due to the availability of earlier data. Rai et al. screened the FDA-approved drug library for repurposing based on novel antileishmanial drug screening, targeting the primase protein (accession id: ldBPK_230850). They identified and utilised the top 30 selected compounds through molecular docking analysis for MD analysis. They also performed in vitro analysis, cloning the target protein, performing purification, and testing it experimentally. Finally, some compounds, such as pioglitazone and mupirocin, were very effective against the *Leishmania* cell culture, and the authors stated that the identified compounds could be utilised in further clinical trials (Rai et al. 2022).

11.6 CONCLUSION AND FUTURE PROSPECTS

VL caused by *L. donovani* can lead to clinical complications in patients; poor clinical management and low surveillance of the parasite are significant obstacles to the timely and accurate detection of this disease. VL and PKDL require early detection, diagnosis, and treatment for better outcomes. Although some medications are available, combining therapy with other diagnostic kits can play a major role in combating this disease. The diverse species and cases verified in different geographical locations require international collaboration to examine the variation in the parasite and its clinical complications. Currently, no successful vaccines are available, so vaccine development could help overcome this disease. The successful treatment of VL may lead to PKDL, indicating a complex clinical syndrome that requires diagnosis, reporting, treatment, and advancement in therapeutic development, considering the most effective approaches that can prevent substantial morbidity and mortality from VL and PKDL.

ACKNOWLEDGEMENT

The authors are thankful to the University of North Bengal for providing the research facility.

FUNDING

None.

DATA AVAILABILITY STATEMENT

Not applicable.

DECLARATION OF COMPETING INTEREST

The authors declare that they have no known competing financial interests or personal relationships that could have appeared to influence the work reported in this paper.

REFERENCES

Abbate, J. M., C. Maia, A. Pereira, F. Arfuso, G. Gaglio, M. Rizzo, G. Caracappa, G. Marino, M. Pollmeier, S. Giannetto, and E. Brianti. 2020. "Identification of trypanosomatids and blood feeding preferences of phlebotomine sand fly species common in Sicily, Southern Italy." *PLoS One* 15 (3):e0229536. doi: 10.1371/journal.pone.0229536.

Alvar, J., S. Yactayo, and C. Bern. 2006. "Leishmaniasis and poverty." *Trends Parasitol* 22 (12):552–557. doi: 10.1016/j.pt.2006.09.004.

Aronson, N., B. L. Herwaldt, M. Libman, R. Pearson, R. Lopez-Velez, P. Weina, E. Carvalho, M. Ephros, S. Jeronimo, and A. Magill. 2017. "Diagnosis and treatment of leishmaniasis: clinical practice guidelines by the Infectious Diseases Society of America (IDSA) and the American Society of Tropical Medicine and Hygiene (ASTMH)." *Am J Trop Med Hyg* 96 (1):24–45. doi: 10.4269/ajtmh.16-84256.

Aronson, N. E., and C. A. Joya. 2019. "Cutaneous Leishmaniasis: Updates in Diagnosis and Management." *Infect Dis Clin North Am* 33 (1):101–117. doi: 10.1016/j.idc.2018.10.004.

Berhe, R., M. Spigt, F. Schneider, L. Paintain, C. Adera, A. Nigusie, Z. Gizaw, Y. A. Tesfaye, D. A. Elnaiem, and M. Alemayehu. 2022. "Understanding the risk perception of visceral leishmaniasis exposure and the acceptability of sandfly protection measures among migrant workers in the lowlands of Northwest Ethiopia: a health belief model perspective." *BMC Public Health* 22 (1):989. doi: 10.1186/s12889-022-13406-3.

Bhattacharjee, M., M. Banerjee, and A. Mukherjee. 2023. "In silico designing of a novel polyvalent multi-subunit peptide vaccine leveraging cross-immunity against human visceral and cutaneous leishmaniasis: an immunoinformatics-based approach." *J Mol Model* 29 (4):99. doi: 10.1007/s00894-023-05503-w.

Bhowmik, D., A. Bhuyan, S. Gunalan, G. Kothandan, and D. Kumar. 2023. "In silico and immunoinformatics based multiepitope subunit vaccine design for protection against visceral leishmaniasis." *J Biomol Struct Dyn* 42:9731–9752. doi: 10.1080/07391102.2023.2252901.

Bhowmik, D., R. Jagadeesan, P. Rai, R. Nandi, K. Gugan, and D. Kumar. 2021. "Evaluation of potential drugs against leishmaniasis targeting catalytic subunit of Leishmania donovani nuclear DNA primase using ligand based virtual screening, docking and molecular dynamics approaches." *J Biomol Struct Dyn* 39 (5):1838–1852. doi: 10.1080/07391102.2020.1739557.

Burki, T. 2022. "Guidelines for visceral leishmaniasis and HIV co-infection." *Lancet Infect Dis* 22 (8):1124–1125. doi: 10.1016/S1473-3099(22)00461-3.

Burza, S., S. L. Croft, and M. Boelaert. 2018. "Leishmaniasis." *Lancet* 392 (10151):951–970. doi: 10.1016/S0140-6736(18)31204-2.

Cavalcante, K. K. S., K. M. O. Borges, F. R. A. Cavalcante, F. G. S. Correia, Cmgd Florencio, and C. H. Alencar. 2022. "Epidemiological aspects and high magnitude of human visceral leishmaniasis in Ceara, Northeast of Brazil, 2007–2021." *Rev Soc Bras Med Trop* 55:e06842021. doi: 10.1590/0037-8682-0684-2021.

Chapman, L. A. C., S. E. F. Spencer, T. M. Pollington, C. P. Jewell, D. Mondal, J. Alvar, T. D. Hollingsworth, M. M. Cameron, C. Bern, and G. F. Medley. 2020. "Inferring transmission trees to guide targeting of interventions against visceral leishmaniasis and post-kala-azar dermal leishmaniasis." *Proc Natl Acad Sci USA* 117 (41):25742–25750. doi: 10.1073/pnas.2002731117.

Chappuis, F., S. Sundar, A. Hailu, H. Ghalib, S. Rijal, R. W. Peeling, J. Alvar, and M. Boelaert. 2007. "Visceral leishmaniasis: what are the needs for diagnosis, treatment and control?" *Nat Rev Microbiol* 5 (11):873–882. doi: 10.1038/nrmicro1748.

Chastonay, A. H. M., and O. J. Chastonay. 2022. "Housing risk factors of four tropical neglected diseases: a brief review of the recent literature." *Trop Med Infect Dis* 7 (7). doi: 10.3390/tropicalmed7070143.

Desjeux, P., R. S. Ghosh, P. Dhalaria, N. Strub-Wourgaft, and E. E. Zijlstra. 2013. "Report of the Post Kala-azar Dermal Leishmaniasis (PKDL) Consortium Meeting, New Delhi, India, 27–29 June 2012." *Parasit Vectors* 6:196. doi: 10.1186/1756-3305-6-196.

Dixit, K. K., R. Singh, and P. Salotra. 2020. "Advancement in molecular diagnosis of Post Kala-Azar dermal leishmaniasis." *Indian J Dermatol* 65 (6):465–472. doi: 10.4103/ijd.IJD_311_19.

Farzan, M., M. Farzan, Y. Mirzaei, S. Aiman, F. Azadegan-Dehkordi, and N. Bagheri. 2023. "Immunoinformatics-based multiepitope vaccine design for the re-emerging monkeypox virus." *Int Immunopharmacol* 123:110725.

Fayaz, S., F. Bahrami, P. Fard-Esfahani, P. Parvizi, G. Bahramali, and S. Ajdary. 2022. "Immunoinformatics evaluation of a fusion protein composed of Leishmania infantum LiHyV and phlebotomus kandelakii apyrase as a vaccine candidate against visceral leishmaniasis." *Iran J Parasitol* 17 (2):145–158. doi: 10.18502/ijpa.v17i2.9530.

Franssen, S. U., Y. Takele, E. Adem, M. J. Sanders, I. Muller, P. Kropf, and J. A. Cotton. 2021. "Diversity and within-host evolution of Leishmania donovani from visceral leishmaniasis patients with and without HIV coinfection in Northern Ethiopia." *MBIO* 12 (3):e0097121. doi: 10.1128/mBio.00971-21.

Gherbi, R., M. Bounechada, M. S. Latrofa, G. Annoscia, V. D. Tarallo, F. Dantas-Torres, and D. Otranto. 2020. "Phlebotomine sand flies and Leishmania species in a focus of cutaneous leishmaniasis in Algeria." *PLoS Negl Trop Dis* 14 (2):e0008024. doi: 10.1371/journal.pntd.0008024.

Ghimire, P. G., P. Ghimire, J. Adhikari, and A. Chapagain. 2019. "A case report of visceral leishmaniasis and malaria co-infection with pancytopenia and splenomegaly - a diagnostic challenge." *BMC Infect Dis* 19 (1):849. doi: 10.1186/s12879-019-4478-1.

Goncalves, Svcb, and C. H. N. Costa. 2018. "Treatment of cutaneous leishmaniasis with thermotherapy in Brazil: an efficacy and safety study." *Ann Bras Dermatol* 93 (3):347–355. doi: 10.1590/abd1806-4841.20186415.

Ibiapina, A. B., F. M. A. Batista, B. G. A. Aguiar, V. J. Mendonca, D. L. Costa, C. H. N. Costa, and C. V. M. Abdala. 2022. "Evidence map of diagnosis, treatment, prognosis, prevention, and control in visceral leishmaniasis." *Rev Panam Salud Publica* 46:e89. doi: 10.26633/RPSP.2022.89.

Jancar, N., F. Sousa Goncalves, J. Duro, I. Pinto, T. Oliveira, and P. Aguiar. 2023. "Kala-Azar: a case report." *Cureus* 15 (2):e34864. doi: 10.7759/cureus.34864.

John, L., G. J. John, and T. Kholia. 2012. "A reverse vaccinology approach for the identification of potential vaccine candidates from Leishmania spp." *Appl Biochem Biotechnol* 167 (5):1340–1350. doi: 10.1007/s12010-012-9649-0.

Johny, A., and S. K. Mishra. 2023. "Pharmacological effects of bioactive compounds from allium sativum." In *Pharmacological Benefits of Natural Agents*, 13–30. IGI Global.

Keshri, A. K., R. Kaur, S. S. Rawat, N. Arora, R. K. Pandey, B. V. Kumbhar, A. Mishra, S. Tripathi, and A. Prasad. 2023. "Designing and development of multiepitope chimeric vaccine against Helicobacter pylori by exploring its entire immunogenic epitopes: an immunoinformatic approach." *BMC Bioinformatics* 24 (1):358. doi: 10.1186/s12859-023-05454-2.

Khan, M. A. A., J. Q. Ami, K. Faisal, R. Chowdhury, P. Ghosh, F. Hossain, A. Abd El Wahed, and D. Mondal. 2020. "An immunoinformatic approach driven by experimental proteomics: in silico design of a subunit candidate vaccine targeting secretory proteins of *Leishmania donovani* amastigotes." *Parasit Vectors* 13(1):196. doi: 10.1186/s13071-020-04064-8.

Khan, M. A. S., M. I. Miah, and S. R. Rahman. 2023. "A comprehensive immunoinformatic analysis of chitin deacetylase's and MP88 for designing multiepitope vaccines against Cryptococcus neoformans." *J Biomol Struct Dyn* 42:10711–10726 . doi: 10.1080/07391102.2023.2258410.

Khatoon, N., R. K. Pandey, and V. K. Prajapati. 2017. "Exploring Leishmania secretory proteins to design B and T cell multiepitope subunit vaccine using immunoinformatics approach." *Sci Rep* 7 (1):8285. doi: 10.1038/s41598-017-08842-w.

Kipp, E. J., M. de Almeida, P. L. Marcet, R. S. Bradbury, T. K. Benedict, W. Lin, E. M. Dotson, and M. Hergert. 2020. "An atypical case of autochthonous cutaneous leishmaniasis associated with naturally infected phlebotomine sand flies in Texas, United States." *Am J Trop Med Hyg* 103 (4):1496–1501. doi: 10.4269/ajtmh.20-0107.

Koley, S., R. K. Mandal, S. Choudhary, and A. Bandyopadhyay. 2013. "Post-kala-azar dermal leishmaniasis developing in miltefosine-treated visceral leishmaniasis." *Indian J Dermatol* 58 (3):241. doi: 10.4103/0019-5154.110848.

Kumar, A., V. K. Singh, R. Tiwari, P. Madhukar, Rajneesh, S. Kumar, V. Gautam, C. Engwerda, S. Sundar, and R. Kumar. 2023. "Post kala-azar dermal leishmaniasis in the Indian sub-continent: challenges and strategies for elimination." *Front Immunol* 14:1236952. doi: 10.3389/fimmu.2023.1236952.

Kumar, P., P. Kumar, N. Singh, S. Khajuria, R. Patel, V. K. Rajana, D. Mandal, and R. Velayutham. 2022. "Limitations of current chemotherapy and future of nanoformulation-based AmB delivery for visceral leishmaniasis: an updated review." *Front Bioeng Biotechnol* 10:1016925. doi: 10.3389/fbioe.2022.1016925.

Kumar, R., and S. Nylen. 2012. "Immunobiology of visceral leishmaniasis." *Front Immunol* 3:251. doi: 10.3389/fimmu.2012.00251.

Kushwaha, V., and S. Kaur. 2021. "Lymphatic filariasis and visceral leishmaniasis coinfection: a review on their epidemiology, therapeutic, and immune responses." *Acta Trop* 224:106117. doi: 10.1016/j.actatropica.2021.106117.

Lage, D. P., P. A. F. Ribeiro, D. S. Dias, D. V. C. Mendonca, F. F. Ramos, L. M. Carvalho, D. de Oliveira, B. T. Steiner, V. T. Martins, L. Perin, A. S. Machado, T. T. O. Santos, G. S. V. Tavares, J. A. Oliveira-da-Silva, J. S. Oliveira, B. M. Roatt, R. A. Machado-de-Avila, A. L. Teixeira, M. V. Humbert, E. A. F. Coelho, and

M. Christodoulides. 2020. "A candidate vaccine for human visceral leishmaniasis based on a specific T cell epitope-containing chimeric protein protects mice against Leishmania infantum infection." *NPJ Vaccines* 5:75. doi: 10.1038/s41541-020-00224-0.

Lari, A., N. Lari, and A. Biabangard. 2022. "Immunoinformatics approach to design a novel subunit vaccine against visceral leishmaniasis." *Int J Pept Res Ther* 28 (1):34. doi: 10.1007/s10989-021-10344-3.

Li, Y., Z. Luo, Y. Hao, Y. Zhang, L. Yang, Z. Li, Z. Zhou, and S. Li. 2022. "Epidemiological features and spatial-temporal clustering of visceral leishmaniasis in mainland China from 2019 to 2021." *Front Microbiol* 13:959901. doi: 10.3389/fmicb.2022.959901.

Mahto, K. K., P. Prasad, M. Kumar, I. Ali, V. Vohra, and D. K. Arya. 2023. "Visceral leishmaniasis: an overview and integrated analysis of the current status, geographical distribution and its transmission." *IntechOpen eBooks*2023: 110567.

Mann, S., K. Frasca, S. Scherrer, A. F. Henao-Martinez, S. Newman, P. Ramanan, and J. A. Suarez. 2021. "A review of leishmaniasis: current knowledge and future directions." *Curr Trop Med Rep* 8 (2):121–132. doi: 10.1007/s40475-021-00232-7.

Maru, H., K. Sharma, and S. K. Mishra. 2023. "A comprehensive review of ethnomedical uses, phytochemical studies, and properties of Holy Basil (Ocimum Sanctum Linn.)." *Pharmacol Benefits Nat Agents* 2023:49–68.

Mishra, S. K., P. Priya, G. P. Rai, R. Haque, and A. Shanker. 2023. "Coevolution based immunoinformatics approach considering variability of epitopes to combat different strains: a case study using spike protein of SARS-CoV-2." *Comput Biol Med* 163:107233. doi: 10.1016/j.compbiomed.2023.107233.

Moafi, M., H. Rezvan, R. Sherkat, and R. Taleban. 2019. "Leishmania vaccines entered in clinical trials: a review of literature." *Int J Prev Med* 10:95. doi: 10.4103/ijpvm.IJPVM_116_18.

Ojha, R., R. K. Pandey, and V. K. Prajapati. 2020. "Vaccinomics strategy to concoct a promising subunit vaccine for visceral leishmaniasis targeting sandfly and leishmania antigens." *Int J Biol Macromol* 156:548–557. doi: 10.1016/j.ijbiomac.2020.04.097.

Oli, A. N., W. O. Obialor, M. O. Ifeanyichukwu, D. C. Odimegwu, J. N. Okoyeh, G. O. Emechebe, S. A. Adejumo, and G. C. Ibeanu. 2020. "Immunoinformatics and vaccine development: an overview." *Immunotargets Ther* 9:13–30. doi: 10.2147/ITT.S241064.

Onile, O. S., F. Musaigwa, N. Ayawei, V. Omoboyede, T. A. Onile, E. Oghenevovwero, and R. T. Aruleba. 2022. "Immunoinformatics studies and design of a potential multiepitope peptide vaccine to combat the fatal visceral Leishmaniasis." *Vaccines (Basel)* 10 (10):1–15. doi: 10.3390/vaccines10101598.

Pacheco-Fernandez, T., G. Volpedo, S. Gannavaram, P. Bhattacharya, R. Dey, A. Satoskar, G. Matlashewski, and H. L. Nakhasi. 2021. "Revival of leishmanization and Leishmanin." *Front Cell Infect Microbiol* 11:639801. doi: 10.3389/fcimb.2021.639801.

Pal, B., S. Kumari, M. Kaur, P. Wadhwa, K. Murti, R. Kumar, K. Pandey, N. A. Siddiqui, S. Dhingra, and S. Padmakar. 2023. "Barriers to the effective management and prevention of post kala-azar dermal leishmaniasis (PKDL) in the Indian subcontinent." *Med J Armed Forces India* 79 (5):500–505. doi: 10.1016/j.mjafi.2023.03.010.

Pasandideh, R., M. Dadmanesh, S. Khalili, M. Mard-Soltani, and K. Ghorban. 2020. "In silico molecular modeling and docking studies on the Leishmania mitochondrial iron transporter-1 (LMIT1)." *Comparative Clinical Pathology* 29:115–125.

Rai, P., H. Arya, S. Saha, D. Kumar, and T. K. Bhatt. 2022. "Drug repurposing based novel anti-leishmanial drug screening using in-silico and in-vitro approaches." *J Biomol Struct Dyn* 40 (21):10812–10820. doi: 10.1080/07391102.2021.1950574.

Ready, P. D. 2013. "Biology of phlebotomine sand flies as vectors of disease agents." *Annu Rev Entomol* 58:227–250. doi: 10.1146/annurev-ento-120811-153557.

Safavi, M., H. Eshaghi, and Z. Hajihassani. 2021. "Visceral Leishmaniasis: Kala-azar." *Diagn Cytopathol* 49 (3):446–448. doi: 10.1002/dc.24671.

Saha, S., S. Vashishtha, B. Kundu, and M. Ghosh. 2022. "In-silico design of an immunoinformatics based multiepitope vaccine against *Leishmania donovani*." *BMC Bioinformatics* 23 (1):319. doi: 10.1186/s12859-022-04816-6.

Santos Junior, C. J. D., M. M. D. Santos, Fcco Lins, J. P. Silva, and K. C. Lima. 2023. "Temporal trend in the incidence of human visceral leishmaniasis in Brazil." *Cien Saude Colet* 28 (9):2709–2719. doi: 10.1590/1413-81232023289.15422022.

Sinha, M., R. Jagadeesan, N. Kumar, S. Saha, G. Kothandan, and D. Kumar. 2022. "In-silico studies on Myo inositol-1-phosphate synthase of Leishmania donovani in search of anti-leishmaniasis." *J Biomol Struct Dyn* 40 (8):3371–3384. doi: 10.1080/07391102.2020.1847194.

Soleymani, S., A. Tavassoli, and M. R. Housaindokht. 2021. "An overview of progress from empirical to rational design in modern vaccine development, with an emphasis on computational tools and immunoinformatics approaches." *Comput Biol Med* 140:105057. doi: 10.1016/j.compbiomed.2021.105057.

Sundar, S., J. Chakravarty, and L. P. Meena. 2019. "Leishmaniasis: treatment, drug resistance and emerging therapies." *Expert Opinion on Orphan Drugs* 7 (1):1–10.

Sundar, S., and O. P. Singh. 2018. "Molecular diagnosis of visceral leishmaniasis." *Mol Diagn Ther* 22 (4):443–457. doi: 10.1007/s40291-018-0343-y.

Tabrez, S., F. Rahman, R. Ali, F. Muhammad, B. M. Alshehri, M. A. Alaidarous, S. Banawas, A. A. B. Dukhyil, and A. Rub. 2021. "Repurposing of FDA-approved drugs as inhibitors of sterol C-24 methyltransferase of *Leishmania donovani* to fight against leishmaniasis." *Drug Dev Res* 82 (8):1154–1161. doi: 10.1002/ddr.21820.

Teixeira, M. C., G. G. Oliveira, P. O. Santos, T. C. Bahiense, V. M. da Silva, M. S. Rodrigues, D. F. Larangeira, W. L. Dos-Santos, and L. C. Pontes-de-Carvalho. 2011. "An experimental protocol for the establishment of dogs with long-term cellular immune reactions to Leishmania antigens." *Mem Inst Oswaldo Cruz* 106 (2):182–189. doi: 10.1590/s0074-02762011000200011.

Thomas, S., A. Abraham, J. Baldwin, S. Piplani, and N. Petrovsky. 2022. "Artificial intelligence in vaccine and drug design." *Methods Mol Biol* 2410:131–146. doi: 10.1007/978-1-0716-1884-4_6.

Vaghasia, V. V., K. Sharma, S. Kumar M., and J. J. Georrge. 2024. "In silico identification of natural product inhibitor for multidrug resistance proteins from selected gram-positive bacteria." In *Nanotechnology and In Silico Tools*, 309–317. Elsevier.

Vakhariya Sakina, S., S. K. Mishra, K. Sharma, and J. J. Georrge. 2023. "Designing of a novel curcumin analogue to inhibit mitogen-activated protein kinase: a cheminformatics approach." *J Phytonanotechnol Pharm Sci* 3 (1):37–47.

Vinjoda, P., S. K. Mishra, K. Sharma, and J. J. Georrge. 2024. "In silico identification of novel drug target and its natural product inhibitors for herpes simplex virus." In *Nanotechnology and In Silico Tools*, 377–383. Elsevier.

Volpedo, G., T. Pacheco-Fernandez, E. A. Holcomb, N. Cipriano, B. Cox, and A. R. Satoskar. 2021. "Mechanisms of Immunopathogenesis in Cutaneous Leishmaniasis And Post Kala-azar Dermal Leishmaniasis (PKDL)." *Front Cell Infect Microbiol* 11:685296. doi: 10.3389/fcimb.2021.685296.

Yadav, S., J. Prakash, H. Shukla, K. C. Das, T. Tripathi, and V. K. Dubey. 2020. "Design of a multiepitope subunit vaccine for immune-protection against Leishmania parasite." *Pathog Glob Health* 114 (8):471–481. doi: 10.1080/20477724.2020.1842976.

Zijlstra, E. E. 2016. "The immunology of post-kala-azar dermal leishmaniasis (PKDL)." *Parasit Vectors* 9 (1):464. doi: 10.1186/s13071-016-1721-0.

12 Review of Existing Control Measures for VL and PKDL

Innovative Approaches to Vector Control, Case Management, Vaccine Development, and Community Education

Prashant Saxena[1]*, *Madhvi Saxena*[2], *Anukrti Mishra*[3], *and Balashankar K.*[4]

[1]Department of Biotechnology Engineering, University Institute of Engineering, Chandigarh University, Mohali, Punjab, India
[2]Department of Computer Science, Vishwakarma University, Pune, Maharashtra, India
[3]School of Biotechnology, IFTM University, Moradabad, Uttar Pradesh, India
[4]Faculty of Bioinformatics, Sathyabama University, Chennai, Tamil Nadu, India
*Corresponding Author: Prashant Saxena, prashant347@gmail.com

12.1 INTRODUCTION

Visceral leishmaniasis (VL), also known as kala-azar, is a serious and potentially fatal disease caused by protozoan parasites of the *Leishmania* species. These parasites are primarily transmitted to humans through the bites of infected female sandflies belonging to the genera *Phlebotomus* in the Old World (Asia, Africa, Europe) and Lutzomyia in the New World (the Americas) (Dinc, 2002; Zijlstra, 2016). The most common species causing VL are *Leishmania donovani*, *Leishmania infantum* (also known as *Leishmania chagasi* in the Americas), and *Leishmania tropica*, which are responsible for distinct geographic distributions of the disease. VL presents a wide range of symptoms, which can include prolonged fever, weight loss, fatigue, enlarged spleen and liver (hepatosplenomegaly), anemia, and a compromised immune system. If left untreated, it can be fatal (Yeşilova et al., 2015).

VL is a significant public health concern in various parts of the world, particularly in regions where the sandfly vectors and *Leishmania* parasites responsible for the disease are endemic. VL is highly endemic in parts of South Asia, with the majority of cases occurring in India, Bangladesh, and Nepal. This region is considered the primary global hotspot for VL. The causative agent is typically *L. donovani*. VL is also prevalent in parts of East Africa, including Sudan, South Sudan, Ethiopia, Kenya, and Somalia, and is caused by both *L. donovani* and *L. infantum*. In South America, *L. infantum* is the primary species responsible for VL, which is endemic in several countries, including Brazil, Colombia, Venezuela, and Peru (Zink, 2006). VL also occurs in other parts of the world, including parts of the Middle East, Central Asia, and the Horn of Africa (Steverding, 2017).

DOI: 10.1201/9781032643120-12

FIGURE 12.1 Global distribution of VL. The size of the red dots refers to the intensity of the cases. The larger the dot, the higher the number of cases, and vice versa

Figure 12.1 (Karimi et al., 2016) illustrates the global distribution of leishmaniasis and its frequency of occurrence worldwide, with red dots indicating the intensity of infection in each area.

The presence and distribution of VL within these regions are influenced by various factors, including environmental conditions, the presence of reservoir hosts (such as infected dogs), and the effectiveness of public health interventions. Efforts to control VL typically involve vector control measures, early diagnosis, treatment with antiparasitic drugs, and, in some cases, the culling of infected animals (e.g., dogs) to reduce reservoirs of infection. Ongoing research focuses on the development of vaccines and improved diagnostic and treatment strategies (Gannavaram et al., 2016).

Preventing VL primarily involves controlling the sandfly vector through insecticide spraying, using bed nets, wearing protective clothing, and avoiding outdoor activities during peak sandfly activity times. Controlling reservoir hosts (such as infected dogs in some regions) can also help reduce transmission (Moafi et al., 2017) (Rezvan et al., 2015).

Diagnosis typically involves clinical evaluation, blood tests, bone marrow aspiration, and/or serological tests to detect antibodies against *Leishmania*. Molecular techniques such as polymerase chain reaction (PCR) can also be used for confirmation (Bezemer et al., 2022). Treatment options for VL include drugs such as pentavalent antimonials (e.g., sodium stibogluconate), amphotericin B, and miltefosine. The choice of treatment depends on the species of *Leishmania* and local drug-resistance patterns (Convit et al., 1999).

Post-kala-azar dermal leishmaniasis (PKDL) is a dermatological condition that can occur as a late complication of VL, typically developing months to years after successful treatment of the initial VL infection. It manifests as hyperpigmented (darkened) or hypopigmented (lightened) skin lesions, often appearing on the face, hands, and other parts of the body. While not life-threatening, PKDL can be disfiguring and significantly impact a person's quality of life (Cook & Zumla, 2008).

The exact mechanism behind the development of PKDL is not fully understood, but it is thought to involve the persistence of the *Leishmania* parasites in the skin even after the apparent cure of the visceral form. PKDL can also serve as a reservoir for the parasite, potentially contributing to the further transmission of leishmaniasis (Kumar et al., 2023).

The treatment for PKDL typically involves antileishmanial medications, such as miltefosine or amphotericin B, administered for several weeks to months, depending on the severity of the condition and the specific medication used. Prompt medical attention and appropriate treatment are essential to manage the skin lesions effectively and prevent further transmission of the disease (Kumar et al., 2023).

PKDL is primarily found in regions where VL is endemic, as both are caused by the same parasite, *L. donovani*. It is most commonly reported in the Indian subcontinent, including parts of India, Nepal, and Bangladesh, with Bihar, India being a well-known endemic region for both VL and PKDL. Cases have also been reported in East African countries such as Sudan, South Sudan, Ethiopia, and Kenya. While PKDL can also be found in other parts of the world where VL is endemic, it is less frequently reported in these areas.

It's important to note that the distribution of PKDL may change over time due to such as climate change, human migration, and vector control efforts. Additionally, the burden of PKDL may be underreported in some areas due to limited healthcare access and surveillance.

Efforts to control and eliminate VL often include strategies to manage PKDL cases, as PKDL can serve as a reservoir for the parasite, potentially leading to VL re-emergence (Costa-Madeira et al., 2022). Public health organizations and researchers strive to better understand and address the distribution of both VL and PKDL to reduce the overall burden of these diseases.

This chapter includes a broad discussion of various issues related to VL and PKDL, including both classical and innovative approaches to vector control, as well as case management systems, community education, diagnosis and treatment methods, and vaccination processes. It offers a roadmap for the research and scientific community to address this infection and develop effective medicines and vaccines for the treatment and benefit of human society.

12.2 VECTOR CONTROL

Vector control is a critical component of public health efforts to prevent the transmission of vector-borne diseases. It involves strategies and measures aimed at reducing or eliminating populations of disease-carrying vectors (usually arthropods like mosquitoes, ticks, and sandflies) or interrupting their ability to transmit pathogens to humans and animals. Common methods for vector control include spraying in vector-prone areas, using insecticide-treated bed nets, and making environmental modifications. However, these primary methods are not too effective in the present-day scenario (Garlapati et al., 2021). To address this, we need more advanced and innovative methods for vector control. This chapter discusses the applications of the Internet of Things (IoT) in vector control (Kim et al., 2019) (Liao et al., 2019).

12.2.1 Methods for Vector Control

12.2.1.1 Remote Monitoring and Surveillance

IoT devices, such as sensors and cameras, can be deployed in vector-prone areas to monitor environmental conditions, including temperature, humidity, and rainfall, which affect vector populations. These devices can also monitor vector presence and behaviour, including the movement of insects like mosquitoes or sandflies. Real-time data from these sensors can be transmitted to central databases or control centres, enabling timely responses to emerging vector threats (Islam MR et. al., 2023).

12.2.1.2 Data Analysis and Predictive Modelling

IoT-generated data can be analyzed using machine learning and data analytics to create predictive models for vector-borne diseases (Oikonomidi, T et al., 2023). These models can forecast vector population dynamics, disease transmission risk, and outbreak potential, helping authorities allocate resources more effectively.

12.2.1.3 Remote Vector Trapping and Monitoring

IoT-enabled traps and monitoring systems can capture vectors and transmit data on their abundance and infection rates. Researchers and control teams can remotely access this data, reducing the need for physical inspections and minimizing risks to field workers.

12.2.1.4 Weather and Climate Data Integration

IoT can integrate weather and climate data into vector control efforts. For example, IoT devices can collect weather information and use it to predict vector breeding sites or disease transmission patterns. Weather data can also help identify conditions conducive to vector proliferation, enabling proactive control measures.

12.2.1.5 Smart Vector Control Devices

IoT can enhance the effectiveness of vector control methods, such as insecticide spraying or larvicide application, by automating processes based on real-time data. For instance, drones equipped with IoT technology can target specific areas for spraying based on sensor data indicating high vector activity (Knapp, J. et al., 2015).

12.2.1.6 Early Warning Systems

IoT data can feed into early warning systems that alert authorities and healthcare providers to potential disease outbreaks. This timely information can lead to rapid response efforts, including the distribution of vaccines, medications, or bed nets to at-risk populations.

12.2.2 Applications of IoT in Vector Control

IoT has various applications in vector control. In contrast to traditional methods, modern approaches are gaining prominence in the developmental scenario (Byul Hur et al., 2021). Figure 12.2 (data collected from various online resources) contains previous data suggesting that technology is making a meaningful impact on vector control. IoT-based methods are being used in pest control, remote monitoring and surveillance, data analysis, and data collection. Another very important role of IoT is analyzing the obtained data and providing better forecasting to implement effective measures

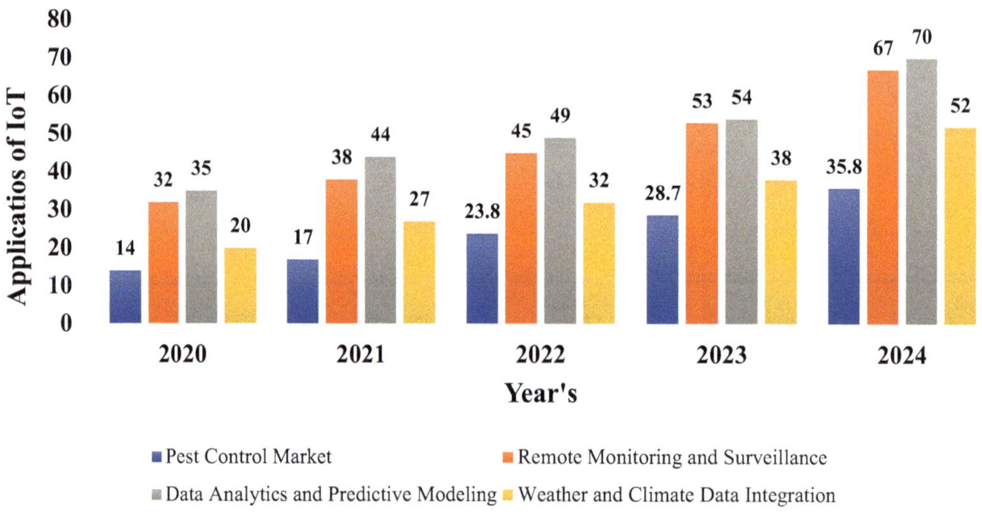

FIGURE 12.2 Role of IoT in various fields of vector control. Blue represents the market for pest control; orange represents remote monitoring and surveillance; gray represents data analytics and predictive modelling; and yellow represents weather and climate data integration. Data is represented in percentages

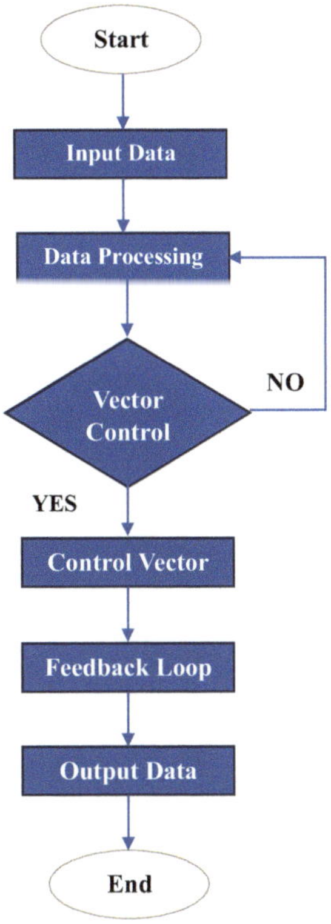

FIGURE 12.3 The top-down approach of controlling vectors using IoT-based methods

for controlling the vectors in infection-prone areas. Figures 12.2 and 12.3 illustrate the increasing demand for such methods in addressing vector control challenges.

12.2.2.1 Case Management

Case management for vector-borne diseases involves a coordinated approach to identifying, diagnosing, treating, and monitoring individuals who have contracted these diseases (Castro et al., 2023). Vector-borne diseases are transmitted to humans through the bites of infected vectors like mosquitoes, ticks, and fleas. The following are some case management methods and strategies commonly used for vector-borne diseases.

12.2.2.2 Early Detection and Diagnosis

The first step in case management is to detect the active cases as early as possible. An effective surveillance system is needed to detect the areas that are more prone or highly susceptible to infection. Additionally, identifying outbreaks helps measure the intensity of the infection. Such surveillance methods help in better understanding the future situation and taking effective measures to manage infections. Effective diagnostic methods are also crucial for identifying infected persons at a very early stage and isolating them from society for a period of time, helping to control the spread of infection. Figure 12.4 describes the detailed process for identifying and managing infection (Dubey et al., 2021).

12.2.2.3 Healthcare Infrastructure

A robust healthcare infrastructure is essential to diagnose and manage infections. Healthcare infrastructure refers to the foundational framework and resources that support the delivery of healthcare services in a particular region, country, or healthcare system (Salari et al., 2022). It encompasses

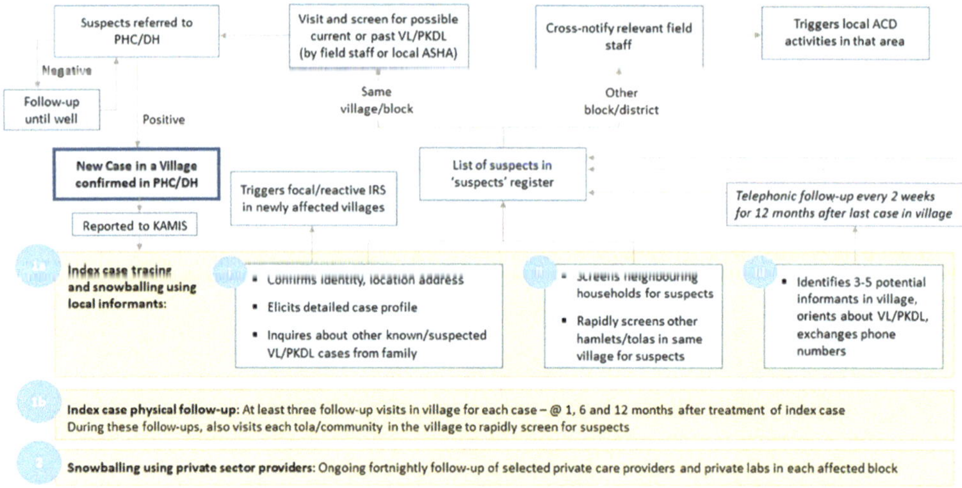

FIGURE 12.4 Schematic representation of the steps involved in the detection and management of active cases of kala-azar

a wide range of elements and components that work together to ensure the provision of medical care to individuals and communities. Healthcare facilities include hospitals, clinics, nursing homes, and other establishments where medical services are provided (de Vries & Schallig, 2022). These facilities vary in size and specialization, ranging from small rural clinics to large urban hospitals with specialized departments. Robust health information systems are essential for tracking patient data, managing healthcare records, and facilitating communication between healthcare providers. Effective planning, organization, and management of healthcare infrastructure are necessary to ensure the efficient and equitable distribution of healthcare resources (Bailey et al., 2019).

The integration of telemedicine and digital health technologies has become increasingly important, especially in providing healthcare services remotely, improving access, and managing healthcare data (Rushton et al., 2012). The quality and accessibility of healthcare infrastructure can vary widely from one country or region to another. Inadequate healthcare infrastructure can lead to disparities in access to care and health outcomes (Pelican et al., 2019). Developing and maintaining a robust healthcare infrastructure is a priority for many governments and organizations to ensure the well-being of their populations. Figure 12.5 (Ghai et al., 2022) depicts the classical framework of healthcare infrastructure functioning and setup to address infectious diseases.

12.2.2.4 Training Programme

Training for case management of vector-borne diseases is essential for effectively preventing, diagnosing, treating, and controlling diseases transmitted by vectors such as mosquitoes, ticks, and flies. These diseases include malaria, dengue fever, Zika virus, Lyme disease, and many others. Case management training typically targets healthcare professionals, public health workers, and community health workers. Understanding the epidemiology of vector-borne diseases, including their distribution, transmission cycles, and seasonal variations, is fundamental. Trainees should learn how to monitor and report cases through disease surveillance systems. Knowledge of vector biology is crucial for understanding disease transmission (Worsley et al., 2016). Training should cover the identification of vectors, their habitats, and strategies for vector control, such as insecticide-treated bed nets and indoor residual spraying (Kothari et al., 2016).

Recognizing the clinical symptoms of vector-borne diseases is vital for early diagnosis and treatment. Trainees should learn how to diagnose these diseases using laboratory tests and clinical evaluations. Familiarity with treatment guidelines is essential, as the management of vector-borne diseases often involves specific medications and therapies (Rowe et al., 2021). Trainees should be

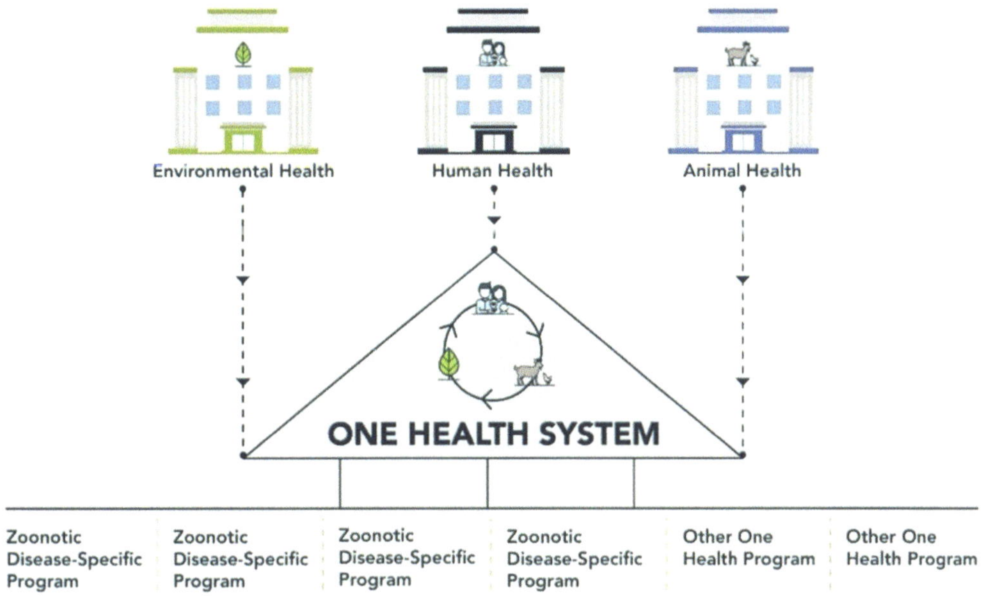

FIGURE 12.5 Organization of a healthy healthcare system mainly for vector-borne infections

up to date on the latest treatment recommendations. Healthcare workers should be trained to educate patients and communities about vector-borne diseases, their prevention, and the importance of seeking care promptly if symptoms develop. Effective data management and reporting systems are crucial for tracking and responding to outbreaks. Training should cover data collection, analysis, and reporting procedures. Training programmes for case management of vector-borne diseases are often provided by government health agencies, nongovernmental organizations, international health organizations, and academic institutions. These programmes may include classroom-based instruction, practical exercises, fieldwork, and ongoing professional development opportunities. Additionally, online courses and resources are increasingly available to help healthcare professionals and public health workers acquire the necessary skills and knowledge to effectively manage vector-borne diseases.

12.2.2.5 Vaccine Development

A safe and effective vaccine is urgently needed to provide long-lasting protective immunity against parasitic diseases. Although it is well recognized that cell-mediated immunity is essential for host defence, the immunological protective mechanism against leishmaniasis is complex. Currently, there is no approved vaccine for human leishmaniasis, but significant efforts are being made to develop a range of promising vaccine modalities worldwide (Dinc, 2022). Some of these candidates are now in the clinical stage, with positive outcomes in terms of safety and immunogenicity. Animal vaccines are expected to be crucial in halting the spread of leishmaniasis to humans in the meantime. Additionally, several vaccine candidates, including those in the first, second, and third generations, are being assessed at various stages of clinical trials (Alvar et al., 2021).

12.2.3 Leishmaniasis Immunological Mechanism

Leishmaniasis has a complex immune protection system. Host immunity heavily depends on cell-mediated immunity. While the innate branch of the immune system initiates the primary immunological response to infection, producing protective innate and adaptive immunity, infection management is primarily handled by an IL-12-driven Th1-type immune response (Srivastava et al.,

2016; Malvolti et al., 2021). T cells shape the host immune system to provide either direct protective immunity or nonprotective immunity.

Patients who recover from leishmaniasis or undergo leishmanization have protective immunity against reinfection, which is unusual for parasitic illnesses (Abdellahi et al., 2022). This finding indicates that immunological mechanisms play a significant part in the development of the disease, and that vaccines may potentially prevent this serious condition (Pacheco-Fernandez et al., 2021). It also offers a blueprint for the creation of effective vaccines that can produce infection-preventive immunity.

12.2.4 Approaches for Leishmaniasis Vaccine Development

There are various possible methods for effectively developing vaccines for infectious diseases. Several strategies have been employed to develop a vaccine for leishmaniasis. Some of them are listed below.

12.2.4.1 First-Generation Vaccine

First-generation vaccines are made of completely killed parasites, partially purified fractions, or parasite excretions. Different techniques, including long-term in vitro culture, heating, pressure, radiation, and chemicals, are used to kill the parasites (Helou et al., 2021). These dead parasite vaccines offer a broad range of parasite antigens and, by simulating real infections, can significantly increase protection against infection. Studies indicate that, compared to live parasites, they elicit a weaker Th1-type immune response, and the results are associated with a mixed effect. These factors have led to the introduction of adjuvants and alternative delivery methods for parasites, such as the mucosal route, in numerous trials.

12.2.4.2 Second-Generation Vaccine

While first-generation vaccines are still being studied, numerous research studies have concentrated on second-generation vaccines. These include purified native protein fractions of parasite antigens, synthetic peptides, genetically modified parasites, and various recombinant proteins produced by genetically modified organisms like viruses and bacteria. Mass immunization is more practical with second-generation vaccines, and their recombinant nature makes them more accessible for high-volume, low-cost production (Velez & Gallego, 2020).

12.2.4.3 Third-Generation Vaccine

Third-generation vaccines use DNA. This relatively new technique involves injecting bare plasmid DNA or DNA enclosed in a viral vector intradermally or intramuscularly (Moafi et al., 2019). DNA vaccines are safe because they do not contain any dangerous organisms that could change their virulence. Additionally, they effectively activate dendritic cells and produce interferon-gamma, protecting against *Leishmania* infection (Aunguldee et al., 2021).

In addition to the aforementioned, scientific advancements have prompted research into novel vaccine production methods, such as the use of nonpathogenic *Leishmania* species and specific bioengineering techniques, including delivery systems and chimeric vaccines.

The use of drug delivery systems in the treatment of infectious and cancerous disorders has received substantial research. Nanoparticles (NP) are thought to be the best vehicle for delivering vaccines (Taslimi et al., 2020). Due to their structure, vaccine delivery systems have several advantages, including site-specific administration, controlled antigen release, and protection from oxidation. They may also act as adjuvants. All these properties of NP carriers increase the bioavailability of antigens, which triggers the immunological response (Cecilio et al., 2017).

The in silico approach is also an advanced and appropriate method for designing small peptide-based vaccines for infectious diseases (Saxena & Mishra, 2021). In this method, antigenic determinants (epitopes) are identified using machine learning and artificial intelligence-based algorithms,

TABLE 12.1
List of Various Undertrial Vaccines for Treating VL and PKDL

S.No.	Name of Vaccine	Action of Mechanism	Effective Outcome
1	Leishvaccine	CD4+, CD+, and B cell activation	Despite being safe, the vaccine did not have the intended impact
2	*L. mexicana* + BCG	T cell activation	Leishmaniasis incidence significantly decreased
3	ALM + BCG +	T cell activation	Decrease in the prevalence of VL
4	Leishmune	T cell activation	High level of protection against canine VL
5	Gentamycin-attenuated *L. infantum*	T cell and B cell activation	Increased defences against canine VL
6	ChAd63KH	Broad activation of CD8+ and T cells	The vaccine provides active immunity against VL and PKDL

TABLE 12.2
Current Position of Different Undertrial Vaccines

S. No.	Name of Vaccine	Phase I	Phase II	Phase III	Reference
1	Leishvaccine			√	Teixeira et al., 2011
2	*L. mexicana* + BCG		√		De Luca, 2016
3	ALM + BCG		√		IRCT, 2015
4	Leishmune			√	Chakravarty et al., 2011
5	Gentamycin-attenuated *L. infantum*	√			USNLAM, 2009
6	ChAd63KH		√		Younis et al., 2021

and the identified epitopes are tested against the human leukocyte antigen (HLA) molecules. The best-matching epitope is confirmed as the best vaccine candidate because it is assumed that this region of the viral protein could provide a better immunological response (Saxena & Mishra, 2020).

Various vaccines have been identified using the above-mentioned methods. Table 12.1 describes their names, mechanisms of action, and outcomes. Table 12.2 contains information about the current trial phases of these vaccines.

From the above tables, we can expect that the efforts of the scientific community will soon yield an approved vaccine for treating VL or PKDL.

12.3 COMMUNITY EDUCATION

Community education is a process of providing learning opportunities and information to individuals and groups within a community to empower them with knowledge and skills that contribute to personal development, community improvement, and social change. It plays a crucial role in promoting awareness, fostering community engagement, and addressing various social, economic, and health-related issues. Community education includes several aspects, such as programme design, community engagement and participation, skill building, healthcare education, and technology-driven methods. These methods help the community better understand infections and possible solutions to cope with them at early stages to protect society from deadly infection diseases (Pang et al., 2022).

12.3.1 Programme Design

Community education programmes are designed to address a wide range of topics and issues, including health, literacy, job training, environmental sustainability, and more. Programme development involves creating curricula, materials, and instructional methods that align with the community's needs. Ensuring that education is accessible to all community members is critical. This includes considering factors like location, transportation, language, and cultural sensitivity to remove barriers to participation (Scott et al., 2020).

12.3.2 Engagement and Participation

Effective community education encourages active engagement and participation from community members. It often involves workshops, seminars, community meetings, and interactive learning experiences. Engaging local community members in the planning, implementation, and evaluation of education programmes fosters a sense of ownership and ensures that programmes are relevant and effective. Community education is not limited to formal schooling; it promotes lifelong learning, encouraging individuals of all ages to continue acquiring new knowledge and skills throughout their lives.

12.3.3 Skill Development

Many community education programmes focus on skill development, such as adult literacy classes, vocational training, financial literacy, and parenting skills workshops. Community education empowers individuals and communities by providing them with the tools and knowledge to make informed decisions and advocate for change. These programmes often align with broader community development initiatives, aiming to improve the overall well-being and quality of life for community members (Smith et al., 2021)

12.3.4 Health Education

Health-related community education programmes provide information on topics such as disease prevention, nutrition, sexual health, mental health, and substance abuse prevention. These education sessions raise awareness about maintaining hygiene in the locality, encourage wearing full clothing during the evening, ensure proper isolation of diseased individuals, and stress taking medication at prescribed times. Such small steps can prevent the spread of infection and protect society from infectious diseases (Pavloff & Labrecque, 2021).

Community education is often delivered through a variety of institutions and organizations, including schools, community centres, nonprofit organizations, government agencies, and grassroots community groups. The approach to community education can vary widely based on the specific goals and needs of the community it serves. Ultimately, it plays a crucial role in promoting social inclusion, civic engagement, and the overall well-being of individuals and communities.

12.4 DISCUSSION

VL and PKDL are re-emerging infectious diseases transmitted via the bites of sandflies carrying protozoan parasites, with dogs serving as reservoirs for the vector. Symptoms of the infection include body rashes, fever, weight loss, and enlargement of the spleen. The symptoms may persist for weeks or longer, and if not treated properly, the infection could be fatal. PKDL is a late-onset complication following VL infection. To date, no direct medication is available to treat the infection; instead, indirect therapies such as antipyretic drugs to lower body temperature and other drugs to treat various symptoms are used. Several vaccine candidates are being tested, suggesting that if the trials are successful, society could have vaccines for both VL and PKDL.

In this chapter, the authors have discussed various aspects of the disease, such as innovative methods for vector control, case management methods, the process of vaccine development and its current status, and community education systems. The authors have discussed IoT-based methods that will help society and key players in the field better understand the situation ahead of time, aiding healthcare practitioners in preparing themselves to deal with this infectious disease effectively.

12.5 CONCLUSION

Leishmaniasis is a complex, vector-borne disease that poses significant challenges worldwide. Globalization, climate change, and population fluctuation are all contributing factors to the spread of this disease. Leishmaniasis control efforts are consistently advancing thanks to the combined efforts of governments and nonprofit organizations, and the commitment of endemic countries is crucial for long-term success. The development of efficient vaccines is the ideal solution for the eradication of leishmaniasis globally. With new investigations into improved therapies, it may be possible to eliminate this disease within a few decades. This chapter describes the current scenario of the disease and will help researchers understand the infection more closely, potentially leading to the eradication of leishmaniasis from society.

ACKNOWLEDGEMENT

The authors are very thankful to the management of IFTM University, Chandigarh University, Vishwakarma University, and Sathyabama Institute of Science and Technology for providing all the necessary resources to complete this chapter.

REFERENCES

Abdellahi, L., Iraji, F., Mahmoudabadi, A., & Hejazi, S. H. (2022). Vaccination in leishmaniasis: a review article. *Iranian Biomedical Journal*, *26*(1), 1.

Alvar, J., den Boer, M., & Dagne, D. A. (2021). Towards the elimination of visceral leishmaniasis as a public health problem in east Africa: reflections on an enhanced control strategy and a call for action. *The Lancet Global Health*, *9*(12), e1763–e1769.

Aunguldee, T., Gerdprasert, O., Tangteerawatana, P., Jariyapongskul, A., Leelayoova, S., & Wongsatayanon, B. T. (2021). Immunogenicity and potential protection of DNA vaccine of Leishmania martiniquensis against Leishmania infection in mice. *The Journal of Infection in Developing Countries*, *15*(09), 1328–1338.

Bailey, F., Mondragon-Shem, K., Haines, L. R., Olabi, A., Alorfi, A., Ruiz-Postigo, J. A., ... & Molyneux, D. H. (2019). Cutaneous leishmaniasis and co-morbid major depressive disorder: a systematic review with burden estimates. *PLoS Neglected Tropical Diseases*, *13*(2), e0007092.

Bezemer, J. M., Meesters, K., Naveda, C. L., Machado, P. R., Calvopiña, M., Leeflang, M. M., ... & de Vries, H. J. (2022). Clinical criteria for Mucosal leishmaniasis diagnosis in rural South America: a systematic literature review. *PLoS Neglected Tropical Diseases*, *16*(8), e0010621.

Byul Hur, et al. (2021). IoT environmental-monitoring system development for mosquito research through capstone project integration in engineering technology. *ASEE Annual Conference*, American Society of Engineering Education.

Castro, M. D. M., Rode, J., Machado, P. R. L., Llanos-Cuentas, A., Hueb, M., Cota, G., Rojas, I. V., Orobio, Y., Oviedo Sarmiento, O., Rojas, E., Quintero, J., Pimentel, M. I. F., Soto, J., Suprien, C., Alvarez, F., Ramos, A. P., Arantes, R. B. D. S., da Silva, R. E., Arenas, C. M., Vélez, I. D., … Alexander, N. (2023). Cutaneous leishmaniasis treatment and therapeutic outcomes in special populations: a collaborative retrospective study. *PLoS Neglected Tropical Diseases*, *17*(1), e0011029. https://doi.org/10.1371/journal.pntd.0011029

Cecilio, P., Pérez-Cabezas, B., Fernández, L., Moreno, J., Carrillo, E., Requena, J. M., ... & Cordeiro-da-Silva, A. (2017). Pre-clinical antigenicity studies of an innovative multivalent vaccine for human visceral leishmaniasis. *PLoS Neglected Tropical Diseases, 11*(11), e0005951.

Chakravarty, J., Kumar, S., Trivedi, S., Rai, V. K., Singh, A., Ashman, J. A., ... & Piazza, F. M. (2011). A clinical trial to evaluate the safety and immunogenicity of the LEISH-F1+ MPL-SE vaccine for use in the prevention of visceral leishmaniasis. *Vaccine, 29*(19), 3531–3537.

Convit, J., Ulrich, M., Zerpa, O., Borges, R., Aranzazu, N., Valera, M., ... & Tomedes, I. (2003). Immunotherapy of American cutaneous leishmaniasis in Venezuela during the period 1990–1999. *Transactions of the Royal Society of Tropical Medicine and Hygiene, 97*(4), 469–472.

Cook, G. C., & Zumla, A. (2008). *Manson's Tropical Diseases*. Elsevier Health Sciences.

Costa-Madeira, J. C., Trindade, G. B., Almeida, P. H., Silva, J. S., & Carregaro, V. (2022). T lymphocyte exhaustion during human and experimental visceral leishmaniasis. *Frontiers in Immunology, 13*, 835711.

De Luca, P. M., & Macedo, A. B. B. (2016). Cutaneous leishmaniasis vaccination: a matter of quality. *Frontiers in Immunology, 7*, 151.

de Vries, H. J., & Schallig, H. D. (2022). Cutaneous leishmaniasis: a 2022 updated narrative review into diagnosis and management developments. *American Journal of Clinical Dermatology, 23*(6), 823–840.

Dinc, R. (2022). New developments in the treatment of cutaneous leishmaniasis. *Asian Pacific Journal of Tropical Medicine, 15*(5), 196–205.

Dubey, P., Das, A., Priyamvada, K., Bindroo, J., Mahapatra, T., Mishra, P. K., ... & Srikantiah, S. (2021). Development and evaluation of active case detection methods to support visceral leishmaniasis elimination in India. *Frontiers in Cellular and Infection Microbiology, 11*, 648903.

Gannavaram, S., Bhattacharya, P., Ismail, N., Kaul, A., Singh, R., & Nakhasi, H. L. (2016). Modulation of innate immune mechanisms to enhance Leishmania vaccine-induced immunity: role of coinhibitory molecules. *Frontiers in Immunology, 7*, 187.

Garlapati, R., Iniguez, E., Serafim, T. D., Mishra, P. K., Rooj, B., Sinha, B., ... & Kamhawi, S. (2021). Towards a sustainable vector-control strategy in the post kala-azar elimination era. *Frontiers in Cellular and Infection Microbiology, 11*, 641632.

Ghai, R.R., Wallace, R.M., Kile, J.C. et al. (2022). A generalizable one health framework for the control of zoonotic diseases. *Scientific Report,* 12, 8588. https://doi.org/10.1038/s41598-022-12619-1

Helou, D. G., Mauras, A., Fasquelle, F., Lanza, J. S., Loiseau, P. M., Betbeder, D., & Cojean, S. (2021). Intranasal vaccine from whole *Leishmania donovani* antigens provides protection and induces specific immune response against visceral leishmaniasis. *PLoS Neglected Tropical Diseases, 15*(8), e0009627. .

Islam, M.R., Kabir, M.M., Mridha, M.F., Alfarhood, S., Safran, M., & Che, D. (2023). Deep learning-based IoT system for remote monitoring and early detection of health issues in real-time. *Sensors, 23*(11), 5204. https://doi.org/10.3390/s23115204

Karimi, A., Alborzi, A., & Amanati, A. (2016). Visceral leishmaniasis: an update and literature review. *Arch Pediatr Infect Dis, 4*(3), e31612. https://doi.org/10.5812/pedinfect.31612.

Kim, K., Hyun, J., Kim, H., Lim, H., & Myung, H. (2019). A deep learning-based automatic mosquito sensing and control system for urban mosquito habitats. *Sensors, 19*(12), 2785.

Knapp, J., Macdonald, M., Malone, D. et al. (2015). Disruptive technology for vector control: the Innovative Vector Control Consortium and the US Military join forces to explore transformative insecticide application technology for mosquito control programmes. *Malar J 14*, 371.

Kothari, M., Maidment, I., Lyon, R., & Haygarth, L. (2016). Medicines reconciliation in comparison with NICE guidelines across secondary care mental health organisations. *International Journal of Clinical Pharmacy, 38*, 289–295.

Kumar, A., Singh, V. K., Tiwari, R., Madhukar, P., Rajneesh, Kumar, S., Gautam, V., Engwerda, C., Sundar, S., & Kumar, R. (2023). Post kala-azar dermal leishmaniasis in the Indian sub-continent: challenges and strategies for elimination. *Front. Immunol. 14*, 1236952. doi: 10.3389/fimmu.2023.1236952

Liao, L. D., Wang, Y., Tsao, Y. C., Wang, I. J., Jhang, D. F., Chu, T. S., ... & Ger, T. R. (2019). Design and validation of a multifunctional android-based smart home control and monitoring system. *IEEE Access, 7*, 163313–163322.

Malvolti, S., Malhame, M., Mantel, C. F., Le Rutte, E. A., & Kaye, P. M. (2021). Human leishmaniasis vaccines: Use cases, target population and potential global demand. *PLoS Neglected Tropical Diseases, 15*(9), e0009742.

Moafi, M., Rezvan, H., Sherkat, R., & Taleban, R. (2019). Leishmania vaccines entered in clinical trials: a review of literature. *International Journal of Preventive Medicine, 10*, 1–6.

Moafi, M., Rezvan, H., Sherkat, R., Taleban, R., Asilian, A., Hamid Zarkesh-Esfahani, S., ... & Ansari, N. (2017). Comparison of pro-inflammatory cytokines of non-healing and healing cutaneous leishmaniasis. *Scandinavian Journal of Immunology*, *85*(4), 291–299.

Murray, H. W., Berman, J. D., Davies, C. R., & Saravia, N. G. (2005). Advances in leishmaniasis. *The Lancet*, *366*(9496), 1561–1577.

Oikonomidi, T., Norman, G., McGarrigle, L., Stokes, J., van der Veer, S. N., & Dowding, D. (2023). Predictive model-based interventions to reduce outpatient no-shows: a rapid systematic review. *Journal of the American Medical Informatics Association*, *30*(3), 559–569.

Pacheco-Fernandez, T., Volpedo, G., Gannavaram, S., Bhattacharya, P., Dey, R., Satoskar, A., ... & Nakhasi, H. L. (2021). Revival of leishmanization and leishmanin. *Frontiers in Cellular and Infection Microbiology*, *11*, 639801.

Pang, X., Zhang, M., & Pang, H. (2022). Analysis of the demand for continuing education of nurses in the department of infectious diseases and its influencing factors. *Evidence-Based Complementary and Alternative Medicine: eCAM*, *2022*, 3743588.

Pavloff, M., & Labrecque, M. E. (2021). Continuing education for home care nurses: an integrative literature review. *Home Healthcare Now*, *39*(6), 310–319.

Pelican, K., Salyer, S. J., Behravesh, C. B., Belot, G., Carron, M., Caya, F., ... & Dutcher, T. (2019). Synergising tools for capacity assessment and One Health operationalisation. *Revue scientifique et technique (International Office of Epizootics)*, *38*(1), 71.

Rezvan, H., Moafi, M., Malekinejad, H., Alizadeh-Tabrizi, N., Ostadi, A., Fink-Gremmels, J., ... & Haghighi, Z. M. S. (2015). Veterinary research forum. *Urmia, Iran: Faculty of Veterinary Medicine, Urmia University.*

Rowe, A. K., Rowe, S. Y., Peters, D. H., Holloway, K. A., & Ross-Degnan, D. (2021). The effectiveness of training strategies to improve healthcare provider practices in low-income and middle-income countries. *BMJ Global Health*, *6*(1), e003229.

Rushton, J., Hasler, B., & De Haan, N. (2012). Economic benefits or drivers of a 'One Health' approach: why should anyone invest?: proceeding. *Onderstepoort Journal of Veterinary Research*, *79*(2), 1–5.

Salari, S., Bamorovat, M., Sharifi, I., & Almani, P. G. N. (2022). Global distribution of treatment resistance gene markers for leishmaniasis. *Journal of Clinical Laboratory Analysis*, *36*(8), e24599.

Saxena, P., & Mishra, S. (2020). Study of the binding pattern of HLA Class I alleles of indian frequency and cTAP binding peptide for chikungunya vaccine development. *International Journal of Peptide Research and Therapeutics*, *26*(4), 2437–2448.

Saxena, P., & Mishra, S. (2021). Chikungunya outbreak from 1952–2019: a prospective to cope-up with this re-emerging disease. *Biointerface Research in Applied Chemistry*, *11*(4), 11724–11736.

Scott, P. N., Lefler, L. L., Jeffs, D., Davis, A., Lear, T., & Haushalter, A. (2020). A tristate education model to prepare nurses to serve on health-promoting boards. *The Journal of Continuing Education in Nursing*, *51*(11), 528–532.

Smith, T. S., Holland, A. C., White, T., Combs, B., Watts, P., & Moss, J. (2021). A distance accessible education model: teaching skills to nurse practitioners. *The Journal for Nurse Practitioners*, *17*(8), 999–1003.

Srivastava, S., Shankar, P., Mishra, J., & Singh, S. (2016). Possibilities and challenges for developing a successful vaccine for leishmaniasis. *Parasites & Vectors*, *9*, 1–15.

Steverding D. (2017). The history of leishmaniasis. *Parasites & Vectors*, *10*(1), 82. https://doi.org/10.1186/s13071-017-2028-5

Taslimi, Y., Zahedifard, F., Taheri, T., Doroud, D., Dizaji, S. L., Saljoughian, N., & Rafati, S. (2020). Comparison of protective potency of DNA and live vaccines expressing A2-CPA-CPB-CTE antigens against visceral leishmaniasis in syrian hamster as preliminary study. *Iranian Journal of Parasitology*, *15*(3), 383.

Teixeira, M. C. A., Oliveira, G. G. D. S., Santos, P. O. M., Bahiense, T. C., Silva, V. M. G. D., Rodrigues, M. S., ... & Pontes-de-Carvalho, L. C. (2011). An experimental protocol for the establishment of dogs with long-term cellular immune reactions to Leishmania antigens. *Memórias do Instituto Oswaldo Cruz*, *106*, 182–189.

Velez, R., & Gallego, M. (2020). Commercially approved vaccines for canine leishmaniosis: a review of available data on their safety and efficacy. *Tropical Medicine & International Health*, *25*(5), 540–557.

Worsley, C., Webb, S., & Vaux, E. (2016). Training healthcare professionals in quality improvement. *Future Hospital Journal*, *3*(3), 207–210. https://doi.org/10.7861/futurehosp.3-3-207

Yeşilova, Y., Aksoy, M., Sürücü, H. A., Uluat, A., Ardic, N., & Yesilova, A. (2015). Lip leishmaniasis: Clinical characteristics of 621 patients. *International Journal of Critical Illness and Injury Science*, *5*(4), 265.

Younis, B. M., Osman, M., Khalil, E. A. G., Santoro, F., Furini, S., Wiggins, R., Keding, A., Carraro, M., Musa, A. E. A., Abdarahaman, M. A. A., Mandefield, L., Bland, M., Aebischer, T., Gabe, R., Layton, A. M., Lacey, C. J. N., Kaye, P. M., & Musa, A. M. (2021). Safety and immunogenicity of ChAd63-KH vaccine in post-kala-azar dermal leishmaniasis patients in Sudan. *Molecular Therapy: The Journal of the American Society of Gene Therapy*, *29*(7), 2366–2377. https://doi.org/10.1016/j.ymthe.2021.03.020

Zijlstra, E. E. (2016). Visceral leishmaniasis: a forgotten epidemic. *Archives of Disease in Childhood*, *101*(6), 561–567.

Zink, A. R., Spigelman, M., Schraut, B., Greenblatt, C. L., Nerlich, A. G., & Donoghue, H. D. (2006). Leishmaniasis in ancient Egypt and Upper nubia. *Emerging Infectious Diseases*, *12*(10), 1616–1617. https://doi.org/10.3201/eid1210.060169

13 Public Health Implications and Challenges of VL and PKDL

*Ritesh Kumar Tiwari, Gaurav Kumar Pandit, and Meenakshi Singh**
Department of Botany, Patna University, Patna, Bihar, India
*Corresponding Author: Meenakshi Singh, Department of Botany, Patna University, Patna, Bihar, India

13.1 INTRODUCTION

Visceral leishmaniasis (VL) and post-kala-azar dermal leishmaniasis (PKDL) pose significant public health implications and challenges in affected regions. VL is a severe and potentially fatal parasitic disease transmitted by sandflies, primarily affecting the liver, spleen and bone marrow. The high morbidity and mortality rates associated with VL can strain healthcare systems and result in economic burdens for affected communities (WHO, 2021a). Leishmaniasis predominantly affects impoverished people residing in remote rural areas with poor housing and little to no access to modern healthcare facilities. Population displacement, malnutrition, lack of awareness and constrained financial resources increase the chances of contracting leishmaniasis. The diagnosis and treatment of leishmaniasis in endemic areas places a huge financial burden on already meagre financial resources at both the individual and community levels. Often, families are obliged to sell their land and livestock or take out loans at heavy interest rates to pay for the diagnosis and treatment of leishmaniasis.

Most infected humans remain asymptomatic, with only about 1–10% developing clinical symptoms, resulting in death if left untreated (Ostyn et al. 2011; Le Rutte et al. 2016). Between 5% and 20% of treated VL cases develop a long-lasting skin condition known as PKDL. Recent studies have identified that individuals with PKDL are equally infectious to sandflies as VL cases, making them an important reservoir of infection (Mondal et al. 2019, Le Rutte et al. 2019). However, the contribution of asymptomatic individuals to transmission has not yet been determined (Mondal et al. 2019; Nagill et al. 2011). After infection, a period of immunity follows, though the duration of this immunity remains debated.

PKDL is a dermatological complication that often occurs after VL treatment. Infected individuals develop skin lesions, which may serve as reservoirs for the transmission of the parasite to sandflies, further complicating control efforts (WHO, 2021b).

13.2 VISCERAL LEISHMANIASIS IN INDIA

According to the WHO, more than 1 billion people reside in areas considered endemic for leishmaniasis infection. Globally, it is estimated that approximately 30,000 and 1 million new cases of VL and cutaneous leishmaniasis (CL) occur annually, respectively.

VL, or kala-azar, is reported to be endemic in 75 countries across Africa, the Americas and Asia. In 2020, India accounted for 18% of the total global burden of kala-azar. It is a major health concern in India, with its presence in 54 districts across four states, namely Bihar (33 out of 38 districts), Uttar Pradesh (6 out of 75 districts), Jharkhand (4 out of 24 districts) and West Bengal

DOI: 10.1201/9781032643120-13

(11 out of 23 districts). According to micro-stratification surveys, 633 blocks in these four states are endemic. Sporadic cases are also reported in other states, including Himachal Pradesh, Assam, Madhya Pradesh, Puducherry, Uttaranchal, Haryana, Gujarat, Sikkim, Kerala and Tamil Nadu.

However, significant improvement has been achieved in recent decades in India. There was a substantial decrease of 98% in kala-azar cases, from 77,102 cases in 1992 (since the intensified activities were started) to 1,275 cases in 2021. To realise the WHO targets for eliminating kala-azar as a public health problem and reach the 2030 Sustainable Development Goals, the incidence of cases should be reduced to less than one case per 10,000 population at the block level. This target aligns with the new roadmap for neglected tropical diseases 2021–2030 (www.who.int/publications/i/item/9789240010352). It is reported that only seven blocks have yet to realise the targets, and almost 98% of blocks have already achieved the WHO elimination threshold by the end of 2021. The national programme for intensifying activities to achieve the targets and ensure the sustainability of kala-azar elimination is well supported by the WHO.

13.3 POST-KALA-AZAR DERMAL LEISHMANIASIS

In East Africa and Southeast Asia, including India, a complication of kala-azar is characterised by hypopigmented (discoloured) macular (flat skin) rash, sometimes in combination with nodular or maculopapular rash, seen in patients who have recovered from the disease. This complication, known as PKDL, typically appears between 6 months and a year after the apparent cure of kala-azar. However, it may also occur earlier or concurrently with kala-azar. PKDL cases often heal spontaneously in Africa but rarely in patients in India.

13.4 IMPLICATIONS

13.4.1 VL-HIV Co-Infection

VL-HIV co-infected people are highly prone to developing full-blown clinical disease, with high relapse and mortality rates. However, antiretroviral treatment has been shown to reduce the progression of the disease and delay relapses. The survival of co-infected patients is also increased. Forty-five countries reported *Leishmania*-HIV co-infection as of 2021. Ethiopia, Brazil and the state of Bihar in India report high *Leishmania*-HIV co-infection rates. According to a report in *The Lancet*, around a third of people with HIV live in VL-endemic areas. It has also been reported that in Bihar, India, around 2–7% of people infected with VL are co-infected with HIV. Northwestern Ethiopia presents even higher rates of around 20–40% of VL-HIV co-infection (Burki. 2022).

According to the WHO (www.who.int/news/item/08-06-2022-visceral-leishmaniasis-and-HIV-coinfection-WHO-publishes-new-guideline-with-region-specific-treatment-recommendations), HIV and *Leishmania* co-infection poses challenges to the control and elimination of VL because people infected with HIV are more vulnerable to the disease. *Leishmania* and HIV appear to reinforce each other, posing a very significant clinical and public health concern. Both conditions impact immunity, resulting in severe morbidity with limited treatment options, a higher relapse rate, exposure to medicines with increased toxicity and a higher mortality rate. This co-infection occurs in 45 countries, with the first reported case in southern Europe in the mid-1980s. A high incidence of co-infection is reported in Ethiopia, Brazil and the state of Bihar in India. This co-infection not only renders people vulnerable to comorbid conditions such as cryptococcal meningitis and tuberculosis but also to issues such as social stigma and human rights concerns.

13.4.2 Clinical Manifestations of VL-HIV Co-Infection

Co-infected patients have similar clinical manifestations as immunocompetent ones (Lindoso et al. 2016; Lindoso et al. 2014). However, co-infected patients may have certain immunological characteristics specific to the co-infection and thus differ from the clinical manifestations of patients not

TABLE 13.1
Comparison of Classical Symptoms in Immune-Competent Patients versus Co-Infected Patients

Classical Symptoms	Symptoms in VL-HIV Co-Infection	Reference
Fever, hepatosplenomegaly, pallor	Weakness, diarrhoea, cough, weight loss and undernourishment. Low frequency of fever and hepatosplenomegaly. More frequent gastrointestinal problems. Haemorrhagic phenomena in almost a third of patients due to enlargement of lymph nodes.	Lindoso et al. 2014; Lindoso et al. 2018; Coutinho et al. 2017; Lima et al. 2013

infected with HIV, making the diagnosis a challenging task (Lindoso et al. 2018) as they reflect similarity with other opportunistic diseases (Table 13.1).

Atypical manifestations are frequently observed when patients with co-infection present to clinical settings, and amastigotes are present at unusual sites, such as pleura, lymph nodes, skin, pericardium, oral and gastrointestinal mucosa, respiratory tract and Kaposi's sarcoma lesions. All these factors contribute to the poor outcome of the co-infection as the diagnosis often gets delayed or misdiagnosed as other opportunistic conditions (Diro et al. 2015; Lindoso et al. 2016). The rate of relapse and lethality are higher in co-infected patients compared to non-HIV patients (Lindoso et al. 2014). In Latin America, the Backspacelethality rate and relapse rate were 8.7–23.5% and 10–56.5% of cases, respectively (Lindoso et al. 2014). Table 13.2 and Figure 13.1 outline the main methods used for diagnosing VL-HIV co-infections based on sensitivity and specificity.

TABLE 13.2
Main Laboratory Techniques Used for the Diagnosis of Visceral Leishmaniasis with HIV Co-Infection (Lindoso et al. 2018)

Method Used in Lab	Specificity	Sensitivity	Brief Description	Comments
PCR	96%	87.2–98%	Molecular method with specific targets	Expensive; limited availability; high specificity and sensitivity
ELISA	90%	50–60%	Enzyme immunoassay using total antigens	Easy to perform; variable specificity and sensitivity based on antigens; may be positive in asymptomatic patients from endemic areas
DAT	90%	81%	Agglutination serological test	Good performance in serological tests
IFAT	93%	51–61%	Immunofluorescence test to detect anti-leishmania antibodies	Easy method; variable specificity and sensitivity based on antigens; may be positive in asymptomatic individuals from endemic areas
Rk39 rapid test	96–100%	46.6–81%	Immunochromatographic test using recombinant K39 antigen	High specificity and sensitivity; easy and rapid method
Microscopic examination (bone marrow aspiration)	100%	93.2%	Microscopic demonstration of amastigote forms using Giemsa staining	High specificity and sensitivity; inexpensive but invasive

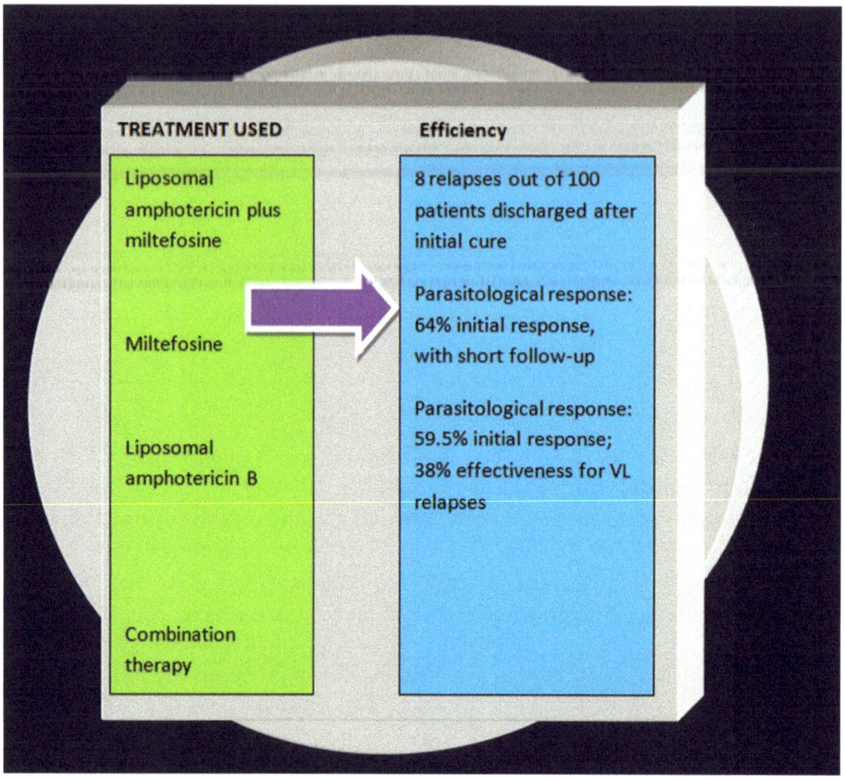

FIGURE 13.1 Commonly used drugs for the treatment of VL-HIV co-infection (Source: Lindoso et al. 2018)

13.5 LEISHMANIASIS AND CANCER

Leishmaniasis and cancer are very prevalent, and the occurrence of these two pathologies in the same tissue is merely coincidental. Innate and adaptive immune responses influence the clinical outcome of both these pathologies, and both are characterised by an impaired Th1 response in the host. As a result, the Th2 microenvironment occurring in progressing leishmaniasis may lead to tumour formation, and vice-versa. Also, the overlapping of clinical aspects of some neoplasms with VL or CL may lead to misdiagnosis (Schwing et al. 2019).

13.6 SOCIOECONOMIC IMPACT

Leishmaniasis is more frequent in the most impoverished parts of developing countries, thus putting a further burden on already strained financial resources. A study conducted in 2006 in Nepal found that the average cost incurred for the treatment of VL was more than the annual household per capita income in Nepal (Rijal et al. 2006). Also, the median cost incurred per household, that is, \$425 USD, was more than the median annual household income of \$405 USD. In the same study, it was found that direct costs, including costs incurred for consultation, laboratory tests and drugs, accounted for 53% of the total costs (75% of the direct costs were incurred even before the treatment of VL was started), while indirect costs were mainly attributed to loss of earnings due to illness. To meet these expenses, most households depleted their savings, while some also had to take high-interest loans to pay for the treatment. Despite the fact that healthcare services are provided for free in general, households still bear a substantial financial burden for the treatment of episodes of leishmaniasis (Chaudhuri et al. 2024).

Catastrophic payment is another aspect of the treatment of VL. Catastrophic payment occurs when healthcare costs exceed a threshold (5–25%), thus diverting consumption from basic needs and pushing households to use their savings. In the study conducted by Singh et al. (2022a), it was projected that 31% and 93% of affected households spent at least 15% and 5%, respectively, of their income on the treatment of leishmaniasis. In another study conducted by Uranw et al. (2013), the catastrophic threshold of 10% of the annual household income was exceeded in almost 51% of the households. It is also noteworthy to mention that without the provision of free drugs, this catastrophic index would have significantly increased from 51% to 74%. This implication is mainly a manifestation of altered treatment-seeking behaviour in the formal healthcare sector and in the informal one. This may be attributed to various factors such as cost, distance, staff, cultural preferences, etc., influencing the timeline of diagnosis and efficiency of treatment (Singh et al. 2023). This further leads to deterioration in the quality of life and the development of chronic lesions (Praveena et al. 2022).

13.7 LEISHMANIASIS AND POVERTY

There is a very strong link between poverty and the development of leishmaniasis. Many factors contribute to this connection (Singh et al. 2022b). Ecological factors such as cracked walls and damp earthen floors enhance the chances of vector (sandfly) survival, thus leading to the development of leishmaniasis. The entry of sandflies into households with poor housing conditions is also facilitated by improper doors (Singh et al. 2020; Singh et al. 2021; Verma et al. 2022). Along with these ecological factors, there are some other socioeconomic factors that highly influence the incidence of leishmaniasis among the poor. Some of these factors are illiteracy, poor garbage collection, unhygienic living conditions and the kind of work they mostly undertake. All these factors culminate in creating a good breeding ground for sandflies, thus facilitating the propagation of leishmaniasis (Ahmed et al. 2023).

A substantial body of evidence of the socioeconomic impact of leishmaniasis can be traced in India through studies on the prevalence of leishmaniasis among different castes. The caste system is a very rigid hierarchy of social structure wherein people are grouped into categories based on various factors, with economic factors being one of them. A study conducted in Bihar, India, found that lower castes infected with leishmaniasis have comparatively delayed access to treatment compared to higher castes infected with leishmaniasis (Rutte et al. 2019).

13.8 GENDER INEQUALITY

Although recent studies suggest that men and women are equally vulnerable to leishmaniasis (Mondal et al. 2019), the implication for females appears much more pronounced due to various socioeconomic factors. While CL has low mortality and morbidity, it leaves infected individuals with disfiguring scars, leading to stigmatisation across genders. The more pronounced effect on women can be attributed to the fact that they have less access to healthcare facilities and, even when available, they hesitate to go to the health centres (). Additionally, they are mostly engaged in unpaid household labour such as cooking, cleaning, taking care of livestock and children, and farming and gardening (Rutte et al. 2019; Mondal et al. 2019) and thus have mostly meagre cash resources. They are very reluctant to spend their already strained cash resources on check-ups and treatment (Mondal et al. 2019).

Gender also differentially affects the experience of illness, the pattern of treatment-seeking behaviour and the care offered by family and society (Mondal et al. 2019). In a study conducted by Mondal et al. (2019), it was found that while 80% of patients attending a CL clinic used alternative therapies, only men self-medicated with the appropriate drug. Women primarily relied on alternative caustic agents and plants.

13.9 PSYCHOSOCIAL EFFECTS

Leishmaniasis has a profound, long-lasting psychosocial impacts, especially on women. The disfiguring scars have many detrimental social and psychological implications for the lives of leishmaniasis patients, particularly affecting women. This leads to social stigmatisation, mental health issues and alienation from community activities. Women also face marital difficulties; single women are unable to marry, and those already married face rejection from their partners, possibly due to shame or fear of contracting the disease (Ahmed et al. 2023; Prasanth et al. 2021). The mental health effects become more pronounced when women are unable to take proper care of their children. Beyond these outward stigmas, perceived stigma is also an issue. Infected women show signs of embarrassment, low self-esteem and shame. Sometimes, the scars become a manifestation of low social status and poverty, further exacerbating the stigma and shame. They are believed to be contagious, and thus the affected individuals are excluded from attending colleges and schools. They also get isolated from communities and families (Panda et al. 2019).

There are other implications as well, which include:

- **Disease control:** Research can help identify high-risk areas, enabling targeted interventions like vector control and early case detection to reduce disease burden.
- **Treatment strategies:** Studying drug efficacy and drug resistance patterns can lead to the development of more effective treatment protocols for VL and PKDL.
- **Diagnostic improvements:** Advancements in diagnostic tools can facilitate timely detection and reduce misdiagnosis, ensuring prompt treatment.
- **Public awareness:** Research findings can increase public awareness, leading to a better understanding and cooperation in disease control efforts.

13.10 CHALLENGES

The public health implications of research on VL and PKDL are significant. Understanding the transmission, epidemiology and treatment of these diseases can lead to improved control measures and better healthcare strategies. However, there are also several challenges associated with this research:

- The lack of effective vaccines and limited treatment options for VL and PKDL contribute to challenges in disease management (Sundar, 2015).
- Inadequate diagnosis and surveillance systems in some regions hinder early detection and timely intervention, leading to delayed or inadequate treatment (Boelaert et al., 2018).
- Socioeconomic factors, such as poverty, lack of education and inadequate access to healthcare, exacerbate the burden of VL and PKDL (Alvar et al., 2012).
- The geographical distribution of the disease varies, with hotspots in specific regions necessitating tailored control strategies (Hirve et al., 2020).
- Vector control measures, such as insecticide-treated bed nets and environmental management, play a crucial role in preventing disease transmission (Alvar et al., 2012).
- Health education and community engagement are essential in raising awareness and promoting preventive measures (WHO, 2021a; Lala et al., 2024).

13.11 CONCLUSION AND FUTURE DIRECTIONS

Addressing these implications and challenges through research can have a profound impact on the prevention, control and management of VL and PKDL, ultimately benefiting public health and affected communities. There are many fronts on which detailed study and analysis should be done to have a better outcome for leishmaniasis, the major areas being:

- **Extensive studies on the national and global impact of leishmaniasis:** Much focus has been given to community-based studies of leishmaniasis, mainly in India and Nepal. However, more comprehensive studies on the socioeconomic impact of the disease are needed, both at the national and global levels, to develop effective plans and strategies for the control and eradication of the disease regionally, nationally and globally (Mondal et al. 2019).
- **Formulation of effective vaccines:** There remains a lack of clarity regarding the factors involved in generating *Leishmania*-specific memory cells and protective immunity. Consequently, there is not yet an effective and specific vaccination strategy for leishmaniasis. A study conducted by Zaph et al. (2004) stated that central memory cells may be more effective in providing protection after a secondary challenge, although both central and effector memory cells are generated after the primary infection with *Leishmania major* in experimental mice. Another study conducted by Peters et al. claimed that memory cells may only be relevant in artificial systems and that recruitment of effector cells to inflammation areas after infection with sandflies was instrumental in providing protection (Peters et al. 2009).

ACKNOWLEDGEMENT

The authors are thankful to Patna University, Patna, Bihar, for providing the necessary support for this research work.

REFERENCES

Ahmed, G., Jamal, F., Tiwari, R. K., Singh, V., Rai, S. N., Chaturvedi, S. K., Pandey, K., Singh, S. K., Kumar, A., Narayan, S., & Vamanu, E. (2023). Arsenic exposure to mouse visceral leishmaniasis model through their drinking water linked to the disease exacerbation via modulation in host protective immunity: a preclinical study. *Scientific Reports 13*(1). https://doi.org/10.1038/s41598-023-48642-z

Alvar, J., Vélez, I. D., Bern, C., Herrero, M., Desjeux, P., Cano, J., Jannin, J., & den Boer, M. (2012). Leishmaniasis worldwide and global estimates of its incidence. *PLoS One,* 7(5): e35671.

Boelaert M, Bhattacharya S, Chappuis F, El Safi, S. H, Hailu A, Mondal, D., Rijal, S., Sundar, S., Wasunna, M., Peeling, R. W., & Alvar, J. (2018). Evaluation of rapid diagnostic tests: visceral leishmaniasis. *Nat Rev Microbiol*, 6(11): 667–671.

Burki, T. (2022). Guidelines for visceral leishmaniasis and HIV co-infection. *The Lancet* 22(8): 1124–1125. https://doi.org/10.1016/S1473-3099(22)00461-3.

Chaudhuri, R., Prasanth, T., Biswas, D., Mandal, S., & Dash, J. (2024). Combating multidrug-resistance in S. pneumoniae: a G-quadruplex binding inhibitor of efflux pump and its bio-orthogonal assembly. *Deleted Journal 1*(2): 1–9. https://doi.org/10.1093/narmme/ugae005

Coutinho, J., Santos, F.S., Ribeiro, R., et al. (2017). Visceral leishmaniasis and leishmaniasis-HIV coinfection: comparative study. *Rev Soc Bras Med Trop* 50(5): 670–674.

Diro, E., van Griensven, J., Mohammed, R., et al. (2015). Atypical manifestations of visceral leishmaniasis in patients with HIV in north Ethiopia: a gap in guidelines for the management of opportunistic infections in resource poor settings. *Lancet Infect Dis* 15(1): 122–129.

Hirve, S., Singh, S. P., Kumar, N., Banjara, M. R., Das, P., Sundar, S., Rijal, S., Joshi, A., & Thakur, C. P. (2020). Effectiveness and feasibility of active and passive case detection in the visceral leishmaniasis elimination initiative in India, Bangladesh, and Nepal. *Am J Trop Med Hygiene* 102(3): 662–676.

Lala, B., Chaudhuri, R., Prasanth, T., Burkhart, I., Schwalbe, H., & Dash, J. (2024). Guanosine-based hydrogel as a supramolecular scaffold for template-assisted macrocyclization. *Chemical Communications 60*(25): 3433–3436. https://doi.org/10.1039/d4cc00129j

Le Rutte, E. A., Coffeng, L. E., Bontje, D. M., Hasker, E. C., Ruiz Postigo, J. A., Argaw, D., et al. (2016). Feasibility of eliminating visceral leishmaniasis from the Indian subcontinent: Explorations with a set of deterministic age-structured transmission models: Quantitative analysis of strategies to achieve the 2020 goals for neglected tropical diseases: wher. *Parasites Vect* 2016: 9.

Le Rutte, E. A., Zijlstra, E. E., & de Vlas, S. J. (2019). Post-Kala-Azar dermal leishmaniasis as a reservoir for visceral leishmaniasis transmission. *Trends Parasitol* 35: 590–592. https://doi.org/10.1016/j.pt.2019.06.007

Lima, I. P., Müller, M. C., Holanda, T. A., et al. (2013). Human immunodeficiency virus/Leishmania infantum in the first foci of urban American visceral leishmaniasis: clinical presentation from 1994 to 2010. *Rev Soc Bras Med Trop* 46(2): 156–160.

Lindoso, J. A., Cota, G. F., da Cruz, A. M., Goto, H., Maia-Elkhoury, A. N., Romero, G. A., de Sousa-Gomes, M. L., Santos-Oliveira, J. R., & Rabello, A. (2014). Visceral leishmaniasis and HIV coinfection in Latin America. *PLoS Negl Trop Dis* 8(9): e3136. https://doi.org/10.1371/journal.pntd.0003136

Lindoso, J. A., Cunha, M. A., Queiroz, I. T., & Moreira, C. H (2016). Leishmaniasis-HIV coinfection: current challenges. *HIV/AIDS (Auckland, N.Z.)* 8: 147–156. https://doi.org/10.2147/HIV.S93789

Lindoso, J. A. L, Moreira C. H. V, Cunha M. A, & Queiroz I. T (2018) Visceral leishmaniasis and HIV coinfection: current perspectives. *HIV/AIDS (Auckland, N.Z.)* 10: 193–201. https://doi.org/10.2147/HIV.S143929

Mondal, D., Bern, C., Ghosh, D., Rashid, M., Molina, R., Chowdhury, R., et al. (2019). Quantifying the infectiousness of Post-Kala-Azar dermal leishmaniasis toward sand flies. *Clin Infect Dis* 69: 251–258.

Nagill, R., & Kaur, S. (2011). Vaccine candidates for leishmaniasis: a review. *Int Immunopharmacol 11*(10): 1464–1488.

Ostyn, B., Gidwani, K., Khanal, B., Picado, A., Chappuis, F., Singh, S.P., et al. (2011). Incidence of symptomatic and asymptomatic *Leishmania donovani* infections in high-endemic foci in India and Nepal: a prospective study. *PLoS Negl Trop Dis* 5: 1–7.

Panda, D., Saha, P., Chaudhuri, R., Prasanth, T., Ravichandiran, V., & Dash, J. (2019). A competitive pull-down assay using G-quadruplex DNA linked magnetic nanoparticles to determine specificity of G-quadruplex ligands. *Analytical Chemistry 91*(12): 7705–7711. https://doi.org/10.1021/acs.analchem.9b00889

Prasanth, T., Chakraborti, G., Mandal, T., Ravichandiran, V., & Dash, J. (2021). Cycloaddition of N-sulfonyl and N-sulfamoyl azides with alkynes in aqueous media for the selective synthesis of 1,2,3-triazoles. *Green Chemistry 24*(2): 911–915. https://doi.org/10.1039/d1gc03340a

Praveena, S. M., Aris, A. Z., & Singh, V. (2022). Quality assessment for methodological aspects of microplastics analysis in soil. *Trends in Environmental Analytical Chemistry 34*: e00159. https://doi.org/10.1016/j.teac.2022.e00159

Rijal, S., Koirala, S., Van der Stuyft, P., & Boelaert, M. (2006). The economic burden of visceral leishmaniasis for households in Nepal. *Trans R Soc Trop Med Hyg 100*(9): 838–841. https://doi.org/10.1016/j.trstmh.2005.09.017

Schwing, A., Pomares, C., Majoor, A., Boyer, L., Marty, P., & Michel, G. (2019). Leishmania infection: misdiagnosis as cancer and tumor-promoting potential. *Acta Tropica* 197: 104855. https://doi.org/10.1016/j.actatropica.2018.12.010

Singh, V., & Mishra, V. (2020). Coronavirus Disease 2019 (COVID-19): Current situation and therapeutic options. *Coronaviruses 2*(4): 481–491. https://doi.org/10.2174/2666796701999201005211854

Singh, V., & Mishra, V. (2021). Environmental impacts of coronavirus disease 2019 (COVID-19). *Bioresource Technology Reports 15:* 100744. https://doi.org/10.1016/j.biteb.2021.100744

Singh, N., Singh, V., Rai, S. N., Mishra, V., Vamanu, E., & Singh, M. P. (2022a). Deciphering the gut microbiome in neurodegenerative diseases and metagenomic approaches for characterization of gut microbes. *Biomedicine & Pharmacotherapy 156*: 113958. https://doi.org/10.1016/j.biopha.2022.113958

Singh, N., Singh, V., & Singh, M. P. (2022b). Microbial degradation of lignocellulosic biomass for bioenergy production: A metagenomic-based approach. *Biocatalysis and Biotransformation 41*(1): 15–25. https://doi.org/10.1080/10242422.2022.2056451

Singh, V., Jain, S. K., & Mishra, V. (2023). A recent update on SARS-CoV-2 transmission and its variants: transmission, pathogenic mechanism, and treatment. *Coronaviruses 4*(1). https://doi.org/10.2174/2666796704666230102121225

Sundar, S. (2015). Drug resistance in Indian visceral leishmaniasis. *Trop Med Intl Health*, 20(10): 1316–1318.

Uranw, S., Meheus, F., Baltussen, R., Rijal, S., & Boelaert, M. (2013). The household costs of visceral leishmaniasis care in South-eastern Nepal. *PLoS Neglected Tropical Diseases* 7(2): e2062. https://doi.org/10.1371/journal.pntd.0002062

Verma, M., Verma, M. K., Singh, V., Singh, J., Singh, V., & Mishra, V. (2022). Advancements in applicability of microbial fuel cell for energy recovery from human waste. *Bioresource Technology Reports 17:* 100978. https://doi.org/10.1016/j.biteb.2022.100978

WHO. (2021a). Visceral leishmaniasis. Retrieved from https://www.who.int/news-room/fact-sheets/detail/visceral-leishmaniasis

WHO. (2021b). Post-kala-azar dermal leishmaniasis. Retrieved from https://www.who.int/news-room/fact-sheets/detail/post-kala-azar-dermal-leishmaniasis

Zaph, C., Uzonna, J., Beverley, S. M., & Scott, P. (October 2004). Central memory T cells mediate long-term immunity to Leishmania major in the absence of persistent parasites. *Nat Med. 10*(10): 1104–1110.

14 Role of Environmental Factors on the Development of PKDL

Moti Lal[1], Neelam[2] and Himanshu Gaur[3]*
[1]Department of Biotechnology, School of Life Sciences, ITM University, Gwalior, Madhya Pradesh, India
[2]Department of Microbiology, School of Life Sciences, ITM University, Gwalior, Madhya Pradesh, India
[3]Department of Food-Tech, School of Life Sciences, ITM University, Gwalior, Madhya Pradesh, India
*Corresponding Author: Dr. Moti Lal, Department of Biotechnology, School of Life Sciences, ITM University, Gwalior, Madhya Pradesh, India

14.1 INTRODUCTION

A century ago, in 1922, an essay by Dr U.N. Brahmachari published in the *Indian Medical Gazette* brought to light a hitherto unidentified form of cutaneous leishmaniasis. It was recognized as a brand-new kind of dermatological manifestation and possibly bridged the gap between cutaneous and visceral leishmaniasis. Post-kala-azar dermal leishmaniasis (PKDL) emerged and made its mark on the natural history of leishmaniasis throughout the ensuing decades, as others began to report comparable cases from different parts of India and the world. At that time, this unknown disease had killed many people worldwide, and no research community in India or the world knew anything about it.

PKDL develops after visceral leishmaniasis (VL) (Ashford, 1992), also called kala-azar. The *Leishmania* genus of protozoa is the source of a range of parasitic disorders known as leishmaniasis, which are spread to humans by sandfly bites (Elnaiem, 1998; Palsson, 1999). PKDL typically develops after the apparent cure of VL; the latter is an extreme and sometimes lethal variant of the illness, impacting internal organs such as the spleen, liver, and bone marrow. PKDL is characterized by the emergence of skin lesions, which can take many different forms, including nodules, papules, plaques, or macules. These lesions usually develop on the face, arms, and upper body but can occur anywhere on the skin.

The exact mechanism that leads to the creation of PKDL is unclear; however, it is thought to be related to the *Leishmania* parasites' continued presence in the skin even after the systemic infection has been treated. In addition to its effect on the patient's quality of life, PKDL is significant because it can serve as a breeding ground for sandfly infections, hypothetically leading to the spread of the illness to other individuals. PKDL can present as a self-limiting condition, with lesions eventually healing on their own, or it may require specific treatment depending on the severity and extent of the lesions. Treatment options for PKDL may include antileishmanial drugs such as miltefosine, paromomycin, or amphotericin B. The choice of treatment depends on various factors, including the patient's overall health, the presence of any underlying conditions, and the drug's availability in the specific region. PKDL is a dermatological complication that can occur after a person has been treated for VL.

 DOI: 10.1201/9781032643120-14

14.2 EPIDEMIOLOGY WORLDWIDE

Prevalence reports rely on estimates of data because the worldwide epidemiology of PKDL has not been fully studied and is currently under investigation. PKDL is primarily seen in East Africa and the Indian landmass. *Leishmania donovani* is the cause of its dissemination. Although the causes of PKDL in endemic regions are not widely understood, they may include the rapid development of VL and inadequate analysis. PKDL has been frequently recorded following sodium stibogluconate treatment for VL. All these regions should be monitored because diseases like VL and PKDL are common there.

14.2.1 Different Social and Environmental Factors in South Sudan and Sudan for the Spread of PKDL

Sudan has the greatest disease load and prevalence. PKDL in Sudan is distinct from that on the Indian subcontinent. In 56–62% of cases in Sudan, PKDL develops following VL. 10% of cases of PKDL present without a history of VL, while 14.99% do so concurrently with VL (a condition known as para-KDL). Patients often get PKDL six months after receiving treatment for VL; however, in rare cases, it may occur as early as two weeks or as late as 13 months after treatment. VL, mainly caused by *L. donovani* and often spread by both *Phlebotomus orientalis* and *P. martini*, is highly endemic in East Africa, with approximately 30,000 VL cases reported annually. Sudan has the largest number of cases, where *P. orientalis* is prevalent within acacia woodlands, particularly along its eastern border with Ethiopia. Other countries with a significant number of cases include South Sudan, Ethiopia, Somalia, Kenya, and Uganda.

The main causes of VL in East African nations are protracted civil war and turmoil. Here, we attempt to establish a link between VL epidemics and civil conflicts. Large-scale and fatal epidemics, including PKDL, were brought on by waves of forced migration caused by civil wars in southern Sudan between 1983 and 2005. Notably, in 2011, with the allocation of more funds for disease treatment and prevention, occurrences of VL decreased, reaching their lowest points since South Sudan's independence. However, 1.5 million internally displaced people have moved to states where VL is extremely endemic, while 800,000 have fled to nearby nations.

14.2.2 Ethiopia

In the lower Omo plains in southwestern Ethiopia, the first case of VL was reported in 1942 (Coles et al., 1942). The illness has now spread throughout the nation and is endemic in several regions, with an estimated yearly incidence of more than 4,000 cases. It is most common in lowland, dry regions. The Humera and Metema plains in the northwest are among the most significant endemic hotspots (Haile & Anderson, 2006). In northern Ethiopia during 1998–1999, PKDL was present in an average of 13% of patients with VL and in 27% of patients who also had HIV. However, these percentages do not accurately reflect the true number of cases because mild PKDL is frequently left untreated and, as a result, is not reported. The prevalence of PKDL appears to be lower in southern Ethiopia.

Parasitic isolates from eastern Sudan often resemble those from northern foci, which are more diverse, while those in the south are more closely related to parasites seen in India. The relevance of this variation for PKDL remains under debate. There are host differences between northern and southern Ethiopia, as well as differences in treatment response. In hyperendemic areas of the north, malnutrition and high coinfection rates are factors, whereas in the south, VL manifests as a microfocal pattern, and people are usually better nourished (Alvar et al., 2007).

Ethiopia is experiencing an increase in this health problem due to constantly expanding endemic areas, where more than 3.2 million people are at risk. There has been a change in the epidemiology of VL in Ethiopia. Nearly all VL cases and outbreaks during the past few decades have occurred in arid and semi-arid regions of the nation; however, recent studies suggest that this disease has spread to the highlands. The VL epidemiology changed because of migration to and from endemic regions, climatic and environmental changes, weakened immunity from HIV/AIDS, and malnutrition. HIV promotes the spread of VL by increasing the probability that an asymptomatic infection will proceed to full VL. On the other hand, VL accelerates the onset of AIDS. Due to the variety of risk factors involved, VL epidemiology in Ethiopia continues to be complex, and its control is becoming an increasingly difficult task (Samson et al., 2014; Alvar et al., 2007).

14.2.3 The Indian Subcontinent

In the Indian subcontinent, between 3% and 21% of people who previously had VL can develop PKDL (Zijlstra et al., 2003). General factors favoring the transmission of VL in anthroponotic foci in Bangladesh, India, and Nepal (excluding Bhutan, Kashmir, Himachal Pradesh, and Uttarakhand in India) include rural locations less than 600 meters above sea level, abundant precipitation each year, a mean humidity of more than 70%, a maximum temperature of 38°C and a minimum of 15°C, with a daily variation of less than 7°C. Although there aren't many cases each year, Thailand and Sri Lanka are also endemic areas. VL and PKDL surveillance should be done in each of these areas. VL is prevalent in agricultural communities with mud walls and earthen floors, where cattle and other livestock are kept close to human habitations (Figure 14.1).

Cases are significantly clustered, according to spatial analysis, both at the household and community levels. Although the proven vector (*P. argentipes*) is more prevalent in cattle sheds than in human dwellings, the vector resourcefully feeds on the blood of both humans and animals. While using bed nets is generally associated with a lower risk of infection, human behavior can increase that risk, such as sleeping on the ground or outside. Living in the same house as a person with PKDL or active VL, as well as factors associated with poverty, such as inadequate shelter and nutrition, are risk factors for VL. The goal of the VL elimination campaign, which Bangladesh, India, and

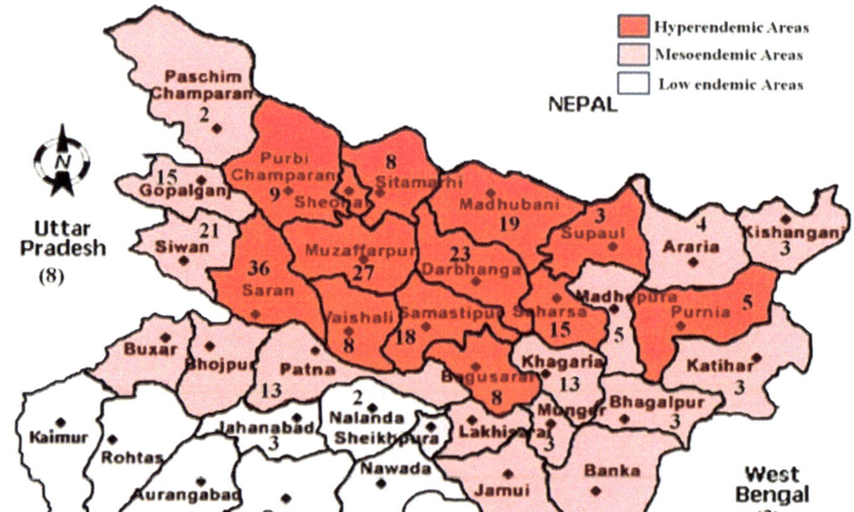

FIGURE 14.1 Cases of PKDL are in Bihar and the surrounding regions, considering high, moderate, and low VL endemicity areas (Sundar & Kumar, 2023). This map illustrates the distribution of PKDL cases in the state of Bihar and its neighboring states, showing the total number of PKDL cases per district (Source: Ramesh et al., 2015)

Nepal agreed to undertake in 2005, is to have less than 1 case per 10,000 population per district or subdistrict by 2015. To accomplish this, it will be essential to address PKDL.

14.3 THE ROLE ENVIRONMENTAL INFLUENCES PLAY IN PKDL DEVELOPMENT

Compared to the factors causing VL, the role of environmental factors in the development of PKDL has not received as much research attention, but several potential interactions between the environment and the disease may influence its development.

14.3.1 Host–Vector Interaction

Socioeconomically disadvantaged residents rely on agriculture and nearby brick kiln industries for a living. People working in these industries typically own cattle, and areas with cow manure serve as sandfly breeding grounds (Kaul et al., 1994). Environmental conditions can influence the behaviors of sandflies and their interaction with human hosts. Factors like deforestation, urbanization, and human migration can alter the ecological balance among sandflies, animal reservoirs, and humans. Changes in land use and human activities can lead to increased contact between sandflies and humans, potentially increasing the risk of PKDL transmission. The vector-borne transmission of PKDL is closely linked to the transmission of VL (Figure 14.2).

14.3.1.1 Sandfly Transmission

The main vector that transmits *Leishmania* parasites is *Phlebotomus argentipes*, better known as the sandfly. Leishmaniasis is a disease caused by these parasites, and it can manifest in different clinical forms, including VL and PKDL. The primary means of transmission for both VL and PKDL is through the bite of infected female sandflies belonging to the genera *Phlebotomus* (Old World) and *Lutzomyia* (New World). These sandflies become infected when they feed on blood from an individual with VL or PKDL, and the parasites multiply within the sandfly. In the Indian subcontinent, the disease is only disseminated by one common parasite species (*Leishmania donovani*) and one species of sandfly (*Phlebotomus argentipes*) (Joshi et al., 2009).

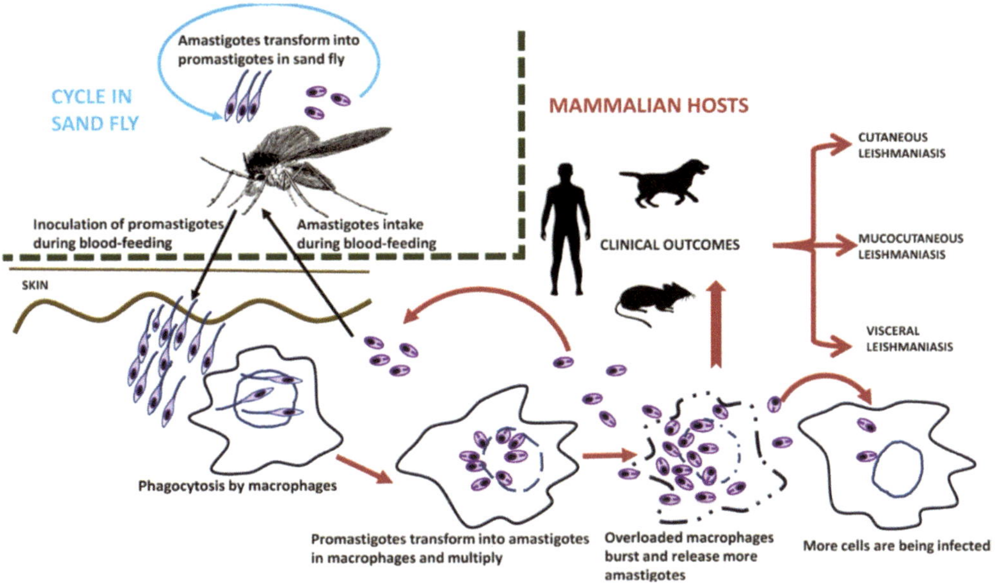

FIGURE 14.2 *Leishmania* has a straightforward life cycle that consists of two phases in both the insect vector and the mammalian host, occurring without sexual reproduction (Source: Virginie et al., 2010)

PKDL presents as painless papulonodular (PN) or macular lesions, or a combination of the two, that harbor parasites that could be transmitted through sandfly bites, potentially playing a significant role in the transmission cycle (Mukhopadhyay & Mishra, 1991). It was observed that after feeding on a patient with depigmented (macular) PKDL lesions, the most common type in the Indian subcontinent, sandflies had a 30% infection rate. In these lesions, parasites are usually scarce in a slit-skin smear or biopsy. In a 1933 study, parasites were found in sandflies that had fed on two patients with nodules (Napier et al., 1993). In a subsequent experiment, 53% of the sandflies that fed on the nodular or nodulo-ulcerative lesions of four PKDL patients became infected. This discovery was thought to be the reason behind the 1980s VL outbreak in southern West Bengal (Addy & Nandy, 1992).

14.3.1.2 Transmission Cycle

The infected sandfly transmits the parasites to a healthy individual when it bites to feed on their blood. During the blood meal, the infected sandfly injects *Leishmania* parasites into the person's skin. These parasites are then taken up by immune cells called macrophages, leading to infection (Carvallo et al., 1981).

1. **Reservoir hosts:** The primary reservoir hosts of the *Leishmania* parasite, which causes both VL and PKDL, are infected humans and certain wild and domestic animals like dogs, rodents, and other mammals (Halder et al., 1983).
2. **Vector:** In the Old and New Worlds, female sandflies belonging to the genera *Phlebotomus* and *Lutzomyia* are the main carriers of the parasite. These sandflies become infected when they feed on an infected reservoir host, taking up the parasites along with their blood meal (Pampiglione et al., 1974).
3. **Transmission to humans:** When an infected sandfly bites a human, it injects the *Leishmania* parasites into the person's bloodstream. This is the primary route of transmission for both VL and, subsequently, PKDL (Rutte et al., 2019).
4. **Visceral leishmaniasis:** The parasites multiply within the human body and can lead to VL, a severe systemic disease characterized by fever, weight loss, enlarged spleen and liver, and other symptoms. If left untreated, VL can be fatal (Pampiglione et al., 1975).

14.3.1.3 Development of PKDL

In some individuals who have had VL, the parasites can persist in the body even after apparent cure, leading to the development of PKDL. This condition can arise months to years after the initial infection and is thought to be the result of incomplete or inadequate treatment for VL (Gasim et al., 2000).

14.3.1.4 Secondary Transmission

Once an individual develops PKDL, they become a potential reservoir for the parasite. If a sandfly bites an individual with PKDL, it can ingest the parasites, thereby continuing the transmission cycle. The infected sandfly can then go on to transmit the parasites to other individuals, potentially causing new cases of VL or PKDL. Controlling the transmission of PKDL and VL involves targeting the sandfly vector through several approaches, such as insecticide-treated bed nets, indoor residual spraying, and personal protective measures. Additionally, to prevent the spread of PKDL and reduce the overall disease transmission, it is imperative that VL cases receive adequate treatment. It is important to be aware that advancements in our understanding of these diseases and their transmission may have occurred since my last knowledge update in September 2021. For the most current information, I recommend consulting recent medical literature or reputable public health sources (Lloyd et al., 2020).

14.3.2 Vector-Borne Transmission

The main way that *Leishmania* parasites are spread is through the bites of infected female sandflies, which serve as vectors for the disease. In the case of PKDL, it is believed that these infected sandflies play a role in transmitting the disease from one individual to another. Evidence suggests that PKDL patients can act as a reservoir for the *Leishmania* parasites, and sandflies that bite an infected PKDL patient can acquire the parasites. When these infected sandflies subsequently bite healthy individuals, they can transmit the parasites, possibly resulting in new PKDL or VL cases. It is important to note that the transmission dynamics of leishmaniasis, including PKDL, can be complex and can vary based on factors such as the specific *Leishmania* species, the sandfly vector species, the host immune response, and environmental factors (Zijlstra, 2012).

14.3.3 Immune Response and Parasite Persistence

The immunological response of the host to *Leishmania* infection is thought to be linked to the development of PKDL. The immune response and the persistence of *Leishmania* parasites in the body following VL treatment can be influenced by environmental factors, including nutritional status, coinfections, and exposure to other pathogens. These elements could influence a person's likelihood of developing PKDL (Kumar et al., 2023).

14.3.4 Geographical Distribution

The distribution of both VL and PKDL is influenced by environmental factors. These diseases are often found in regions with specific ecological conditions that support the lifecycle of the parasite and its vector. Changes in environmental conditions due to factors like climate change, deforestation, and urban expansion could potentially expand the geographical range of both VL and PKDL (Figure 14.3).

FIGURE 14.3 Geographic distribution of unusual presentations in leishmaniasis by region. CL: cutaneous leishmaniasis; PKDL: post-kala-azar dermal leishmaniasis; VL: visceral leishmaniasis (Source: Yadav et al., 2023)

14.3.5 Prevalence of Malnutrition

The phrase "malnutrition" refers to excesses, deficiencies, or imbalances in an individual's energy or nutrient consumption in all its forms. This broad definition includes undernutrition (wasting, stunting, and underweight), inadequate vitamin and mineral intake, overweight, obesity, and the resulting diet-related noncommunicable diseases. Malnutrition and infection have historically been closely related (Millward, 2017). As more research is conducted on the pathophysiology of this relationship, the role of nutrition in preventing illness and reducing morbidity and death rates, particularly in people living with infectious diseases including tuberculosis, leishmaniasis, HIV, and malaria, has become evident (Singh et al., 2022a).

According to joint estimates of malnutrition published by UNICEF in 2018, based on data from the World Bank Group and the World Health Organization (WHO), 462 million adults (over the age of 18) are underweight, and 1.9 billion are overweight or obese (body mass index [BMI]: healthy weight, 18.5–24.9 kg/m^2; overweight, 25–29.9 kg/m^2; obese, >30 kg/m^2) (Singh et al., 2022b). In addition, of the 52 million wasted children, 41 million are overweight or obese, and 155 million children under the age of five are stunted. Undernutrition is linked to 45% of deaths in children under five, alongside the rise in childhood overweight and obesity rates, which are predominantly seen in low- and middle-income countries. Over half of the global population lives in countries where obesity and overweight cause over three million deaths annually—three times more than famine. A recent analysis found that between 1975 (4%) and 2016 (18%), the prevalence of obesity increased more than fourfold worldwide (Reedy et al., 2014). Undernourishment remains a significant problem in developing nations. A nutritious diet can help prevent nutrition-related diseases, as well as support those who already have these conditions to better manage their symptoms, maximize the effectiveness of their medications, and improve their quality of life (Reedy et al., 2014).

14.3.5.1 Leishmaniasis and Malnutrition

Multiple species of *Leishmania* cause the parasitic disease leishmaniasis, which remains the deadliest neglected tropical disease globally, particularly in developing countries where chronic starvation coexists with it. Typically, this parasite lives in sandflies, and humans can contract it from infected sandflies (Akhoundi et al., 2016). Malnutrition plays a significant role in determining the severity of the infection and manifests as a variety of physical abnormalities, some of which have common etiologic pathways in undernutrition and overnutrition: poor growth and development in children, immunological dysfunction, diminished epithelial integrity, childhood undernutrition that increases the risk of obesity in adulthood, altered metabolic rate, chronic inflammation, and altered food absorption are all contributing factors (Weger-Lucarelli et al., 2018; Bresnahan & Tanumihardjo, 2014). Unquestionably, frequent exposure to infectious diseases causes appetite loss, failure to eat enough, or inability to make appropriate food choices; this causes starvation and infection to enter a vicious cycle. Considering the links between the host, pathogens, and the types of foods that an individual, an animal, or a community regularly consumes, the intricacy of the connections between viral diseases and nutritional status becomes evident (Tanner et al., 2014).

14.3.5.2 Effect of Nutrition on Leishmania Infection

There are claims that the host immune system, dietary state, and, to a lesser extent, the type of sandfly species involved in transmission all play a role in determining the severity of *Leishmania* (Ferro et al., 2015). It is now widely accepted that malnourished people living in poverty are more prone to infections and certain diseases because of deficits in macronutrients and micronutrients. The importance of these nutrients in enhancing immune system activity to ward off specific infections, inflammation, and potentially even some malignancies has only recently come to light. Changes in immunity changes, which are strongly linked to nutritional status, significantly increase the hosts' susceptibility to infectious agents or the severity of infection. The connection between the prevalence/endemicity of the *Leishmania* parasite and malnutrition makes it imperative to understand more about the interaction between *Leishmania* and host and/or vector nutrition (Nweze et al., 2020).

Since seroprevalence remains the primary method of linking nutrition to *Leishmania* infection and infection susceptibility in both human and animal hosts, there are few prospective studies on patient nutritional status and susceptibility to leishmaniasis. Most research has demonstrated a direct correlation between obesity and excessive body weight and a history of *Leishmania* infection. In southern Ethiopia, a study showed that individuals with a BMI higher than 18 kg/m^2 were more likely to have *Leishmania* infection (Ali, 1997). Malnourished people were found to have a roughly three times higher risk of contracting VL in the North Gondar zone, Amhara region, of Ethiopia (Ali, 1997). However, another study in the same Ethiopian location found that VL can increase the risk of malnutrition and lower BMI. Malnutrition can increase one's susceptibility to illness, and infection can worsen malnutrition, creating a vicious cycle.

According to a study conducted in the Bangladeshi district of Mymensingh, dogs with poor nutritional status had a seroprevalence of *Leishmania* approximately 13 times higher than dogs with adequate nutritional status (Singh et al., 2022c). Furthermore, malnutrition can impair the host's immunological response, leaving people more susceptible to infections. Several studies have described how various immune cells and leishmaniasis interact. Additional nutritional parameters that reflect intrauterine and maternal nutrition, including the frequency of breastfeeding and birth weight, also play a significant role in an infant's increased risk of contracting *Leishmania* spp. infection. According to a study by Maciel, patients with active VL had considerably lower blood levels of vitamin A (serum retinol) than control groups (Nweze et al., 2020)

14.3.5.3 Therapeutic Action of Micronutrients for Leishmaniasis

Leishmaniasis is a disease that the WHO has recognized as requiring innovative medicines. The available medications need intravenous administration, have frequent and substantial adverse effects, and are not always totally successful. Miltefosine is the only oral medication that successfully treats leishmaniasis. However, it has many drawbacks, including drug resistance, high cost, and unavailability in some endemic regions (de Souza et al., 2022). It has been demonstrated that micronutrient deficiencies worsen VL illness, which in turn accelerates the progression of malnutrition. Given that VL frequently interferes with quality of life in developing nations where nutritious diets are scarce, providing additional nutrients to people receiving VL treatments could enhance their micronutrient status and, consequently, their well-being (Custodio et al., 2018).

A study conducted in a region of northeastern Brazil where VL is widespread has found that vitamin A may have a dual role in immune response control. Numerous studies have shown that zinc may be an effective and promising anti-cutaneous leishmaniasis agent. Studies examining the effectiveness of zinc sulfate in vivo or in vitro revealed that it can be used as a treatment or preventative measure for cutaneous leishmaniasis (Kumar et al., 2023).

Iron deficiency inhibits *Leishmania* proliferation by enhancing the host's oxidative defense mechanism to the maximum extent possible as a result of an oxidative process that modifies the signaling cascade and produces substances harmful to the parasite. Calcium, as a micronutrient, performs two important functions: inhibiting the growth of the *Leishmania* parasite and boosting the host's immunity to the disease. Vitamin C functions as an antioxidant and slows the advancement of the disease by balancing the levels of oxidants and antioxidants, which are brought on by elevated lipid peroxidation and low antioxidant defense. Additionally, vitamin C has a preventative effect and lowers the parasite load in animal models. Furthermore, it regulates free radicals that damage *Leishmania* species and inhibit them (Kumar et al., 2023).

As a potent immunomodulator, vitamin D triggers the production of the antimicrobial peptide LL-37, which inhibits *Leishmania* growth and kills promastigote leishmaniasis. Skin lesions in mice treated with vitamin D were reduced. Both the effectiveness of treatment and the development of VL are impacted by vitamin D deficiency. Vitamin D further suppresses IFN-mediated

TABLE 14.1
Therapeutic Actions of Different Micronutrients for Leishmaniasis

Vitamins/Minerals	Therapeutic Effect on Leishmaniasis	References
Vitamin A	VL: Control of IL-10, IL-17, Treg and monocytes, IL-10 production, and T-cells	(Stephensen et al., 2004)
Vitamin B	Causes changes to the nucleic acids that pathogens and white blood cells contain	(Kumar et al., 2023)
Vitamin C	Enhances immune response, tissue regeneration, wound healing, and tissue growth and development, and impedes the growth of *Leishmania donovani*	(Kumar et al., 2023)
Vitamin D	Robust immunomodulator that triggers the production of LL-37, an antimicrobial peptide that inhibits *Leishmania* growth and kills the promastigote form of the parasite	(Kumar et al., 2023)
Zinc	Zinc sulfate was found to lead to high VL cure rates	(Prasad, 2007)
Calcium	Slows the *Leishmania* parasite's growth and boosts the host's immunity to the disease	(Millward, 2017)
Iron	Crucial for the antioxidant activity of iron superoxide dismutase (Fe-SOD) and acts as a precursor to the parasite's production of cofactor Fe-S clusters	(Singh et al. 2023)

macrophages, which increases the Arginase1 enzyme and lowers NO, reducing the host's ability to defend against *Leishmania major* (Table 14.1) (Kumar et al., 2023).

14.3.6 Access to Healthcare

Environmental factors can indirectly affect the development of PKDL by influencing access to healthcare services. Remote or impoverished areas with limited access to healthcare facilities may experience delays in diagnosing and treating VL, which, in turn, could contribute to the development of PKDL.

Access to healthcare for PKDL, especially in resource-limited areas like many parts of the Indian subcontinent, can be a complex issue. Here are some key points to consider:

- **Diagnosis:** Proper diagnosis of PKDL requires skilled medical personnel and appropriate diagnostic tools. Skin lesions need to be examined, and sometimes a biopsy or other laboratory tests are required to confirm the presence of the *Leishmania* parasite. In areas with limited access to medical facilities, diagnosis might be delayed or inaccurate due to the lack of trained healthcare professionals and diagnostic resources.
- **Healthcare infrastructure:** Access to healthcare facilities, especially in rural or economically disadvantaged areas, can be a challenge. People with PKDL may need to travel great distances to reach a medical facility where they can be properly diagnosed and treated. In some cases, adequate medical facilities may not be available, making it difficult for individuals to receive timely and appropriate care.
- **Awareness and education:** A lack of awareness and education about PKDL can lead to underreporting and delayed treatment. If people are not informed about the potential consequences of PKDL and its connection to VL, they might not seek medical attention when skin lesions appear. Health education campaigns and community outreach are essential to raise awareness about the condition and its implications.
- **Treatment availability:** PKDL can be treated with medications, including antimonial drugs, miltefosine, and amphotericin B. However, the availability of these drugs might be

limited in certain regions. Stockouts of essential medications can further hinder the proper treatment and management of PKDL.

- **Cost and affordability:** Even if healthcare facilities are accessible, the cost of diagnosis and treatment can be a significant barrier, especially for individuals from low-income backgrounds. The financial burden of seeking medical care, along with potential transportation costs, can prevent people from getting the help they need.
- **Stigma and social factors:** Skin conditions like PKDL can be stigmatized in some communities. This social stigma might discourage individuals from seeking medical care due to fear of discrimination or social isolation.
- **Government and NGO efforts:** Government health programs and non-governmental organizations (NGOs) play a crucial role in improving access to healthcare services, providing training to healthcare workers, and raising awareness about neglected tropical diseases like PKDL. Collaborative efforts can help address the various challenges associated with PKDL diagnosis and treatment.

Improving access to healthcare for PKDL requires a multifaceted approach that involves strengthening healthcare infrastructure, increasing awareness, reducing financial barriers, and ensuring the availability of necessary medications. It is important for governments, NGOs, healthcare professionals, and the international community to work together to address these challenges and provide effective care for individuals with PKDL.

14.4 CONCLUSION

Post-kala-azar dermal leishmaniasis (PKDL) is a dermatological sequel of visceral leishmaniasis caused by *Leishmania donovani* parasites. This chapter has explored the multifaceted role of environmental factors in the development of PKDL. The formation of PKDL and the dynamics of *Leishmania* parasite transmission are significantly influenced by environmental conditions, including temperature, humidity, and ecological shifts. Socioeconomic factors, including poverty and malnutrition, also contribute to the prevalence of PKDL. Moreover, the impact of environmental pollution, such as arsenic contamination in water sources, has been identified as a possible PKDL risk factor. Understanding the interplay between environmental factors and PKDL is crucial for designing effective control strategies. Integrated vector management (IVM) approaches, focusing on environmental management and community engagement, are essential for reducing the burden of PKDL. Additionally, public health interventions, including health education and improved access to healthcare, are critical for the early detection and management of PKDL cases. In conclusion, the development of PKDL is significantly influenced by environmental factors, highlighting the need for multidisciplinary approaches that address environmental, social, and healthcare aspects to effectively control and eliminate PKDL.

ACKNOWLEDGMENT

The authors are thankful to ITM University Gwalior for the necessary support for this work.

REFERENCES

Addy, M., Nandy, A. (1992). Ten years of kala-azar in west Bengal, Part I. Did post-kala-azar dermal leishmaniasis initiate the outbreak in 24-Parganas? *Bull World Health Organ.* 70: 341–346.

Akhoundi, M., Kuhls, K., Cannet, A., Votypka, J., Marty, P., Delaunay, P., & Sereno, D. (2016). A historical overview of the classification, evolution, and dispersion of leishmania parasites and sandflies. *PLoS Negl Trop Dis.* 10: 1–40.

Ali, A. (1997). Visceral leishmaniasis in Southern Ethiopia II. Nutritional risk factors. *Int J. Health Dev.* 11: 03–30.

Alvar, J., Bashaye, S., Argaw, D., Cruz, I., Aparicio, P., Kassa, A., Orfanos, G., Parreño, F., Babaniyi, O., Gudeta, N., Cañavate, C., Bern C. (2007). Kala-azar outbreak in Libo Kemkem, Ethiopia: epidemiologic and parasitologic assessment. *Am J Trop Med Hyg.* 77: 275–282.

Ashford, R.W. (1992). Epidemic visceral leishmaniasis in southern Sudan: identity and systematic position of the parasites from patients and vectors. *Trans R Soc Trop Med Hyg.* 86: 379–380.

Bresnahan, K.A., Tanumihardjo, S. A. (2014). Undernutrition, the acute phase response to infection, and its effects on micronutrient status indicators. *In Adv Nutrit.* 5: 702–711.

Carvallo, E.M., Teixeira, R.S., Johnoson, W.D. (1981). Cell mediated immunity in American visceral leishmaniasis; reversible immunosuppression during acute infection. *Infect Immun.* 33: 498–502.

Coles, A., Consgrove, P., Robinson, G. (1942). A preliminary report of an outbreak of kala-azar in a battalion of the King's African Rifles. *Trans Roy Soc Trop Med Hyg* 36: 25–34.

Custodio, E., López-Alcalde, J., Herrero, M., Bouza, C., Jimenez, C., Storcksdieck, G., Bonsmann, S., Mouratidou, T., Lopez-Cuadrado, T., Benito, A., Alvar, J. (2018). Nutritional supplements for patients being treated for active visceral leishmaniasis. *Cochrane Database Systematic Rev.* 3: 1–26.

Elnaiem, D.E. (1998). Evidence for active transmission of visceral leishmaniasis within a village in eastern Sudan. *Acta Trop.* 71: 305–309.

Ferro, C., Lopez, M., Fuya, P., Lugo, L., Cordovez, J.M., González, C. (2015). Spatial distribution of sand fly vectors and eco-epidemiology of cutaneous leishmaniasis transmission in Colombia. *PLoS One*, 10: 1–16.

Gasim, S., Elhassan, M.A., Kharazmi, A., Khalil, E.A.G., Ismail, A., Theander, T.G. (2000). The development of post-kala-azar dermal leishmaniasis (PKDL) is associated with acquisition of *Leishmania* reactivity by peripheral blood mononuclear cells (PBMC). *Clin Exp Immunol.* 119: 523–529.

Haile, T., Anderson, S. (2006). Visceral leishmaniasis in northern Ethiopia. *East Afr Med J.* 83: 392.

Halder, J.P., Ghose, S., Saha, K.C., Ghose, A.C. (1983). Cell-mediated immune response in Indian Kala-azar and post kala-azar dermal leishmaniasis. *Infect Immun.* 42: 702.

Joshi, A.B., Das, M.L., Akhter, S., Chowdhury, R., Mondal, D., Kumar, V., Das, P., Kroeger, A., Boelaert, M., Petzold, M. (2009). Chemical and environmental vector control as a contribution to the elimination of visceral leishmaniasis on the Indian subcontinent: cluster randomized controlled trials in Bangladesh, India and Nepal. *BMC Med.* 7: 54.

Kaul, S.M., Sharma, R.S., Borgohain, B.K., Das, N.S., Verghese, T. (1994). Absence of *Phlebotomus argentipes* Ann & Brun. (Diptera: Psychodidae) the vector of Indian kala-azar from Kamrup district, Assam. *J Commun Dis.* 26: 68–74.

Kumar, A., Singh V.K., Tiwari, R., Madhukar, P., Rajneesh, Kumar S., Gautam, V., Engwerda, C., Sundar, S., Kumar, R. (2023). Post kala-azar dermal leishmaniasis in the Indian sub-continent: challenges and strategies for elimination. *Front Immunol* 14: 1236952.

Kumar, V.U., Kt, M.F., Sharma, A., Bisht, P., Dhingra, S., Ravichandiran, V., Ramesh, M., Murti, K. (2023). The possible role of selected vitamins and minerals in the therapeutic outcomes of leishmaniasis. *In Biol Trace Element Res.* 201: 1672–1688).

Lloyd, A.C., Chapman, S.E.F., Spencer, T.M., Pollington, C.P., Jewell, D., Mondal, J., Alvar, T.D., Hollingsworth, M.M., Cameron, C. (2020). Inferring transmission trees to guide targeting of interventions against visceral leishmaniasis and post–kala-azar dermal leishmaniasis. *Proc Natl Acad Sci USA.* 117: 25742–25750.

Millward, D.J. (2017). Nutrition, infection, and stunting: the roles of deficiencies of individual nutrients and foods, and of inflammation, as determinants of reduced linear growth of children. *Nutrit Res Rev.* 30: 50–72.

Mukhopadhyay, A.K., Mishra, R.N. (1991). Development of *Leishmania donovani* in *Phlebotomus argentipes* & Ph. papatasi fed on kala-azar patients in Bihar. *Indian J Med Res.* 93: 152–154.

Napier, L.E., Smith, R.O., Das, A., Gupta, C.R., Mukherjee, S. (1993). The infection of *Phlebotomus argentipes* from dermal leishmanial lesions. *Indian J Med Res.* 11: 173–177.

Nweze, J.A., Nweze, E.I., Onoja, U.S. (2020). Nutrition, malnutrition, and leishmaniasis. *Nutrition.* 73: 110712.

Palsson, K. (1999). Plant products used as mosquito repellents in Guinea Bissau, West Africa. *Acta Trop.*72: 39–52

Pampiglione, S (1974). Studies on Mediteranean leishmaniasis. An outbreak of visceral leishmaniasis in northern Italy. *Trans R Soc Trop Med Hyg.* 68, 349–359.

Pampiglione, S., Manson-Bahr, P.E., Placa, M.L., Borgatti, M.A., Musumeci, S. (1975). Studies in Mediterranean leishmaniasis. 3. The leishmanin skin test in kala-azar. *Trans R Soc Trop Med Hyg.* 69: 60–68.

Prasad, A.S. (2007). Zinc: mechanisms of host Defense 1–3. *In J Nutr.*137: 353–357.

Ramesh, V., Kaushal, H., Mishra, A. K., Singh, R., Salotra, P. (2015). Clinico-epidemiological analysis of Post kala-azar dermal leishmaniasis (PKDL) cases in India over last two decades: a hospital based retrospective study. *BMC Public Health, 15*(1): 1092. https://doi.org/10.1186/s12889-015-2424-8

Reedy, J., Krebs-Smith, S.M., Miller, P.E., Liese, A.D., Kahle, L.L., Park, Y., Subar, A.F. (2014). Higher diet quality is associated with decreased risk of all-cause, cardiovascular disease, and cancer mortality among older adults. *J Nutrit.* 144: 881–889.

Rutte, E.A.L., Zijlstra, E.E., Vlas, S.J.D. (2019). Post-Kala-Azar dermal leishmaniasis as a reservoir for visceral leishmaniasis transmission. *Trends Parasitol.* 35: 590–592.

Samson, L., Thi, H., Thanh, D., Frehiwot, M., Gezahegn, A. (2014). Visceral leishmaniasis in Ethiopia: an evolving disease. *PLoS Negl Trop Dis.* 8: 3131.

Singh, N., Singh, V., Rai, S. N., Mishra, V., Vamanu, E., Singh, M. P. (2022a). Deciphering the gut microbiome in neurodegenerative diseases and metagenomic approaches for characterization of gut microbes. *Biomedicine & Pharmacotherapy 156*: 113958. https://doi.org/10.1016/j.biopha.2022.113958

Singh, N., Singh, V., Rai, S. N., Vamanu, E., Singh, M. P. (2022b). Metagenomic analysis of garden soil-derived microbial consortia and unveiling their metabolic potential in mitigating toxic hexavalent chromium. *Life 12*(12): 2094. https://doi.org/10.3390/life12122094

Singh, N., Singh, V., Singh, M. P. (2022c). Microbial degradation of lignocellulosic biomass for bioenergy production: A metagenomic-based approach. *Biocatalysis and Biotransformation 41*(1): 15–25. https://doi.org/10.1080/10242422.2022.2056451

Singh, V., Abhishek, K., Rai, S. N., Singh, S. K., Vamanu, E., Kumar, A. (2023). Source of Cr(VI) in the aquatic ecosystem, its genotoxic effects and microbial removal from contaminated water. *Green Chemistry Letters and Reviews 16*(1). https://doi.org/10.1080/17518253.2023.2267079

Souza, M.L., dos, S., Sousa, W.M.D., Ferraz, A.L.M.D.L., Silva, Neto, P. J. (2022). Cutaneous leishmaniasis: new oral therapeutic approaches under development. *Int J Dermatol.* 61: 89–98.

Stephensen, C.B., Jiang, X., Freytag, T. (2004). Nutritional immunology vitamin A deficiency increases the in vivo development of IL-10-positive Th2 cells and decreases development of Th1 cells in mice. *J Nutr.* 134: 2660–2666.

Sundar, S., Kumar, R. (2023). Post kala-azar dermal leishmaniasis in the Indian sub-continent: challenges and strategies for elimination. *Front Immunol.* 14: 123–129.

Tanner, S., Leonard, B., Godoy, R., Reyes-Garcia, V., Huanca, T., Conde, E. (2014). Health and disease: exploring the relation between parasitic infections, child nutrition status, and markets. *Am J Phys Anthropol.* 155: 221–228.

Virginie, R., Thierry, D.M., Sandrine, K.O., Mallorie, H., Anne-Laure, B. (2010). Everything you always wanted to know about sex (but Were Afraid to Ask)" in Leishmania after two decades of laboratory and field analyses. *PLoS Pathogens.* 6: 1–4.

Yadav, P., Azam, M., Ramesh, V., Singh, R. (2023). Unusual observations in leishmaniasis—an overview. *Pathogens* 12: 297.

Zijlstra, E.E. (2012). *The Post Kala-azar Dermal Leishmaniasis (PKDL) Atlas A Manual for Health Workers.* WHO Library Cataloguing-in-Publication Data.

Zijlstra, E., Musa, A., Khalil, E., El-Hassan, I., El-Hassan, A. (2003). Post-kala-azar dermal leishmaniasis. *Lancet Infect Dis.* 3: 87–98.

15 Arsenic and Antimony Toxicity in Drinking Water and Their Roles in the Development of PKDL

*Manisha Verma, Mahesh S. Chivate, Vishal Singh, and Vishal Mishra**

School of Biochemical Engineering, IIT (BHU), Varanasi, Uttar Pradesh, India

*Corresponding Author: Dr. Vishal Mishra, Assistant Professor, School of Biochemical Engineering, IIT (BHU), Varanasi, Uttar Pradesh, India

15.1 INTRODUCTION

Visceral leishmaniasis (VL), prevalent in tropical regions, poses a significant threat to global health (Sinha et al., 2006). In India, the state of Bihar has the highest incidence of VL, accounting for nearly 90% of all cases (V. N. R. Das et al., 2012). Post-kala-azar dermal leishmaniasis (PKDL), which develops after VL recovery, is a skin condition (Ganguly, 2010). The parasite *Leishmania donovani* is to blame. PKDL is most prevalent in South Asia (Bangladesh, India, and Nepal) and East African countries such as South Sudan, Kenya, Uganda, Somalia, Eritrea, and Tanzania (V. N. R. Das et al., 2009; Musa et al., 2002).

Africa and South Asia exhibit distinctly different PKDL manifestations (Berman, 2006). South Asia is more likely to be affected by macules, papulonodules, and nodules, whereas Africa is more likely to be affected by papular or nodular lesions (Berman, 2006). Sudan has a higher PKDL incidence (Zijlstra, 2003; Zijlstra & El-Hassan, 2001), with 50–60% of VL patients potentially developing PKDL, in contrast to India's 5–10% (Zijlstra, 2003; Zijlstra & El-Hassan, 2001; Perry et al., 2011). In contrast to the Indian scenario, PKDL in Africa may naturally subside, coexist with VL, or emerge during VL treatment. Indian PKDL is characterised by pale patches, rashes, and nodules, frequently accompanied by facial reddening and hardening. The debate regarding PKDL's role in VL transmission continues, necessitating additional research (Perry et al., 2011). Due to PKDL's crucial role in the global eradication of VL, extensive research is required to identify its risk factors (Perry et al., 2011).

Groundwater is essential for drinking, irrigation, and industry, but its quality decline due to geological and anthropogenic activities is cause for concern (Chowdhury et al., 2001). Due to interactions between water, geological materials, the atmosphere, and surface water bodies, groundwater naturally contains dissolved inorganic chemicals in varying concentrations (Chowdhury et al., 2001). Arsenic, an odourless and colourless toxic element, is present in various environments, including aerial particles, food, water, and soil (Chowdhury et al., 2001). In terms of toxicity, inorganic arsenic dominates organic arsenic in groundwater. Chronic arsenic toxicity (CAT) may result from prolonged exposure to arsenic, which has numerous adverse health effects (Mitra et al., 2004). Long-term exposure to arsenic is linked to diabetes, skin conditions, different cancers, chronic coughing, and adverse effects on numerous systems (Chakraborty & Saha, 1987; Datta & Kaul, 1976). CAT causes skin lesions and systemic effects such as chronic lung disease, liver disease, vascular problems, neuropathy, hypertension, heart disease, oedema, diabetes, weakness,

 DOI: 10.1201/9781032643120-15

and anaemia. Common arsenic poisoning symptoms include thickening of the palms and soles (hyperkeratosis) and skin darkening (hyperpigmentation). CAT-induced keratosis is characterised by symmetrically distributed palm and sole thickening, possibly with nodules (Chen et al., 2006; Vahter et al., 2006).

Access to clean potable water is an essential human right. Arsenic pollution of drinking water is regarded as one of the most pressing environmental health issues on a global scale, including in India. Arsenic from minerals such as arsenopyrite, realgar, orpiment, and iron oxyhydroxide can contaminate groundwater, releasing arsenic into the water (Abernathy et al., 2003). Natural processes release arsenic from these minerals, including oxidation, desorption, dissolution, concentration, leaching, and microbial activity. Notably, arsenopyrite can release arsenic into groundwater under certain conditions (Shi et al., 2004). Rapid industrialisation, insufficient water treatment, sewage management, monsoons, and flooding have exacerbated India's groundwater arsenic problem (Nordstorm, 2002).

Poisoning from arsenic at elevated levels is a significant concern in the Gangetic plain, particularly in Bihar, Uttar Pradesh, and West Bengal, which have all been documented as having drinking water contaminated with arsenic. Since the initial report in the district of Bhojpur in June 2002, groundwater contamination in Bihar's districts has reached alarming proportions. 1861 ppb of arsenic exceeds the WHO's permissible limit of 10 ppb. Contamination of agricultural land through irrigation with arsenic-tainted water exacerbates the problem, heightening exposure risks via consumed crops. Because arsenic and antimony, which are both common in treatments for VL, share chemical properties, patients with VL in the Indian state of Bihar who were exposed to arsenic developed antimonial resistance (Perry et al., 2011).

The development of PKDL is intrinsically linked to antimony-based VL therapy. The emergence of PKDL may be influenced by environmental factors, such as persistent exposure to arsenic-contaminated groundwater. This study explores the role of prolonged exposure to arsenic-contaminated groundwater as a contributor to the development of PKDL in VL-recovered patients (Perry et al., 2011).

15.2. POST-KALA-AZAR DERMAL LEISHMANIASIS

PKDL is a dermatological disorder that develops after the effective treatment of VL, also known as kala-azar. These diseases are caused by various species of the protozoan parasite *Leishmania*. Kala-azar is a potentially fatal form of leishmaniasis that primarily affects the spleen, liver, and bone marrow (Ganguly, 2010). Without intervention, it has the potential to be fatal. However, antiparasitic medications can be used to effectively treat kala-azar with judicious medical intervention. In contrast, PKDL is a potential complication following successful kala-azar treatment (Ganguly, 2010). It manifests as skin lesions within months to years following the initial infection. PKDL is believed to be caused by VL treatment-residant parasites in the skin, although the precise mechanism remains incompletely understood (Chappuis et al., 2007).

Macules, papules, nodules, and plaques are typical skin lesions observed in patients with PKDL. These lesions could be hypopigmented, hyperpigmented, or erythematous (red and inflamed). The severity and extent of PKDL can vary considerably between individuals (Chappuis et al., 2007). Unlike VL, PKDL is typically not fatal, but it can cause significant cosmetic and social problems for those affected (Musa et al., 2002). It is most prevalent in VL-endemic regions, which include parts of Africa, the Indian subcontinent, and the Middle East. *Leishmania* parasites are typically detected through clinical presentation and laboratory tests, such as microscopic examination of skin smears or polymerase chain reactions. Similar antiparasitic drugs, such as miltefosine or sodium stibogluconate, are used for treatment, as they are for VL. The duration and choice of treatment rely on the severity of the disease and the species of *Leishmania* involved (V. N. R. Das et al., 2009).

Given that PKDL has the potential to contribute to leishmaniasis transmission via sandfly vectors, it becomes a crucial factor in efforts to control and eradicate leishmaniasis. Effective management requires preventative measures, prompt diagnosis, and appropriate treatment, targeting both VL and potential complications such as PKDL (V. N. R. Das et al., 2009).

15.2.1 Chronic Arsenic Exposure and Microbial Drug Resistance

In areas where VL is endemic, such as Bihar, India, alarmingly high levels of arsenic contamination in drinking water from groundwater sources have been detected, reaching up to 1861 ppb (Ahmed et al., 2019). Consequently, residents of these regions are susceptible to arsenic poisoning. Even though most arsenic is rapidly eliminated from the bloodstream within hours, prolonged exposure to elevated levels can lead to its accumulation in vital organs such as the liver, spleen, and bone marrow (Ahmed et al., 2019). These organs may contain higher concentrations of arsenic than the bloodstream, where the virulent form of the *Leishmania donovani* parasite resides during infection (Ahmed et al., 2019).

The shared use of microbial aquaglyceroporin (AQP1) channels for transporting arsenicals and antimonials leads to the development of resistance in *Leishmania* parasites following arsenic exposure (Marquis et al., 2005). Previous research has shown elevated thiol levels in arsenic-exposed *Leishmania* parasites that are resistant to antimony. These thiol compounds bind to intracellular vesicles and accumulate there (Maciaszczyk-Dziubinska et al., 2012). Mitochondria play a crucial role in producing adenosine triphosphate (ATP), which is vital for energy (Légaré et al., 2001). Arsenic directly affects mitochondria, disrupting the Krebs cycle and reducing its rate while altering the expression of several mitochondrial enzymes (Bergquist et al., 2009; Leu & Mohassel, 2009). This disruption leads to membrane potential depolarisation and impaired oxidative phosphorylation impairment (Bergquist et al., 2009).

Furthermore, arsenic exposure induces immunotoxic effects, evidenced by reduced T cell proliferation and altered excretion of proinflammatory cytokine, including TNF-α, IFN-β, IL-2, IL-4, IL-5, IL-6, IL-10, and IL-12 (Dangleben et al., 2013; Soto-Peña et al., 2006). Additionally, arsenic impairs macrophage defence mechanisms by inhibiting the production of reactive oxygen species (ROS) and nitric oxide (NO) (Banerjee et al., 2009). This decline weakens the host's defences against invading pathogens, increasing susceptibility. While arsenic has specific impacts on the immune system and its role in infection, further investigation is needed to clarify how arsenic-induced changes in *Leishmania* biology modulate the immune responses of healthy host cells and influence infection patterns (Ahmed et al., 2019). Therefore, an in vitro study aims to explore how these arsenic-induced changes contribute to alterations in immune function and infection dynamics within healthy cells (Ahmed et al., 2019).

The emergence of antimicrobial drug resistance, particularly in infections caused by parasitic protists, poses a significant threat to managing infectious diseases (Mcconville & Ralph, n.d.). Pathogens responsible for diseases such as leishmaniasis are of substantial medical importance, highlighting the urgent need to address this issue. In India during the 1970s and 1980s, the rise in cases of antimonial resistance among *Leishmania donovani* indicated clear signs of prolonged exposure. *L. donovani* is the pathogen responsible for VL, a disease that can be fatal if left untreated (Mcconville & Ralph, n.d.). Perry et al. (2013) propose a compelling link between increased antimonial resistance in *Leishmania* in India and high concentrations of arsenic in drinking water. Environmental factors can contribute to developing drug resistance among microorganisms, which has implications for developing novel treatments (Perry et al., 2013).

Pentostam and various pentavalent antimonials have been successfully employed in treating VL since their introduction in the 1920s. However, resistance started to emerge in the 1970s, primarily due to the discontinuation of vector control initiatives in Bihar. As a result, lipid amphotericin B and miltefosine gradually replaced them as the preferred treatment options starting in 2002, driven by declining success rates and the need for higher drug dosages along with prolonged treatment durations. While improper usage is often cited as a reason for antimonial resistance in Bihar, it's essential to recognise that this issue isn't universally applicable. The absence of similar resistance levels in other regions suggests that unique contributing factors exist within Bihar.

Perry et al. (2013) discovered a notable correlation between the surge in antimonial resistance in Bihar during the 1970s and a simultaneous increase in arsenic concentrations in drinking water. This rise was attributed to the widespread adoption of shallow tube wells as sources of safe water. Notably, both antimony and arsenic belong to group 15 of the periodic table and share various chemical properties. Intriguingly, historical records also indicate that organic arsenic compounds were used in treating leishmaniasis; moreover, arsenicals continue to serve as essential therapeutic agents for managing human African trypanosomiasis (Perry et al., 2013).

Perry et al. (2013) advanced the hypothesis that the emergence of antimonial resistance may be attributed to prolonged exposure of *L. donovani* to sublethal concentrations of arsenic within the human host combined with adaptive changes (see Figure 15.1). To support this assertion, the researchers provide compelling evidence. They conducted experiments where *L. donovani* strains, initially susceptible to antimonials, were systematically cultivated in mice exposed to low levels of arsenic in their drinking water, replicating conditions in Bihar. Remarkably, these strains displayed heightened resistance to pentostam (Perry et al., 2013) (Figure 15.1).

The emergence of antimonial resistance coincided with elevated arsenic levels detected in the liver and persisted even after repeated passages through animals not exposed to arsenic (Mcconville & Ralph, n.d.). Despite the lack of drug selection pressure, these antimonial-resistant parasites displayed increased virulence compared to their antimonial-susceptible counterparts. This indicates that parasites resistant to both antimonials and arsenic could potentially persist and spread into neighbouring regions not contaminated by arsenic (Mcconville & Ralph, n.d.).

Pentavalent antimonials Sb(V) traverse both host and parasite membranes before being converted to their active Sb(III) form. As illustrated in Figure 15.2, this transformation occurs either within host cells or during intracellular stages within parasites.

It is widely postulated that Sb(III) specifically targets essential parasite enzymes responsible for maintaining reduced thiol compounds. A prime example of this phenomenon is observed with trypanothione, a distinctive combination of bis-glutathione and spermidine (Croft et al., 2006).

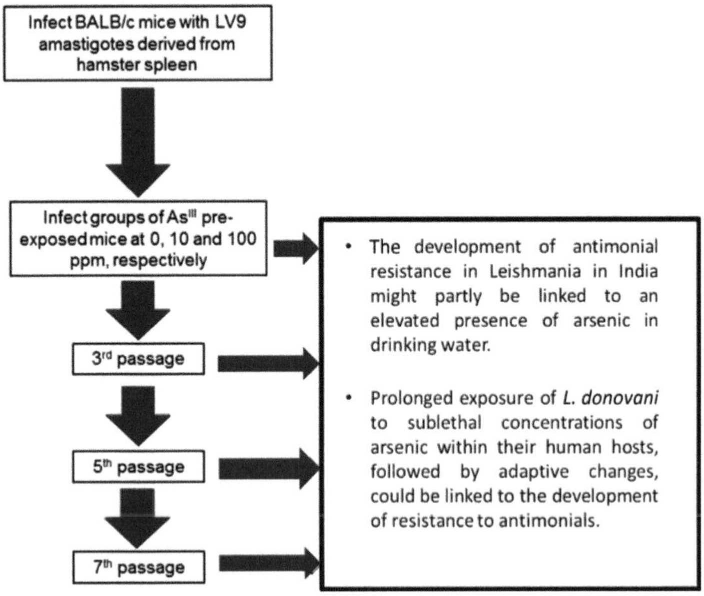

FIGURE 15.1 In vivo generation of an antimonial-resistant line by oral arsenic exposure (Perry et al. 2013)

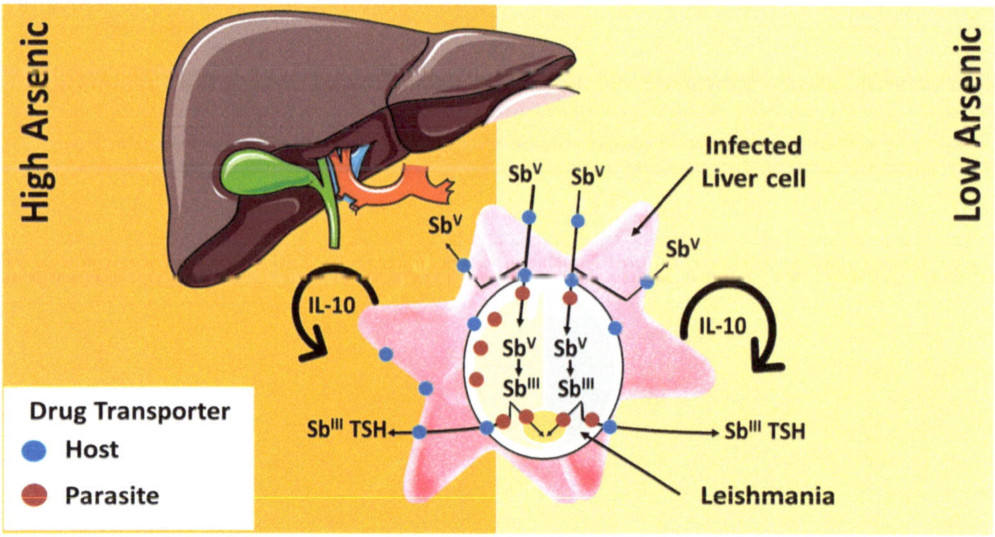

FIGURE 15.2 Mechanisms of antimonial resistance in *Leishmania* that may be induced by chronic arsenic exposure. Arsenic (As) accumulation in the liver

The development of resistance mechanisms may involve an upsurge in the expression of parasite transporter proteins. These proteins have the dual function of either expelling trypanothione-Sb(III) conjugates from the parasite or shuttling them into intracellular organelles, ultimately reducing the accumulation of cytotoxic Sb(III) within the cell (Valentine et al., 1979).

Furthermore, resistance might be associated with substantial alterations in parasite metabolism, leading to the establishment of a distinct physiological state that enables parasites to endure within mammalian macrophages. This adaptive state is imperative for parasite survival (Perry et al., 2011). These findings align with those reported by Perry and colleagues in 2013 (Perry et al., 2013), suggesting that prolonged exposure to arsenic could elevate parasite levels in animals vulnerable to its effects.

Additionally, evidence suggests that antimonial-resistant *Leishmania donovani* strains may alter the rate at which infected host cells take up or release antimonial compounds and modify the host's immune system. Notably, resistant parasite lines may activate host signalling pathways that lead to an increase in IL-10 release, a potent anti-inflammatory cytokine that encourages *Leishmania* development within the host. Therefore, persistent arsenic exposure may not only activate mechanisms of drug resistance against antimonials but may also disrupt the equilibrium of host-parasite interactions. This could be advantageous to the pathogen by improving its fitness and shaping the immune response in a manner that promotes parasite proliferation (Mcconville & Ralph, n.d.).

Evidence suggests that persistent arsenic exposure has contributed to the emergence of antimony resistance in regions beyond India. Antimonial resistance has been documented in Peru, although to a lesser degree than in India. In Peru, the cutaneous variants of the disease are still treated with antimonials. Recent research suggests a correlation between arsenic contamination and the emergence of antimonial resistance in multiple regions of Peru where drinking and irrigation water contain elevated concentrations of arsenic. These findings raise the question of whether prolonged arsenic exposure may also trigger resistance to other drugs used to treat leishmaniasis. One such drug is miltefosine, an unrelated alkyl-lipid compound that has replaced antimonials as the primary treatment for VL in India and is now widely used for various forms of leishmaniasis worldwide (Ahmed et al., 2019).

Using an arsenic dosage ranging from 0.5 to 1.5 mg/L, Ghufran Ahmed et al. (2019) produced a standard laboratory culture of *Leishmania donovani* parasites. This concentration was determined to be optimal for modulating the immune response and promoting infection establishment (Ahmed et al., 2019). Arsenic exposure resulted in antimony resistance in *Leishmania donovani* parasites. The arsenic-exposed *Leishmania* strain exhibited resistance to the drug sodium stibogluconate (SAG) at concentrations ranging from 20 to 150 µg/mL. The increased production of reduced thiol compounds may contribute to the drug's efflux (Ahmed et al., 2019).

Ahmed and colleagues (2019) compiled a substantial body of evidence regarding the deleterious impacts of arsenic on the facilitation of *Leishmania* infection. Their research underscores that exposure of parasites to arsenic leads to heightened infectivity within macrophages derived from BALB/c mice compared to cells not exposed to arsenic. Previous investigations have proposed that the parasite bolsters its survival within the host by modulating the trypanothione synthase (TSH2) pathway. This modulation, in turn, affects the production of crucial reduced thiol compounds and the regulated release of ROS (Ahmed et al., 2019).

Compared to *Leishmania* parasites that were not exposed to arsenic, *Leishmania* parasites that were exposed to arsenic significantly increased their production of reduced thiol compounds by 1.7-fold. This finding supports the hypothesis that decreased reactive nitrogen and oxygen intermediates (NO, ROS) would increase the parasite's infectiousness when exposed to arsenic (Ahmed et al., 2019). However, as a result of the decrease in mitochondrial membrane potential, ATP was utilised for energy production. This was demonstrated by the significant depolarisation of the membrane potential in arsenic-exposed *Leishmania* (LdAS) versus unexposed *Leishmania* (Ld). This result is consistent with reports indicating that arsenic causes depolarisation of the mitochondrial membrane potential (Ahmed et al., 2019).

Because arsenite's entry into the glycolytic pathway affects cellular energy metabolism, this discovery raises concerns about increased infections caused by arsenic (S. Das et al., 2016). The accumulation of lactate, indicating a shift from pyruvate to acetyl-CoA, is a possible argument in favour of this hypothesis (Ahmed et al., 2019). It is likely that the latter substance can't efficiently enter the TCA cycle because of the depolarisation of the mitochondrial membrane potential. The excessive production of lactate could then be converted back to glucose via gluconeogenesis, as evidenced by the high glucose ingestion in *Leishmania* parasites exposed to arsenic (Ahmed et al., 2019). This suggests that the rate of glycolysis could be significantly accelerated, followed by the Warburg effect, ultimately increasing infectiousness (Ahmed et al., 2019).

The effects of arsenic exposure from contaminated groundwater on patients with VL who received SSG or other treatments for their VL episode symptoms and the likelihood of acquiring PKDL are clear. Daily ingestion of water contaminated with arsenic results in a medical condition marked by high amounts of sodium arsenite in different body areas (S. Das et al., 2016). Hyperkeratosis and hyperpigmentation are prominent examples of the visible skin effects resulting from arsenic exposure that reflect its toxicity. Studies on the dangers of arsenic reveal a strong correlation between the clothing choices of individuals and the development of health risks, which are exacerbated by exposure to ultraviolet rays. Notably, PKDL manifests a variety of skin-related symptoms, especially in sun-exposed body regions (S. Das et al., 2016). These patterns of skin lesions correlate with a person's clothing habits and UV radiation exposure. Sunlight has been observed to induce rapid immunological changes in the skin and peripheral blood, and it can also contribute to immunosuppression by decreasing specific dermal dendritic cell subsets in psoriasis patients (S. Das et al., 2016).

Given that the study population consists primarily of VL patients who work in agriculture and are therefore susceptible to arsenic exposure and excessive sunlight, their vulnerability is increased. Another study demonstrates that glass factory workers are exposed to higher levels of arsenic, which are five times higher than those of the control group (S. Das et al., 2016). This increased exposure is associated with DNA damage in leukocytes. Given the importance of leukocytes to the treatment and cure of VL, excessive arsenic exposure may impair their capacity to eliminate parasites during treatment. Consequently, parasites are able to evade immune killing mechanisms.

In conclusion, groundwater contamination with arsenic contributes to the development of PKDL in Bihar through these risk factors (V. N. R. Das et al., 2012).

15.3 CONCLUSION

In regions of Bihar, India, where VL is endemic, humans are exposed to arsenic contamination in groundwater via their drinking water. This exposure can lead to arsenic accumulation in the tissues of patients with VL. As a result, these patients may harbour intracellular amastigotes, contributing to the parasite's resistance to chemotherapy. This resistance appears to affect the host's immune response. This theory is corroborated by laboratory findings showing that arsenic-exposed parasites exhibit depolarised mitochondrial membrane potential, increased reduced thiol and lactate production, and increased glucose consumption. This situation could induce immunosuppression in host cells. Reduced expression of proinflammatory cytokines and impaired anti-leishmanial function in macrophages support this notion. In contrast, the presence of *Leishmania* parasites that have been exposed to arsenic encourages the generation of Th2 cytokines, which further contributes to the pathophysiology of VL. The dysfunction in parasite mitochondria and the increase in lactate production indicate the potential involvement of the Warburg effect, leading to a rise in energy production and possibly enhanced virulence. In summary, the exposure of *Leishmania donovani* to arsenic has the potential to alter the immune response and infection dynamics, thereby compromising the anti-leishmanial effect in VL.

Furthermore, the widespread resistance to antimonial drugs commonly used to treat VL corresponds with the widespread contamination of drinking water with arsenic. The hypothesis suggests parasite resistance may develop due to exposure to trivalent arsenic, resulting in cross-resistance with antimony, a related metalloid found in antimonial drugs. *Leishmania donovani* was passed in mice exposed to environmentally relevant levels of arsenic in their drinking water to test this hypothesis. Mice exposed to arsenic produced parasites with complete resistance to antimonial drugs, while parasites from control mice remained susceptible. This resistance was also observed in vivo when antimonial drugs were administered to infected mice. These findings suggest that arsenic contamination may play a significant role in the evolution of antimonial drug resistance in *Leishmania* parasites, particularly in regions such as Bihar, and may have implications beyond its borders.

ACKNOWLEDGEMENT

The authors thank the Indian Institute of Technology (BHU) Varanasi, Varanasi, for extending their technical and financial support.

CONFLICT OF INTEREST

The authors have declared no conflict of interest.

ABBREVIATIONS

ADP	Adenosine diphosphate
AQP1	Aquaglyceroporin
ATP	Adenosine triphosphate
CAT	Chronic arsenic toxicity
IL	Interleukin
PKDL	Post-kala-azar dermal leishmaniasis
ROS	Reactive oxygen species

TCA cycle	Tricarboxylic acid cycle, or Krebs cycle
TNF	Tumour necrosis factor
SSG	Sodium stibogluconate
VL	Visceral leishmaniasis

REFERENCES

Abernathy, C., Thomas, D., & Calderon, R. (2003). Health effects and risk assessment of arsenic. *J Nutr, 133*, 1536S–1538S.

Ahmed, G., Thakur, A. K., Chaturvedi, S. K., Shivam, P., Jamal, F., Singh, M. K., Bimal, S., Singh, S. K., Choudhary, S. K., Das, P., & Narayan, S. I. (2019). Modulation of the immune response and infection pattern to *Leishmania donovani* in visceral leishmaniasis due to arsenic exposure: An in vitro study. https://doi.org/10.1371/journal.pone.0210737

Banerjee, N., Banerjee, S., Sen, R., Bandyopadhyay, A., Sarma, N., Majumder, P., Das, J. K., Chatterjee, M., Kabir, S. N., & Giri, A. K. (2009). Chronic arsenic exposure impairs macrophage functions in the exposed individuals. *J Clin Immunol, 29*(5), 582–594. https://doi.org/10.1007/S10875-009-9304-X

Bergquist, E., Fischer, R.J., Sugden, K.T., & Martin, B.D. (2009). Inhibition by methylated organoarsenicals of the respiratory 2-oxo-acid dehydrogenases. *Organomet Chem. 694*, 973–980. https://www.sciencedirect.com/science/article/pii/S0022328X08008401

Berman, J. (2006). Visceral leishmaniasis in the New World & Africa. *Indian J Med Res, 123*, 289–294.

Chakraborty, A., & Saha, K. (1987). Arsenical dermatosis from tubewell water in West Bengal. *Indian J Med Res, 85*, 326–334.

Chappuis, F., Sundar, S., Hailu, A., Ghalib, H., Rijal, S., Peeling, R., Alvar, J., & Boelaert, M. (2007). Visceral leishmaniasis: What are the needs for diagnosis, treatment and control? *Nat Rev Microbiol, 5*(11), 873–882.

Chen, Y., Graziano, J. H., Parvez, F., Hussain, I., Momotaj, H., Van Geen, A., Howe, G. R., & Ahsan, H. (2006). Modification of risk of arsenic-induced skin lesions by sunlight exposure, smoking, and occupational exposures in Bangladesh. *Epidemiology, 17*(4), 459–467. https://doi.org/10.1097/01.ede.0000220554.50837.7f

Chowdhury, U., Rahman, M., Mandal, B., Paul, K., Lodh, D., & Biswas, B. (2001). Groundwater arsenic contamination and human suffering in West Bengal, India and Bangladesh. *Environ Sci, 8*, 393–397.

Croft, S. L., Sundar, S., & Fairlamb, A. H. (2006). Drug resistance in leishmaniasis. *Clinical Microbiology Reviews, 19*(1), 111–126. https://doi.org/10.1128/CMR.19.1.111-126.2006

Dangleben, N. L., Skibola, C. F., & Smith, M. T. (2013). Arsenic immunotoxicity: A review. *Environ Health: Global Access Sci Source, 12*(1). https://doi.org/10.1186/1476-069X-12-73

Das, S., Mandal, R., Rabidas, V. N., Verma, N., Pandey, K., Ghosh, A. K., Kesari, S., Kumar, A., Purkait, B., Lal, C. S., & Das, P. (2016). Chronic arsenic exposure and risk of post kala-azar dermal Leishmaniasis development in India: A retrospective cohort study. *PLoS Negl Trop Dis, 10*(10), e0005060. https://doi.org/10.1371/JOURNAL.PNTD.0005060

Das, V. N. R., Pandey, K., Verma, N., Lai, C. S., Bimal, S., Topno, R. K., Singh, D., Siddiqui, N. A., Verma, R. B., & Das, P. (2009). Short report: Development of post-kala-azar dermal leishmaniasis (PKDL) in miltefosine-treated visceral leishmaniasis. *Am J Trop Med Hyg, 80*(3), 336–338. https://doi.org/10.4269/ajtmh.2009.80.336

Das, V. N. R., Ranjan, A., Pandey, K., Singh, D., Verma, N., Das, S., Lal, C. S., Sinha, N. K., Verma, R. B., Siddiqui, N. A., & Das, P. (2012). Short report: Clinical, epidemiologic profile of a cohort of post-kala-azar dermal leishmaniasis patients in Bihar, India. *Am J Trop Med Hyg, 86*(6), 959–961. https://doi.org/10.4269/ajtmh.2012.11-0467

Datta, D., & Kaul, M. (1976). Arsenic content of drinking water in villages in northern India. A concept of arsenicosis. *J Assoc Physicians India, 24*, 599–604.

Ganguly, S. (2010). Post-kala-azar dermal leishmaniasis—an overview. *Int. J. Dermatol, 49*(8), 921–931.

Légaré, D., Richard, D., Mukhopadhyay, R., Stierhof, Y.D., Rosen, B.P., Haimeur, A., Papadopoulou, B., & Ouellette, M. (2001). The Leishmania ATP-binding cassette protein PGPA is an intracellular metal-thiol transporter ATPase. *J Biol Chem., 276*(28), 26301–26307. Retrieved August 26, 2023, from https://www.jbc.org/article/S0021-9258(19)30189-9/abstract

Leu, L., & Mohassel, L. (2009). Arsenic trioxide as first-line treatment for acute promyelocytic leukaemia. *Am J Health-System Pharm, 66*(21), 1913–1918. https://doi.org/10.2146/ajhp080342

Maciaszczyk-Dziubinska, E., WWawrzycka, D., & Wysocki, R. (2012). Arsenic and antimony transporters in eukaryotes. *International J Molecular Sci, 13*(3), 3527–3548. https://doi.org/10.3390/ijms13033527

Marquis, N., Gourbal, B., Rosen, B. P., Mukhopadhyay, R., & Ouellette, M. (2005). Modulation in aquaglyceroporin AQP1 gene transcript levels in drug-resistant Leishmania. *Mol Microbiol*, *57*(6), 1690–1699. https://doi.org/10.1111/J.1365-2958.2005.04782.X

Mcconville, M. J., & Ralph, S. A. (2013). Chronic arsenic exposure and microbial drug resistance. *Proc Natl Acad Sci USA*. *110*(49), 19666–19667. https://doi.org/10.1073/pnas.1319659110

Mitra, S. R., Guha Mazumder, D. N., Basu, A., Block, G., Haque, R., Samanta, S., Ghosh, N., Hira Smith, M. M., von Ehrenstein, O. S., & Smith, A. H. (2004). Nutritional factors and susceptibility to arsenic-caused skin lesions in West Bengal, India. *Environ Health Perspect*, *112*(10), 1104–1109. https://doi.org/10.1289/ehp.6841

Musa, A. M., Khalil, E. A. G., Raheem, M. A., Zijlstra, E. E., Ibrahim, M. E., Elhassan, I. M., Mukhtar, M. M., & El Hassan, A. M. (2002). The natural history of Sudanese post-kala-azar dermal leishmaniasis: Clinical, immunological and prognostic features. *Anna Trop Med Parasitol*, *96*(8), 765–772. https://doi.org/10.1179/000349802125002211

Nordstorm, D. K. (2002). Worldwide occurences of arsenic in ground water. *Science*, *269*(5576), 2143–2145.

Perry, M. R., Wyllie, S., Prajapati, V. K., Feldmann, J., Sundar, S., Boelaert, M., & Fairlamb, A. H. (2011). Visceral leishmaniasis and arsenic: An ancient poison contributing to antimonial treatment failure in the Indian subcontinent? *PLoS Negl Trop Dis*, *5*(9), 1–5. https://doi.org/10.1371/journal.pntd.0001227

Perry, M. R., Wyllie, S., Raab, A., Feldmann, J., & Fairlamb, A. H. (2013). Chronic exposure to arsenic in drinking water can lead to resistance to antimonial drugs in a mouse model of visceral leishmaniasis. *Proc Nat Acad Sci USA*, *110*(49), 19932–19937. https://doi.org/10.1073/pnas.1311535110

Shi, H., Shi, X., & Liu, K. J. (2004). Oxidative mechanism of arsenic toxicity and carcinogenesis. *Molecular Cellular Biochem*, *255*(1–2), 67–78. https://doi.org/10.1023/B:MCBI.0000007262.26044.e8

Sinha, P., Ranjan, A., Singh, V., Das, V., Pandey, K., Kumar, N., Verma, N., Lal, C., Sur, D., Manna, B., & Bhattacharya, S. (2006). Visceral leishmaniasis (kala-azar)—the Bihar (India) perspective. *J Infect*, *53*(1), 60–64.

Soto-Peña, G. A., Luna, A. L., Acosta-Saavedra, L., Conde-Moo, P., López-Carrillo, L., Cebrián, M. E., Bastida, M., Calderón-Aranda, E. S., & Vega, L. (2006). Assessment of lymphocyte subpopulations and cytokine secretion in children exposed to arsenic. *FASEB J*, *20*(6), 779–781. https://doi.org/10.1096/FJ.05-4860FJE

Vahter, M., Li, L., Nermell, B., Rahman, A., Arifeen, S. El, & Rahman, M. (2006). Arsenic, a global public health problem. *Toxicol Lett*, *164S*, S45–S46.

Valentine, J., Kang, H., & Spivey, G. (1979). Arsenic levels in human blood, urine and hair in response to exposure via drinking water. *Environ. Res*, *20,* 24–32.

Zijlstra, E. E. (2003). Post kala-azar dermal leishmaniasis. *Lancet Infect. Dis*, *3*.

Zijlstra, E. E., & El-Hassan, A. M. (2001). Leishmaniasis in Sudan. 3. Visceral leishmaniasis. *Trans R Soc Trop Med Hyg*, *95*, 27–58. https://doi.org/10.1016/S0035-9203(01)90218-4

16 Government Policies toward Providing Safe Drinking Water to the Public

Priyanka Yadav[1], Kusum Yadav[1], Mayank Saxena[2], and Isha Singh[2]*

[1]Department of Biochemistry, University of Lucknow, Lucknow, Uttar Pradesh, India
[2]Department of Biotechnology and Department of Computer Science and Engineering, IILM University, Greater Noida, Uttar Pradesh, India
*Corresponding Author: Priyanka Yadav, Department of Biochemistry, University of Lucknow, Lucknow, Uttar Pradesh, India

16.1 INTRODUCTION

Government approaches to providing safe and clean drinking water to the public vary across nations. However, the common goal is to ensure universal access to safe, clean drinking water (1, 2). Below are some general policies and approaches commonly implemented by governments.

16.1.1 Water Quality Standards

Governments establish specific standards as well as regulations for drinking water quality. These standards specify the maximum allowable levels of various contaminants to ensure that public water supplies meet or exceed these criteria. Regular testing and monitoring are conducted to assess compliance (3).

16.1.2 Water Treatment

Governments invest in water treatment infrastructure and technologies aimed at removing or reducing contaminants from drinking water sources. Common treatment processes include filtration and disinfection. In some cases, advanced techniques like reverse osmosis or ultraviolet treatment are used. The primary goal is to eliminate harmful bacteria, viruses, chemicals, and other pollutants (4). Figure 16.1 illustrates the typical drinking water treatment process.

16.1.3 Water Source Protection

Governments implement strategies to prevent pollution and depletion of water resources including rivers, lakes, reservoirs, and groundwater. The strategies often involve establishing rules and enforcing regulations to control industrial discharges, agricultural runoff, and other activities that can contaminate water supplies (5, 6).

DOI: 10.1201/9781032643120-16

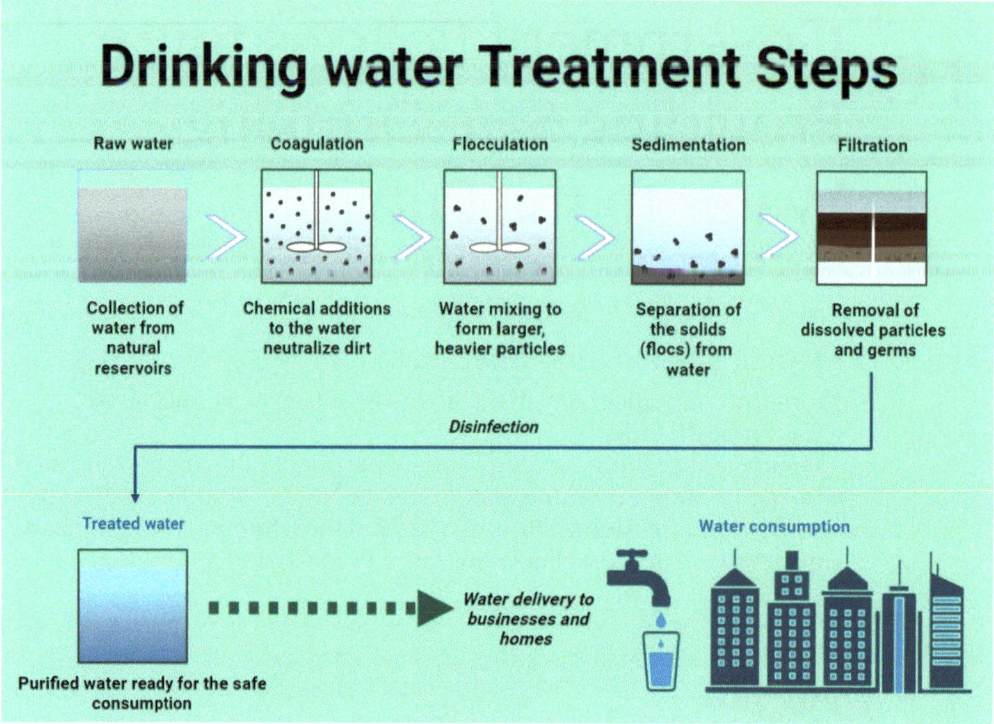

FIGURE 16.1 Drinking water treatment process

16.1.4 Water Infrastructure Development

Governments invest in building, maintaining, and upgrading the infrastructure needed for water delivery, including pipelines, reservoirs, pumping stations, and treatment facilities. This investment aims to ensure access to clean, safe water for all populations, with a particular focus on underserved or remote areas (7).

16.2 PUBLIC HEALTH AND EDUCATION

Governments support public awareness programs to educate people about the importance of clean water and good hygiene. These programs provide information on waterborne diseases, proper water storage, and household water treatment methods, particularly for communities without access to purified water (8).

Government policies regarding safe drinking water can vary significantly depending on the country and its specific circumstances. For the most up-to-date information on regulations, it is advisable to consult official government websites, health departments, or relevant regulatory agencies in your country. These sources typically offer information about current regulations, rules, and recent government initiatives aimed at guaranteeing public access to clean drinking water.

Further research is needed on the increasing consumption of bottled water and its implications. Such studies could inform policies to enhance public water delivery systems and support initiatives that boost consumer confidence in tap water. The United Nations Development Program's Human Development Reports highlight that more than 1 billion people worldwide, particularly in the so-called "developing countries," lack access to clean drinking water. This ongoing crisis is a daily catastrophe that affects millions worldwide, underscoring the fundamental human right to safe drinking water.

The right to clean water is a fundamental human right, enshrined in the Universal Declaration of Human Rights and other legal instruments, alongside rights to life, liberty, education, health, and shelter. Protecting this right is both an aim and an end in itself. Governments must implement effective measures to ensure people's right to high-quality tap water, including public policies that prioritize population health over economic interests.

Some researchers advocate for effective strategies to promote public understanding, communication, and confidence in water supply quality. These efforts aim to enable informed decision-making about drinking water choices and foster an understanding of potential social and environmental repercussions. A study of Canadian print media found that news stories informed readers about the water supply situation, provided crucial advice such as boiling water to reduce health risks, and reported complaints and breaking news alerts.

The collective subject discourse method, based on social representation theory, was used to analyze media narratives on bottled and tap water in Canada. The study found that ambiguous information could mislead the public and change consumption habits. Media channels that raise awareness and encourage reflection could also promote research on the subject of drinking water. Government agencies, the general public, academics, and media professionals are all collaborating to ensure that water commodification does not threaten the fundamental right to access clean water. Canada is considered privileged in terms of its water resources. This chapter aims to provide a better grasp of the existing state of affairs to facilitate future research rather than exhaustively cover the topic (9).

16.3 BACKGROUND

Despite 71% of the Earth's surface being covered by water, access to clean drinking water remains a global challenge (10). The development of government policies for providing safe drinking water originated from the recognition of its critical importance for public health and well-being. Throughout history, communities have faced various issues related to accessing clean water, necessitating government intervention through policies and regulations to address these issues. In ancient India, wells were a common source of clean drinking water. With the increment in the population growth as well as significant alterations to the natural water systems following colonization, the quality of water began to deteriorate. Some societies developed systems to transport clean water to cities through aqueducts and wells. The emergence of waterborne diseases in the early 20th century prompted the development of government policies regarding safe drinking water. In response to these challenges, governments implemented measures to improve water treatment and supply systems while establishing standards for water quality.

India's first government policy on drinking water, the National Water Policy, was launched in 1987 and adopted by the National Resource Water Council in September of that year. The policy's mission statement declared, "Every person should have access to safe and adequate water for drinking, cooking, bathing and other domestic needs."

The main objectives of the National Water Policy were:

a) Ensuring universal access to adequate water for basic needs
b) Developing and managing water resources in a sustainable manner
c) Protecting water from pollution
d) Integrating water resource management with sectors such as industry, agriculture, and energy

16.4 SAFE DRINKING WATER ACCESS

One of the key Millennium Development Goals was to reduce the number of people without access to reliable sources of clean water by half. This ambitious target led to extraordinary efforts in the water industry, improving the lives of millions of people. However, while the goal has theoretically

FIGURE 16.2 Source of safe drinking water

been achieved, the results have been overstated due to inappropriate standards. Indicators exaggerate improvements in water supplies while ignoring the fact that many remain dirty, unreliable, or prohibitively expensive. The alleged success needs to be reassessed in order to prevent misunderstanding, maintain investment in the water sector, and provide more accurate estimates of water availability.

Approximately one billion people live in urban slums worldwide, despite the United Nations' assertion that access to clean water is a fundamental human right. Slum dwellers typically require multiple water sources to meet their needs. Important factors to consider include water quality, availability, dependability, and cost. These factors are not static; they fluctuate across various timescales (from sub-daily to annual and beyond) in response to changes in water sources and household consumption capacity. For example, a slum household's ability to pay for water may change due to fluctuations in income, while the cost of water may vary with seasonal availability.

Current national and international monitoring of safe water access, including Sustainable Development Goal 6, does not account for these temporal dynamics of water supply, quality, and health risks in slums. This chapter presents a study design for investigating the potential health impacts of temporal variations in drinking water access and quality in urban slums. It is proposed that temporal dynamics should be considered across at least three interrelated timescales to improve access and health in urban slums: short-term (from sub-daily to weekly), medium-term (monthly to seasonal), and long-term (annual). The report concludes by recommending additional research into the temporal dynamics of drinking water and health in slums (11) (Figure 16.2).

16.4.1 Basic Principles of Safe Drinking Water Supplies

While more than two-thirds of the Earth's surface is covered by water, much of it is salty and unsuitable for use (12). Water providers worldwide are concerned about climate change, but so far, only qualitative shifts in boundary conditions have been observed, rather than quantifiable impacts on water delivery systems. For instance, extreme precipitation during dry spells may cause short-term alterations in the quality of surface water. Longer, hotter spells without rain can lead to medium-term quality changes in surface waters due to reduced dilution. Additionally, poor groundwater management can lead to water shortages and long-term quality changes, especially during periods of high demand. As the anticipated effects on supply systems depend heavily on specific circumstances, a one-size-fits-all adaptation strategy is unlikely to be effective (13).

16.4.2 Government Policies toward Providing Safe Drinking Water to the Public

The Sustainable Development Goals (SDGs) replaced the Millennium Development Goals, with SDG 6 aiming to ensure that everyone has unrestricted access to safe, affordable drinking water and sanitation. This study surveyed 400 households using a stratified sampling method, distributing 700 well-structured questionnaires with both open- and closed-ended questions. The inquiries aligned with Water and Sanitation Hygiene recommendations for household surveys on drinking water and sanitation. The study examined accessibility, availability, price, and acceptability of access to clean water in Ado-Odo Ota, Ogun State, Nigeria. Personal interviews were conducted to gauge respondents' emotional reactions to the topic. Cross-tabulation and a multinomial logit model were used to analyze the relationships between variables affecting water access.

The data revealed that water access is largely limited to private sources due to quality and accessibility issues. The research highlights a significant decline in the importance given to public water facilities. The report also emphasized the need to review government policies and incorporate subsidies, cost recovery mechanisms, and rainwater harvesting as effective strategies for enhancing and promoting equitable water access (14).

16.5 CONCLUSION

Water security has become a global concern. Water is essential for all life-sustaining activities, including food production, potable drinkable water supply, economic development, and sanitation. Water security ensures that every person has access to sufficient, safe water at affordable rates to lead a clean, healthy, and productive life while preserving and improving the natural environment, from the household to the global level. Although water is a renewable resource, excessive consumption has led to scarcity. Achieving water security is particularly challenging in populous in countries like India, which has more than 18% of the world's population but only 4% of the planet's freshwater resources.

India's population, metropolitan areas, water-dependent industries, and agriculture are all expanding quickly alongside its economy. The country already faces problems with pollution, scarcity, groundwater depletion, inequitable distribution, and unavailability. The Indian government has developed a National Water Policy to address these challenges. An effective water policy must acknowledge and address concerns about current and foreseeable water availability. Sustainable water use is crucial for India's water supply and economic growth.

The government must investigate technological solutions for managing drinking water, such as establishing water treatment facilities at storage sites to remove dissolved solids and harmful inorganic contaminants. India cannot provide its citizens with access to safe and pure drinking water at the expense of depleting its resources, as this would be counterproductive. At the same time, India cannot afford to wait until water supplies are exhausted due to population growth. Therefore, it is imperative that the government take action to conserve water while pursuing the goal of universal access to clean drinking water.

ACKNOWLEDGMENT

The authors are grateful to the University of Lucknow for providing the necessary support.

REFERENCES

1. National Water Policy, 1987. Ministry of Water Resources, Government of India. {#written by Isha}.
2. Martínez-Santos, P., 2017. Does 91% of the world's population really have "sustainable access to safe drinking water"? *International Journal of Water Resources Development*, *33*(4), pp. 514–533.
3. World Health Organization, 2022. *Guidelines for Drinking-Water Quality: Incorporating the First and Second Addenda*. World Health Organization, Geneva, Switzerland..
4. World Bank, 2007. *Making the Most of Scarcity: Accountability for Better water Management in the Middle East and North Africa*. The World Bank, Washington, D.C., USA..
5. Anju, A., Ravi S, P. and Bechan, S., 2010. Water pollution with special reference to pesticide contamination in India. *Journal of Water Resource and Protection*, *2010*, pp. 1–17..

6. Sarker, B., Keya, K.N., Mahir, F.I., Nahiun, K.M., Shahida, S. and Khan, R.A., 2021. Surface and ground water pollution: Causes and effects of urbanization and industrialization in South Asia. *Scientific Review*, *7*(3), pp. 32–41.
7. Blomquist, W., Schlager, E. and Heikkila, T., 2010. *Common Waters, Diverging Streams: Linking Institutions and Water Management in Arizona, California, and Colorado.* Routledge, New York, NY..
8. Cutler, D. and Miller, G., 2005. The role of public health improvements in health advances: the twentieth-century United States. *Demography*, *42*(1), pp. 1–22.
9. Queiroz, J.M., Rosenberg, M.W., Heller, L., Zhouri, A.M. and Silva, S.R., 2012. News about tap and bottled water: Can this influence people's choices? *Journal of Environmental Protection*, *3*(4), p. 324.
10. Gao, M., Zhu, L., Peh, C.K. and Ho, G.W., 2019. Solar absorber material and system designs for photothermal water vaporization towards clean water and energy production. *Energy & Environmental Science*, *12*(3), pp. 841–864.
11. Price, H., Adams, E. and Quilliam, R.S., 2019. The difference a day can make: the temporal dynamics of drinking water access and quality in urban slums. *Science of the Total Environment*, *671*, pp. 818–826.
12. Dinka, M.O., 2018. Safe drinking water: concepts, benefits, principles and standards. *Water Challenges of an Urbanizing World*, *163*, pp. 163–181.
13. Staben, N., Nahrstedt, A. and Merkel, W., 2015. Securing safe drinking water supply under climate change conditions. *Water Science and Technology: Water Supply*, *15*(6), pp. 1334–1342.
14. Madrigal, R., Alpízar, F. and Schlüter, A., 2011. Determinants of performance of community-based drinking water organizations. *World Development*, *39*(9), pp. 1663–1675.

17 Factors Influencing the Development of Post-kala-azar Dermal Leishmaniasis

Himanshu N. Panchal[1], Divya J. Tandel[1], and Sapna Desai[2]*

[1]Department of Pharmacology, PIP, Parul University, Vadodara, Gujarat, India

[2]Department of Pharmacology at Pioneer Pharmacy College, Vadodara, Gujarat, India

*Corresponding Author: Dr. Himanshu N Panchal, Assistant Professor, Department of Pharmaceutical Chemistry, PIP, Parul University, Vadodara, Gujarat, India

17.1 INTRODUCTION

Leishmaniasis is a protozoal disease caused by parasites of the *Leishmania* genus. Post-kala-azar dermal leishmaniasis is a consequence of visceral leishmaniasis (VL) caused by *Leishmania donovani,* transmitted by female sandflies (1). It is manifested by nodules, papules, and macules on the face that spread to other parts of the body (Figures 17.1 and 17.2) (2, 3). In some cases, PKDL lesions predominantly develop in scars that remit following PKDL therapy or become more pronounced (Köbner phenomenon) (4). PKDL cases are primarily found in a few northern states of India and African countries like Sudan and Kenya (5).

In 5–10% of cases in India, PKDL follows VL. The first instances of PKDL in India were recorded by Brahmachari in 1922, who described the emergence of skin patches and eruptions with *Leishmania* parasites in patients who had previously received treatment for VL (6, 7). The incidence of both VL and PKDL decreased in India in the early 1960s, after decades of high transmission, primarily in Bihar. This was likely due to DDT dusting used in the National Malaria Eradication Program, which affected sandflies in addition to malaria mosquitoes. Bihar and West Bengal saw a significant VL outbreak following the end of pesticide treatment (Figures 17.1 and 17.2) (8, 9).

After Kirk and Drew's initial study in 1938, reports on PKDL in Sudan were still limited. Kirk and Sati reported that epidermal infection occurred in more than 50% of VL patients, but they did not differentiate between PKDL and leishmaniaisis. Two hospital-based investigations identified sporadic cases (10–12). A severe VL epidemic with incidence rates of 20–30/1000 persons per year in one community occurred in the endemic region of eastern Sudan in the 1990s. PKDL patients had a similar gender and age distribution to VL cases, with 56% of PKDL patients progressing from VL cases (2, 13). During that one decade, more than 400 cases were reported in the same area. A recently conducted survey suggested the average period of PKDL occurrence was 9.7 months (4, 14).

17.2 LATENCY PERIOD OF PKDL FROM VL

Patients with VL are known to develop PKDL after treatment; however, the temporal lag varies between research groups and case series. The interval between VL and PKDL shows a distinct

DOI: 10.1201/9781032643120-17

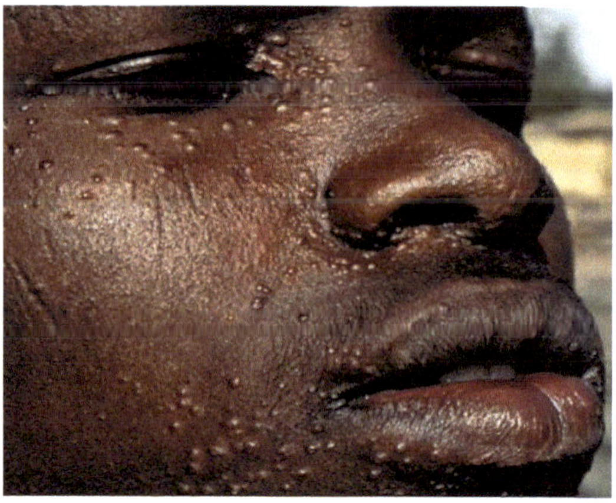

FIGURE 17.1 Lesions spread to other parts of the face

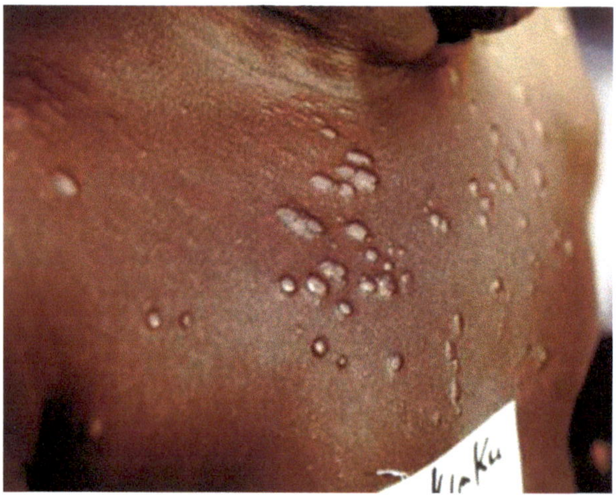

FIGURE 17.2 Nodular lesions

pattern in India and Sudan, and this period is known as the incubation or latent period. In contrast to India, where PKDL often appears 2–7 years after an episode of VL, PKDL in Sudan typically appears concurrently with or within 3–6 months of VL. However, recent research has shown that there is also a decreasing discrepancy between VL and PKDL in India, with up to 36% of cases presenting within a year after VL (15). The age of the patient and any previous VL treatment appear to affect the time between VL and PKDL. Research conducted in India found that when VL was treated with sodium stibogluconate, amphotericin B deoxycholate, and miltefosine, the mean intervals recorded were 36 months, 48 months, and 21 months, respectively (16).

17.3 FACTORS INFLUENCING THE DEVELOPMENT OF PKDL AND VL

Several factors are responsible for PKDL development. Some important factors responsible for influencing VL and PKDL development are listed in Figure 17.3.

17.3.1 Reinfection

The persistence of pathogens after a clinical cure is a characteristic of many disorders, including tuberculosis (TB), viral diseases like herpes, and protozoal infections like trypanosomiasis (17).

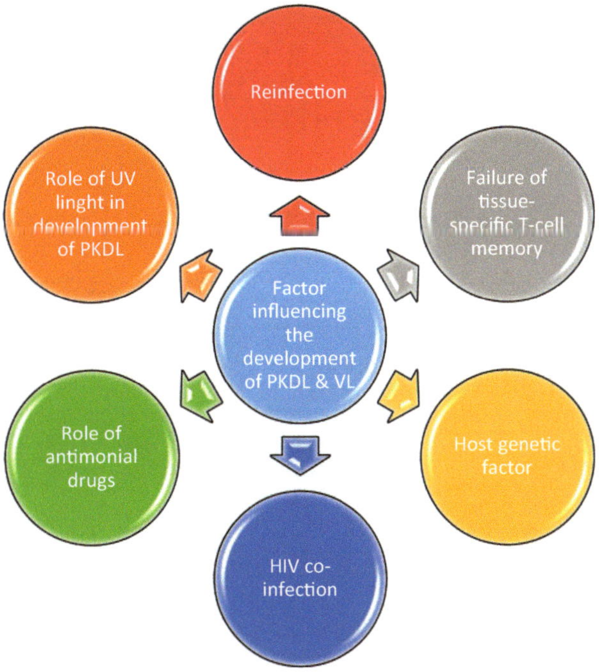

FIGURE 17.3 Factors influencing the development of PKDL and VL

Leishmaniasis shows indications of parasite persistence even after clinical treatment, and recurrence has been linked to parasite persistence and/or reinfection in endemic areas. The best approach to characterize these parasites and compare them to the parental type would be through genomic and/or proteomic studies. This would determine if the parasites originated from the disease's visceral form or from a reinfection. However, investigations in humans are logistically challenging due to inadequate parasite density for the creation of parasite isolates, though one study was carried out in a mouse model and observed that surviving parasites were genetically identical to the parent clone (Figure 17.4) (17–19).

The evidence seems to be biased in favor of reinfection because strains obtained from people with VL and PKDL in India display genetic variability (20). On the other hand, it is possible to hypothesize that during the latency stage, parasites survive within a different cell type, such as fibroblasts or keratinocytes, and upon reactivation, induce PKDL (20, 21). For this issue to be resolved, parasites would need to be isolated from a patient with VL and compared with those recovered from the same person when PKDL manifests. However, this is practically very challenging due to the lengthy and unpredictable lag time between VL and PKDL. Nevertheless, similar trials are possible in Sudan due to the substantially quicker onset of PKDL after VL there (22). Patients who were cured of VL in Sudan still developed PKDL even after moving to a non-endemic region, suggesting that PKDL is likely caused by parasite persistence. One unresolved question is why different strains of *Leishmania* cause PKDL but do not cause a return of VL. Various studies on the pathogenesis of PKDL suggest that reinfection occurred due to the failure of tissue-specific immunity, which was generated by the parasite during the previous infection, allowing the parasite to proliferate inside the body (23) (Figure 17.4).

17.3.2 Failure of Tissue-Specific T Cell Memory

Recovery from *Leishmania* infection is frequently accompanied by the emergence of a robust immune response with long-lasting protection against reinfection in humans. This infection-induced immunity, however, may be compromised in some circumstances, such as immunosuppression,

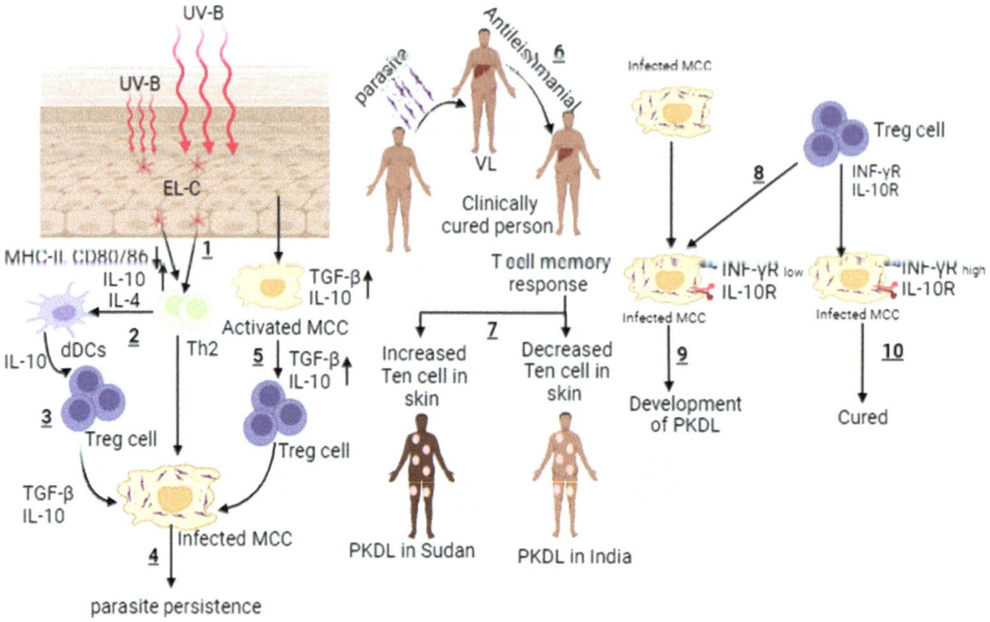

FIGURE 17.4 Contribution of environmental and host factors in the development of PKDL from VL: 1) Ultraviolet B radiation causes morphological changes in epidermal Langerhans cells (E-LCs), suppressing the production of major histocompatibility complex (MHC) class II and co-stimulatory substances while increasing IL-10 levels. 2) Th2 cells produce IL-4, stimulating dermal dendritic cells (dDCs). 3) This increases the cytokine pool by releasing more IL-10, which promotes regulatory T (Treg) cell growth and generates an immunosuppressive atmosphere. 4) These conditions promote parasite survival. 5) Activated macrophages secrete TGF-β, which activates Treg cells and encourages further TGF-β release, aiding in parasite retention. 6) Insufficient antileishmanial therapy for VL may lead to PKDL development. 7) PKDL patients in the Indian subcontinent have lower titers of effector memory T cells, resulting in a weak immune response and non-healing disease. In contrast, self-healing patients in Sudan show increased levels of effector memory T cells, indicating an enhanced immune reaction in the skin. 8) T cells release both IFN-γ and IL-10. 9) Inadequate IFN-γ signaling in individuals with polymorphism in the IFN-γ receptors causes immunological suppression and IL-10 prevalence. 10) Conversely, individuals with high IFN-γ receptor expression are least likely to develop PKDL (89)

making the host vulnerable to infection and/or the recurrence of dormant parasites (23, 24). The numerous tissue microenvironments where parasites are present in VL, such as the liver and spleen, qualitatively and quantitatively influence many different elements of host immunity (25).

In this context, PKDL presents an intriguing challenge because skin-specific immune deficits exist even though systemic protective immunity has been acquired, as shown by the recall cytokine or proliferating T cell responses measured using peripheral blood mononuclear cells or whole blood assays. This was demonstrated by a local increase in FoxP3-expressing Tregs and a predominance of CD8+ T cells (26–28). The course of PKDL in Sudan and South Asian countries differs significantly: lesions in Sudanese PKDL resolve on their own, while in South Asian PKDL, prolonged therapy is necessary (22). This discrepancy may be related to the length of the disease's latent period, as in Sudan, individuals develop PKDL much more quickly after being cured of VL, and their immune responses may resemble post-VL characteristics, such as active immune memory. In South Asia, there is a significant lag time, immunological responses vary widely, and immune anergy, or the inability to trigger an antigen-specific immune response, is evident (26).

Central memory T cells and effector memory T cells possess the important property known as immune memory, which becomes more useful in later periods (29). In Indian PKDL, the disease develops a non-self-healing character as a result of the lack of T helper (Th) cells in the dermal lesions, whereas self-healing PKDL in Sudan is linked to higher concentrations of dermal CD4+ T cells. This hypothesis would be valid if comparison experiments with Sudanese PKDL and non-self-healing Indian PKDL revealed distinct memory responses.

Antigen-specific recall responses, as determined by circulating CD4+ T cell responses to *Leishmania* antigen or phytohemagglutinin (PHA) activation, were intact in Indian PKDL, although CD8+ T cells had functional limitations (26, 28). Thus, CD8+ functionality is important. The expression of CD25 (the IL-2 receptor), which is located on the regulatory T cell surface, may also be a significant contributor to the different pathophysiology of Indian and Sudanese PKDL (30). Evidence from biopsies conducted in patients from Sudan showed a decreased presence of CD25+ T cells, while in Indian patients, it was significantly increased (30, 31).

Increased cutaneous lymphocyte antigen (CLA) expression is necessary for memory CD4+ T cell homing, as seen in localized self-resolving cutaneous leishmaniasis (CL), whereas the absence of CLA expression in diffuse CL (DCL) corresponds with a non-healing outcome (32). It is hypothesized that CD4+ EM T cells cannot migrate to cutaneous lesions due to abnormal CLA expression. As observed in CL, granuloma development in the skin has been considered to protect against leishmaniasis (33). Indeed, the limited granuloma formation in PKDL and the significant decline in the percentage of CD4+ T cells in Indian PKDL support the idea that poor tissue-specific immunity underlies the persistence of the illness (27, 28, 33, 34).

17.3.3 Host Genetic Factor

All the above-mentioned factors have focused on various facets of host defense, wound healing, and regeneration of the immunological memory response, all of which may be influenced by host genetics. In-depth genetic research has been conducted on a number of leishmaniasis strains in humans, but the information on patients with PKDL is still preliminary and is only available from three studies (35–37). The mystery of why some but not all patients with VL develop PKDL, however, remains unanswered because all three studies were conducted in Sudan. A significant correlation has been found between the interferon-γ receptor and the emergence of PKDL (38, 39).

This correlation is crucial, as multiple investigations have found that the lesions contain elevated levels of both pro-inflammatory and anti-inflammatory cytokines (27, 40). Although IFN-γ is elevated, IFN-γR polymorphisms are linked to lower production and function, which leads to decreased responsiveness (Figure 17.4). Leishmaniasis requires a delicate balance between cells that secrete IFN-γ and those that secrete IL-4 and/or IL-10; however, there is no connection between IFN-γ or IL-10 promoter polymorphism and susceptibility to the disease (39, 41, 42). Furthermore, a damaged IFN-γR1 would make it easier for parasites to proliferate and multiply if the body didn't respond to IFN-γ during an active infection. In fact, compared to individuals without Indian PKDL, IFN-γR1 expression is lower in PKDL patients from India (40).

Patients with VL in Sudan tend to have higher plasma concentrations of C-reactive protein (CRP), although it is still unclear if genetic variants in the CRP genes and promoter regions are associated with those who are more likely to develop PKDL. Increased susceptibility to CL is linked to genes involved in innate immunity, especially those associated with complement stimulation, like FCN-2 encoding ficolin-2. The FCN-2 gene's lower expression and greater susceptibility to CL are both correlated with polymorphisms in its promoter regions (43–45). Mannose-binding lectin 2 (MBL2) was found to be more expressed in VL, and elevated blood levels of MBL2, as well as genetic variations at the exons and promoter of MBL2, were linked to disease susceptibility (46). The solute carrier family 11a (SLC11A) gene, also known as NRAMP1, is a significant candidate gene for leishmaniasis. SLC11A1 is a proton/divalent cation antiporter that is found on the phagocytes' endosomal membrane. SLC11A1 regulates IL-1β, iNOS, class II MHC molecules, TNF-α, nitric oxide (NO) production, oxidative stress, and antimicrobial properties, among other macrophage activation-related processes (47). SLC11A1 exon and promoter polymorphisms render it functionally deficient and are linked to an elevated risk of VL (48, 49).

17.3.4 HIV Coinfection

Immunosuppression brought on by concomitant infections such as measles, malaria, TB, and HIV increases the chances of recurrence of VL and PKDL in approximately 25% of patients. Patients with HIV coinfection are known to relapse frequently, and when it happens, they typically do not respond to antimonials. Such patients are very infectious to sandflies in Europe, especially if their CD4+ cell count is low. This decreases the T cell tissue-specific memory, which is the main cause of reinfection (50–52).

17.3.5 Additional Clinical Observations

Patients don't have a fever and are normally healthy (53). In the endemic area of eastern Sudan, where citizens are particularly conscious of the illnesses, which they refer to as "kala-azar," and where many have lost children as a result of this disease, the occurrence of a PKDL rash following therapy is regarded as a positive sign that indicates that the illness has manifested and that the child is likely to survive (54). In additional clinical observations of patients who experience PKDL shortly after VL, there is a gradual change from VL, in which the patient is ill but does not have a skin rash, to PKDL, in which the patient is healthy, the rash has appeared, and parasites are no longer visible anywhere other than the skin. Some patients present at an intermediate stage; one study found that 15% of 416 PKDL patients had parasites in their lymph nodes or bone marrow aspirates, indicating a widespread infection. These patients, who have both PKDL and concurrent VL, may be better described as para-kala-azar dermal leishmaniasis cases because they fall somewhere between the two conditions; they have spleens that are larger than those of VL patients (mean 99 cm) but smaller than those of PKDL patients (mean 41 cm) (55). At least seven cases from India have been reported with comparable simultaneous VL and PKDL; five more cases with VL and likely PKDL were found in Iran, and one case was seen in France (56–58).

17.3.6 The Role of Antimonial Drugs

Previous clinical and epidemiologic studies have suggested a possible connection between sodium antimony gluconate (SAG) and the subsequent emergence of PKDL (59). This association was observed in Sudan, Bangladesh, and Nepal, where all PKDL patients had received SAG treatment (22, 60). In India, approximately 75% of individuals with PKDL who were monitored for 9 years following the cure of VL had been treated with SAG for VL (61). Very few patients developed PKDL after being treated with amphotericin B, AmBisome, miltefosine, miltefosine–amphotericin B, or paromomycin, respectively (62–64). A small subset of 15–20% of individuals with PKDL reported no previous history of VL, possibly due to subclinical illness, and had not taken any medications during that time.

However, SAG is still being used to treat PKDL, albeit at a higher dose (20 mg/kg per day) and for a longer duration (4 months) than is needed to treat VL (20 mg/kg per day for 3 weeks). It is also conceivable that variations in antileishmanial medication concentrations between the systemic circulation and skin allow parasites to persist in the latter. More information is needed about the prevalence of PKDL in treated VL patients receiving other recently introduced antileishmanial treatment regimens before SAG can be implicated in the emergence of PKDL.

Miltefosine and amphotericin B have long been used to treat patients with VL in India due to the increase in antimony resistance (61, 65). As previously mentioned, Thakur et al. found that utilizing amphotericin B at 20 mg/kg for the treatment of VL significantly reduced the development of PKDL, while PKDL was reported in patients treated with lower doses of amphotericin B at

15 mg/kg (66). PKDL has rarely developed in patients who received miltefosine, paromomycin, and an amphotericin B–miltefosine combination treatment (67–69).

Another theory holds that antileishmanial medications, when administered at lower doses and for a shorter period, eliminate parasites from the viscera but not from the skin, which requires a higher dose. Therefore, PKDL may not only be drug-related but may also be a consequence of a dose-related phenomenon (Hypothesis 1). It is clear that a more rigorous understanding of drug pharmacokinetics and pharmacodynamics is necessary. Immunological data also support this hypothesis, which is corroborated by epidemiological data, as levels of the immunoregulatory cytokines transforming growth factor β (TGF-β) and interleukin-10 (IL-10) remained high even after the completion of treatment with SAG but not with amphotericin B and miltefosine (70). Additionally, the treatment of THP1 macrophages with SAG led to an increase in two anti-inflammatory, disease-sustaining molecules, notably glutathione and heme oxygenase 1 (HO-1) (71). Another contributing factor is the inability of SAG to restore the host's peroxisome function (72). Furthermore, the inability of SAG to provide a sterile cure strongly suggests that drug-induced genetic changes may exist in resistant parasites; these parasites are thought to be more "fit" than non-resistant parasites (73). Another interesting finding was that PKDL-causing strains expressed lower levels of amastigote antigen 2 (A2), which is associated with increased viscerotropism, and higher amounts of promastigote surface antigen (PSA2) and glycoprotein 63 (gp63), molecules linked to dermatotropism (21, 74).

17.3.7 The Role of Ultraviolet Light in PKDL Development

PKDL lesions frequently develop in sunlight-exposed areas, especially on the face, earlobes, limbs, and chest, as opposed to areas that are not directly exposed to the sun. This supports the idea that exposure to ultraviolet (UV) radiation contributes to the etiology of PKDL. The occurrence of photosensitive reactions, a hallmark of PKDL patients, further emphasizes the involvement of UV radiation in these patients' conditions. UV radiation has a strong immunosuppressive effect due to its capacity to harm epidermal Langerhans cells, which present antigens, and to reduce alloantigen and contact allergic reactions (75). The immunosuppressive effects of UVB radiation between 280 and 320 nm can be mediated via the chromophore cis-urocanic acid or by altering vitamin D3 levels (76, 77). Trans-urocanic acid is generated from histidine, which undergoes trans-to-cis UV-induced photoisomerization during keratinization. All dermal cell types are affected, notably E-LCs (Figure 17.4), whose number is decreased and whose shape, particularly the distinctive dendritic pattern, is altered; collectively, this weakens the ability of skin cells to present antigens (78, 79). MHC class II production is downregulated in these E-LCs from UV-irradiated skin, and co-stimulatory molecules CD80 and CD86 are lost, impairing the presentation of antigens (80, 81).

Additionally, cis-urocanic acid modifies a wide spectrum of cytokines, from pro-inflammatory tumor necrosis factor-α (TNF-α) to anti-inflammatory and immunosuppressive IL-10, by acting on keratinocytes and lymphocytes. Investigations on lesional patterns frequently reflect people's attire choices, which clearly implies a connection between UV light exposure and the infectious nature of PKDL (82). However, no comparable research has been carried out in India, although it is noteworthy that photoprotected sites are always spared.

Findings from Indian PKDL demonstrated a downregulation of co-stimulatory molecules at the lesional site, comparable to that caused by UV irradiation, and an upregulation of the cytokines TNF-α and IL-10 (40, 83, 84). These findings support the notion that UV radiation contributes to immunosuppression and the onset of disease. UV light enhances the production of vitamin D3 and dihydroxylated vitamin D3 (1a, 25(OH) 2D3), which have strong immunomodulatory effects (85). After absorbing UV rays, 7-dehydrocholesterol undergoes a sequence of enzymatic and nonenzymatic processes that result in vitamin D3. Vitamin D3 in the blood binds to the vitamin D-binding protein (VBP), which then allows it to enter cells like monocytes and macrophages.

The active 1a,25(OH)2D3 is created inside the cells after being further hydroxylated by the mitochondrial enzyme CYP27B1, which then forms a complex with the vitamin D receptor (VDR). After binding to the vitamin D response elements (VDREs), the complex enters the cell's nucleus, where it induces the production of antimicrobial peptides and immunomodulatory substances such as cathelicidin (LL-37), TGF-β, and arginase (86, 87).

The relationship between vitamin D3 and the immune system seems to be contradictory. On the one hand, it stimulates macrophage synthesis of potentially host-protective antimicrobial peptides, while on the other hand, it inhibits the activation of macrophages caused by the toll-like receptor, downregulates co-stimulatory molecules and pro-inflammatory cytokines, and stimulates TGF β and IL-1. This ability to activate the arginase pathway suggests that vitamin D3 or 1a, 25(OH) 2D3 promotes alternative activation of macrophages (88). Further, an increase in regulatory T cells (Tregs) encourages the release of IL-10, which inhibits the local immune response and aids in parasite survival. Forkhead box P3 (FoxP3), a crucial molecular biomarker of Tregs, has been shown to be more prevalent in Indian PKDL (26).

REFERENCES

1. Desjeux P. Leishmaniasis: public health aspects and control. *Clinics in Dermatology.* 1996 Sep 1;14(5):417–23.
2. Zijlstra EE, El-Hassan AM, Ismael A. Endemic kala-azar in eastern Sudan: post-kala-azar dermal leishmaniasis. *The American Journal of Tropical Medicine and Hygiene.* 1995 Apr 1;52(4):299–305.
3. World Health Organization. The post kala-azar dermal leishmaniasis (PKDL) atlas: a manual for health workers. *World Health Organization.* 2012;2012:37–45.
4. Zijlstra EE, Khalil EA, Kager PA, El-Hassan AM. Post-kala-azar dermal leishmaniasis in the Sudan: clinical presentation and differential diagnosis. *British Journal of Dermatology.* 2000 Jul 1;143(1):136–43.
5. Napier LE, Gupta CD. Further clinical observations on post-kala-azar dermal leishmaniasis. *The Indian Medical Gazette.* 1934 Mar;69(3):121.
6. Brahmachari UN. A new form of cutaneous leishmaniasis—dermal leishmanoid. *The Indian Medical Gazette.* 1922 Apr;57(4):125.
7. Ramesh V, Mukherjee A. Post-kala-azar dermal leishmaniasis. *International Journal of Dermatology.* 1995 Feb;34(2):85–91.
8. Thakur CP. Epidemiological, clinical and therapeutic features of Bihar kala-azar (including post kala-azar dermal leishmaniasis). *Transactions of the Royal Society of Tropical Medicine and Hygiene.* 1984 Jan 1;78(3):391–8.
9. Addy M, Nandy A. Ten years of kala-azar in west Bengal, Part I. Did post-kala-azar dermal leishmaniasis initiate the outbreak in 24-Parganas? *Bulletin of the World Health Organization.* 1992;70(3):341
10. Kirk R, Drew CB. Preliminary notes on dermal leishmaniasis in the Anglo-Egyptian Sudan. *Transactions of the Royal Society of Tropical Medicine and Hygiene.* 1938;32(2), 501.
11. Kirk R, Sati MH. Studies in Leishmaniasis in the Anglo-Egyptian Sudan. IV. A punctate rash in treated oases. *Transactions of the Royal Society of Tropical Medicine and Hygiene.* 1940;34(2), 489–92.
12. El-Hassan AM, Ahmed MA, Rahim AA, Satir AA, Wasfi A, Kordofani AA, Mustafa MD, Wasfi S, Bella H, Karrar MO. Visceral leishmaniasis in the Sudan: clinical and hematological features. *Annals of Saudi Medicine.* 1990 Jan;10(1):51–6.
13. Khalil EA, Zijlstra EE, Kager PA, El Hassan AM. Epidemiology and clinical manifestations of *Leishmania donovani* infection in two villages in an endemic area in eastern Sudan. *Tropical Medicine and International Health.* 2002 Jan;7(1):35–44.
14. Musa AM, Khalil EA, Raheem MA, Zijlstra EE, Ibrahim ME, Elhassan IM, Mukhtar MM, El Hassan AM. The natural history of Sudanese post-kala-azar dermal leishmaniasis: clinical, immunological and prognostic. *Annals of Tropical Medicine and Parasitology.* 2002;96(8):765–72.
15. Zijlstra EE, Alves F, Rijal S, Arana B, Alvar J. Post-kala-azar dermal leishmaniasis in the Indian subcontinent: a threat to the South-East Asia Region Kala-azar Elimination Programme. *PLoS Neglected Tropical Diseases.* 2017 Nov 16;11(11):e0005877.
16. Ramesh V, Kaushal H, Mishra AK, Singh R, Salotra P. Clinico-epidemiological analysis of Post kala-azar dermal leishmaniasis (PKDL) cases in India over last two decades: a hospital based retrospective study. *BMC Public Health.* 2015 Dec;15:1–8.

17. Mendonça MG, de Brito ME, Rodrigues EH, Bandeira V, Jardim ML, Abath FG. Persistence of Leishmania parasites in scars after clinical cure of American cutaneous leishmaniasis: is there a sterile cure?. *The Journal of Infectious Diseases.* 2004 Mar 15;189(6):1018–23.
18. Aebischer TO, Moody SF, Handman EM. Persistence of virulent Leishmania major in murine cutaneous leishmaniasis: a possible hazard for the host. *Infection and Immunity.* 1993 Jan;61(1):220–6.
19. Schubach A, Marzochi MC, Cuzzi-Maya T, Oliveira AV, Araujo ML, Oliveira AL, Pacheco RS, Momen H, Conceicao-Silva F, Coutinho SG, Marzochi KB. Cutaneous scars in American tegumentary leishmaniasis patients: a site of Leishmania (Viannia) braziliensis persistence and viability eleven years after antimonial therapy and clinical cure. *The American Journal of Tropical Medicine and Hygiene.* 1998 Jun;58(6):824–7.
20. Schubach A, Marzochi MC, Cuzzi-Maya T, Oliveira AV, Araujo ML, Oliveira AL, Pacheco RS, Momen H, Conceicao-Silva F, Coutinho SG, Marzochi KB. Cutaneous scars in American tegumentary leishmaniasis patients: a site of Leishmania (Viannia) braziliensis persistence and viability eleven years after antimonial therapy and clinical cure. *The American Journal of Tropical Medicine and Hygiene.* 1998 Jun;58(6):824–7.
21. Salotra P, Duncan RC, Singh R, Raju BS, Sreenivas G, Nakhasi HL. Upregulation of surface proteins in *Leishmania donovani* isolated from patients of post kala-azar dermal leishmaniasis. *Microbes and Infection.* 2006 Mar 1;8(3):637–44.
22. Zijlstra EE, Musa AM, Khalil EA, El Hassan IM, El-Hassan AM. Post-kala-azar dermal leishmaniasis. *The Lancet Infectious Diseases.* 2003 Feb 1;3(2):87–98.
23. Gollob KJ, Antonelli LR, Dutra WO. Insights into CD4+ memory T cells following Leishmania infection. *Trends in Parasitology.* 2005 Aug 1;21(8):347–50.
24. Desjeux P. Global control and Leishmania HIV co-infection. *Clinics in Dermatology.* 1999 May 1;17(3):317–25.
25. Engwerda CR, Kaye PM. Organ-specific immune responses associated with infectious disease. *Immunology Today.* 2000 Feb 1;21(2):73–8.
26. Ganguly S, Mukhopadhyay D, Das NK, Chaduvula M, Sadhu S, Chatterjee U, Rahman M, Goswami RP, Guha SK, Modak D, Mallik S. Enhanced lesional Foxp3 expression and peripheral anergic lymphocytes indicate a role for regulatory T cells in Indian post-kala-azar dermal leishmaniasis. *Journal of Investigative Dermatology.* 2010 Apr 1;130(4):1013–22.
27. Ismail A, El Hassan AM, Kemp K, Gasim S, Kadaru AE, Møller T, Kharazmi A, Theander TG. Immunopathology of post kala-azar dermal leishmaniasis (PKDL): T-cell phenotypes and cytokine profile. *The Journal of Pathology.* 1999 Dec;189(4):615–22.
28. Rathi SK, Pandhi RK, Chopra P, Khanna N. Lesional T-cell subset in post-kala-azar dermal leishmaniasis. *International Journal of Dermatology.* 2005 Jan;44(1):12–3.
29. Okwor I, Uzonna J. Persistent parasites and immunologic memory in cutaneous leishmaniasis: implications for vaccine designs and vaccination strategies. *Immunologic Research.* 2008 Jun;41:123–36.
30. Katara GK, Ansari NA, Verma S, Ramesh V, Salotra P. Foxp3 and IL-10 expression correlates with parasite burden in lesional tissues of post kala azar dermal leishmaniasis (PKDL) patients. *PLoS Neglected Tropical Diseases.* 2011 May 31;5(5):e1171.
31. Ismail A, Gadir AF, Theander TG, Kharazmi A, El Hassan AM. Pathology of post-kala-azar dermal leishmaniasis: a light microscopical, immunohistochemical, and ultrastructural study of skin lesions and draining lymph nodes. *Journal of Cutaneous Pathology.* 2006 Dec;33(12):778–87.
32. Diaz NL, Zerpa O, Ponce LV, Convit J, Rondon AJ, Tapia FJ. Intermediate or chronic cutaneous leishmaniasis: leukocyte immunophenotypes and cytokine characterisation of the lesion. *Experimental Dermatology.* 2002 Feb;11(1):34–41.
33. Tuon FF, Fernandes ER, Pagliari C, Duarte MI, Amato VS. The expression of TLR9 in human cutaneous leishmaniasis is associated with granuloma. *Parasite Immunology.* 2010 Nov;32(11–12):769–72.
34. Rathi SK, Pandhi RK, Chopra P, Khanna N. Post-kala-azar dermal leishmaniasis: a histopathological study. *Indian Journal of Dermatology, Venereology and Leprology.* 2005 Jul 1;71:250.
35. Lipoldová M, Demant P. Genetic susceptibility to infectious disease: lessons from mouse models of leishmaniasis. *Nature Reviews Genetics.* 2006 Apr;7(4):294–305.
36. Foote SJ, Handman E. Genetics of murine leishmaniasis. *Briefings in Functional Genomics.* 2005 Nov 1;4(3):270–6.

37. LeishGEN Consortium, Wellcome Trust Case Control Consortium 2, Fakiola M, Strange A, Cordell HJ, Miller EN, Pirinen M, Su Z, Mishra A, Mehrotra S, Monteiro GR. Common variants in the HLA-DRB1–HLA-DQA1 HLA class II region are associated with susceptibility to visceral leishmaniasis. *Nature Genetics.* 2013 Feb;45(2):208–13.
38. Mohamed HS, Ibrahim ME, Miller EN, Peacock CS, Khalil EA, Cordell HJ, Howson JM, El Hassan AM, Bereir RE, Blackwell JM. Genetic susceptibility to visceral leishmaniasis in The Sudan: linkage and association with IL4 and IFNGR1. *Genes and Immunity.* 2003 Jul;4(5):351–5.
39. Salih MA, Ibrahim ME, Blackwell JM, Miller EN, Khalil EA, ElHassan AM, Musa AM, Mohamed HS. IFNG and IFNGR1 gene polymorphisms and susceptibility to post-kala-azar dermal leishmaniasis in Sudan. *Genes and Immunity.* 2007 Jan;8(1):75–8.
40. Ansari NA, Ramesh V, Salotra P. Interferon (IFN)–γ, tumor necrosis factor–α, interleukin-6, and IFN-γ receptor 1 are the major immunological determinants associated with post–kala azar dermal leishmaniasis. *The Journal of Infectious Diseases.* 2006 Oct 1;194(7):958–65.
41. Alexander J, Brombacher F. T helper1/t helper2 cells and resistance/susceptibility to leishmania infection: is this paradigm still relevant?. *Frontiers in Immunology.* 2012 Apr 17;3:80.
42. Farouk S, Salih MA, Musa AM, Blackwell JM, Miller EN, Khalil EA, Elhassan AM, Ibrahim ME, Mohamed HS. Interleukin 10 gene polymorphisms and development of post kala-azar dermal leishmaniasis in a selected sudanese population. *Public Health Genomics.* 2010 Aug 1;13(6):362–7.
43. Gasim S, Theander TG, ElHassan AM. High levels of C-reactive protein in the peripheral blood during visceral leishmaniasis predict subsequent development of post kala-azar dermal leishmaniasis. *Acta Tropica.* 2000 Feb 25;75(1):35–8.
44. Kilpatrick DC, Chalmers JD. Human L-ficolin (ficolin-2) and its clinical significance. *Journal of Biomedicine and Biotechnology.* 2012 Oct;2012, 138797..
45. Assaf A, Hoang TV, Faik I, Aebischer T, Kremsner PG, Kun JF, Velavan T. Genetic evidence of functional ficolin-2 haplotype as susceptibility factor in cutaneous leishmaniasis. *PloS One.* 2012 Mar 23;7(3):e34113.
46. Peres Alonso D, Ferreira AF, Ribolla PE, de Miranda Santos IK, do Socorro Pires e Cruz M, De Carvalho FA, Abatepaulo AR, Lamounier Costa D, Werneck GL, Farias TJ, Soares MJ. Genotypes of the mannan-binding lectin gene and susceptibility to visceral leishmaniasis and clinical complications. *The Journal of Infectious Diseases.* 2007 Apr 15;195(8):1212–7.
47. Blackwell JM, Goswami T, Evans CA, Sibthorpe D, Papo N, White JK, Searle S, Miller EN, Peacock CS, Mohammed H, Ibrahim M. SLC11A1 (formerly NRAMP1) and disease resistance. *Cellular Microbiology.* 2001 Dec;3(12):773.
48. Mohamed HS, Ibrahim ME, Miller EN, White JK, Cordell HJ, Howson JM, Peacock CS, Khalil EA, El Hassan AM, Blackwell JM. SLC11A1 (formerly NRAMP1) and susceptibility to visceral leishmaniasis in The Sudan. *European Journal of Human Genetics.* 2004 Jan;12(1):66–74.
49. Blackwell JM, Mohamed HS, Ibrahim ME. Genetics and visceral leishmaniasis in the Sudan: seeking a link. *Trends in Parasitology.* 2004 Jun 1;20(6):268–74.
50. Molina R, Lohse JM, Pulido F, Laguna F, Lopez-Velez R, Alvar J. Infection of sand flies by humans coinfected with Leishmania infantum and human immunodeficiency virus. *The American Journal of Tropical Medicine and Hygiene.* 1999 Jan;60(1):51–3.
51. Faraut-Gambarelli F, Piarroux R, Deniau M, Giusiano B, Marty P, Michel G, Faugère B, Dumon H. In vitro and in vivo resistance of Leishmania infantum to meglumine antimoniate: a study of 37 strains collected from patients with visceral leishmaniasis. *Antimicrobial Agents and Chemotherapy.* 1997 Apr;41(4):827–30.
52. Nandy A, Addy M, Maji AK, Guha SK, Banerjee D, Chaudhuri D. Recurrence of kala-azar after PKDL: role of co-factors. *Tropical Medicine and International Health.* 1998 Jan;3(1):76–8.
53. El Hassan AM, Ghalib HW, Zijlstra EE, Eltoum IA, Satti M, Ali MS, Ali HM. Post kala-azar dermal leishmaniasis in the Sudan: clinical features, pathology and treatment. *Transactions of the Royal Society of Tropical Medicine and Hygiene.* 1992 May 1;86(3):245–8.
54. Zijlstra EE, El-Hassan AM. Leishmaniasis in Sudan. 3. Visceral leishmaniasis. *Transactions of the Royal Society of Tropical Medicine and Hygiene.* 2001 Apr 1;95(Supplement_1):S27–58.
55. Zijlstra EE, Khalil EA, Kager PA, El-Hassan AM. Post-kala-azar dermal leishmaniasis in the Sudan: clinical presentation and differential diagnosis. *British Journal of Dermatology.* 2000 Jul 1;143(1):136–43.
56. Singh NK, Agrawal SK, Jha TK. Visceral kala-azar associated with post-kala-azar dermal leishmaniasis. *The Journal of the Association of Physicians of India.* 1989 Feb 1;37(2):191.
57. Kumar PV, Sadeghi E, Torabi S. Kala azar with disseminated dermal leishmaniasis. *The American Journal of Tropical Medicine and Hygiene.* 1989 Feb 1;40(2):150–3.

58. Frances C, Merle Beral H, Franceschini PH, Lessana-Leibowitch M, Escande JP. Kala-azar chez l'immunodéprimé: à propos d'un casrévélé par des signescutanés. *La Presse médicale* (1983). 1984;13(40):2433–6.
59. Gedda MR, Singh B, Kumar D, Singh AK, Madhukar P, Upadhyay S, Singh OP, Sundar S. Post kala-azar dermal leishmaniasis: a threat to elimination program. *PLoS Neglected Tropical Diseases.* 2020 Jul 2;14(7):e0008221.
60. Mondal D, Khan MG. Recent advances in post-kala-azar dermal leishmaniasis. *Current Opinion in Infectious Diseases.* 2011 Oct 1;24(5):418–22.
61. Das VN, Ranjan A, Pandey K, Singh D, Verma N, Das S, Lal CS, Sinha NK, Verma RB, Siddiqui NA, Das P. Clinical epidemiologic profile of a cohort of post–kala-azar dermal leishmaniasis patients in Bihar, India. *The American Journal of Tropical Medicine and Hygiene.* 2012 Jun 6;86(6):959.
62. Mondal D, Khan MG. Recent advances in post-kala-azar dermal leishmaniasis. *Current Opinion in Infectious Diseases.* 2011 Oct 1;24(5):418–22.
63. Rahman KM, Islam S, Rahman MW, Kenah E, Galive CM, Zahid MM, Maguire J, Rahman M, Haque R, Luby SP, Bern C. Increasing incidence of post-kala-azar dermal leishmaniasis in a population-based study in Bangladesh. *Clinical Infectious Diseases.* 2010 Jan 1;50(1):73–6.
64. Uranw S, Ostyn B, Rijal A, Devkota S, Khanal B, Menten J, Boelaert M, Rijal S. Post-kala-azar dermal leishmaniasis in Nepal: a retrospective cohort study (2000–2010). *PLoS Neglected Tropical Diseases.* 2011 Dec 20;5(12):e1433.
65. Sundar S, Jha TK, Thakur CP, Engel J, Sindermann H, Fischer C, Junge K, Bryceson A, Berman J. Oral miltefosine for Indian visceral leishmaniasis. *New England Journal of Medicine.* 2002 Nov 28;347(22):1739–46.
66. Thakur CP, Kumar A, Mitra G, Thakur S, Sinha PK, Das P, Bhattacharya SK, Sinha A. Impact of amphotericin-B in the treatment of kala-azar on the incidence of PKDL in Bihar, India. *Indian Journal of Medical Research.* 2008 Jul 1;128(1):38–44.
67. Das VN, Pandey K, Singh D, Forwood C, Lal CS, Das P. Development of post-kala-azar dermal leishmaniasis in AmBisome treated visceral leishmaniasis: a possible challenge to elimination program in India. *Journal of Postgraduate Medicine.* 2013 Jul 1;59(3):336–338.
68. Kumar D, Ramesh V, Verma S, Ramam M, Salotra P. Post-kala-azar dermal leishmaniasis (PKDL) developing after treatment of visceral leishmaniasis with amphotericin B and miltefosine. *Annals of Tropical Medicine and Parasitology.* 2009 Dec 1;103(8):727–30.
69. Pandey K, Das VN, Singh D, Das S, Lal CS, Verma N, Bimal S, Topno RK, Siddiqui NA, Verma RB, Sinha PK. Post-kala-azar dermal leishmaniasis in a patient treated with injectable paromomycin for visceral leishmaniasis in India. *Journal of Clinical Microbiology.* 2012 Apr;50(4):1478–9.
70. Saha S, Mondal S, Ravindran R, Bhowmick S, Modak D, Mallick S, Rahman M, Kar S, Goswami R, Guha SK, Pramanik N. IL-10-and TGF-β-mediated susceptibility in kala-azar and post-kala-azar dermal leishmaniasis: the significance of amphotericin B in the control of *Leishmania donovani* infection in India. *The Journal of Immunology.* 2007 Oct 15;179(8):5592–603.
71. El Fadili K, Imbeault M, Messier N, Roy G, Gourbal B, Bergeron M, Tremblay MJ, Légaré D, Ouellette M. Modulation of gene expression in human macrophages treated with the anti-leishmania pentavalent antimonial drug sodium stibogluconate. *Antimicrobial Agents and Chemotherapy.* 2008 Feb;52(2):526–33.
72. Gupta S, Raychaudhury B, Datta SC. Host peroxisomal properties are not restored to normal after treatment of visceral leishmaniasis with sodium antimony gluconate. *Experimental Parasitology.* 2009 Oct 1;123(2):140–5.
73. Vanaerschot M, De Doncker S, Rijal S, Maes L, Dujardin JC, Decuypere S. Antimonial resistance in *Leishmania donovani* is associated with increased in vivo parasite burden. *PloS One.* 2011 Aug 1;6(8):e23120.
74. McCall LI, Matlashewski G. Involvement of the *Leishmania donovani* virulence factor A2 in protection against heat and oxidative stress. *Experimental Parasitology.* 2012 Oct 1;132(2):109–15.
75. Ismail A, Khalil EA, Musa AM, Hassan IE, Ibrahim ME, Theander TG, Hassan AE. The pathogenesis of post kala-azar dermal leishmaniasis from the field to the molecule: does ultraviolet light (UVB) radiation play a role? *Medical Hypotheses.* 2006 Jan 1;66(5):993–9.
76. Amerio P, Carbone A, Auriemma M, Varrati S, Tulli A. UV induced skin immunosuppression. *Anti-Inflammatory and Anti-Allergy Agents in Medicinal Chemistry (Formerly Current Medicinal Chemistry-Anti-Inflammatory and Anti-Allergy Agents).* 2009 Mar 1;8(1):3–13.
77. Hart PH, Gorman S, Finlay-Jones JJ. Modulation of the immune system by UV radiation: more than just the effects of vitamin D?. *Nature Reviews Immunology.* 2011 Sep;11(9):584–96.

78. Noonan FP, De Fabo EC. Immunosuppression by ultraviolet B radiation: initiation by urocanic acid. *Immunology Today*. 1992 Jan 1;13(7):250–4.
79. Gibbs NK, Norval M. Urocanic acid in the skin: a mixed blessing?. *Journal of Investigative Dermatology*. 2011 Jan 1;131(1):14–7.
80. Weiss JM, Renkl A, Denfeld RW, Schöpf E, Simon JC. UVB-radiation perturbs the functional expression of B7-1 and B7-2 costimulatory molecules on human langerhans cells (LC). *Journal of Investigative Dermatology*. 1995;6(105):863.
81. Simon JC, Edelbaum D, Bergstresser PR, Cruz Jr PD. Distorted antigen-presenting function of Langerhans cells induced by tumor necrosis factor alpha via a mechanism that appears different from that induced by ultraviolet B radiation. *Photodermatology, Photoimmunologyand Photomedicine*. 1991 Oct 1;8(5):190–4.
82. Musa AM, Khalil EA, Raheem MA, Zijlstra EE, Ibrahim ME, Elhassan IM, Mukhtar MM, El Hassan AM. The natural history of Sudanese post-kala-azar dermal leishmaniasis: clinical, immunological and prognostic. *Annals of Tropical Medicine and Parasitology*. 2002;96(8):765–72.
83. Mukhopadhyay D, Das NK, Roy S, Kundu S, Barbhuiya JN, Chatterjee M. Miltefosine effectively modulates the cytokine milieu in Indian post kala-azar dermal leishmaniasis. *Journal of Infectious Diseases*. 2011 Nov 1;204(9):1427–36.
84. Ismail A, Gadir AF, Theander TG, Kharazmi A, El Hassan AM. Pathology of post-kala-azar dermal leishmaniasis: a light microscopical, immunohistochemical, and ultrastructural study of skin lesions and draining lymph nodes. *Journal of Cutaneous Pathology*. 2006 Dec;33(12):778–87.
85. Baeke F, Takiishi T, Korf H, Gysemans C, Mathieu C. Vitamin D: modulator of the immune system. *Current Opinion in Pharmacology*. 2010 Aug 1;10(4):482–96.
86. Griffin MD, Xing N, Kumar R. Vitamin D and its analogs as regulators of immune activation and antigen presentation. *Annual Review of Nutrition*. 2003 Jul;23(1):117–45.
87. Whitcomb JP, DeAgostino M, Ballentine M, Fu J, Tenniswood M, Welsh J, Cantorna M, McDowell MA. The role of vitamin D and vitamin D receptor in immunity to Leishmania major infection. *Journal of Parasitology Research*. 2009 Apr 2;2012, 1–10..
88. Martinez FO, Helming L, Gordon S. Alternative activation of macrophages: an immunologic functional perspective. *Annual Review of Immunology*. 2009 Apr 23;27:451–83.
89. Kumar A, Singh VK, Tiwari R, Madhukar P, Rajneesh, Kumar S, Gautam V, Engwerda C, Sundar S, Kumar R. Post kala-azar dermal leishmaniasis in the Indian sub-continent: challenges and strategies for elimination. *Frontiers in Immunology*. 2023 Aug 11;14:1236952,05–06.

Index

Q

R

S

T

V